# neutrino astronomy

## current status, future prospects

*Editors*

**Thomas Gaisser**
University of Delaware, USA

**Albrecht Karle**
University of Wisconsin–Madison, USA

# neutrino astronomy

## current status, future prospects

**World Scientific**

NEW JERSEY · LONDON · SINGAPORE · BEIJING · SHANGHAI · HONG KONG · TAIPEI · CHENNAI · TOKYO

*Published by*

World Scientific Publishing Co. Pte. Ltd.

5 Toh Tuck Link, Singapore 596224

*USA office:* 27 Warren Street, Suite 401-402, Hackensack, NJ 07601

*UK office:* 57 Shelton Street, Covent Garden, London WC2H 9HE

**Library of Congress Cataloging-in-Publication Data**

Names: Gaisser, Thomas K., editor. | Karle, Albrecht, editor.

Title: Neutrino astronomy : current status, future prospects / edited by: Thomas Gaisser
   (University of Delaware, USA), Albrecht Karle (University of Wisconsin-Madison, USA).

Description: Singapore ; Hackensack, NJ : World Scientific, [2017] |
   Includes bibliographical references and index.

Identifiers: LCCN 2016053883| ISBN 9789814759403 (hard cover ; alk. paper) |
   ISBN 9814759406 (hard cover ; alk. paper)

Subjects: LCSH: Neutrino astrophysics. | Astronomy--Observations.

Classification: LCC QB464.2 .N46 2017 | DDC 522/.686--dc23

LC record available at https://lccn.loc.gov/2016053883

**British Library Cataloguing-in-Publication Data**

A catalogue record for this book is available from the British Library.

Desk Editor: Ng Kah Fee

Typeset by Stallion Press

Email: enquiries@stallionpress.com

# Contents

# Preface

The idea of using a large volume of water as a target for detecting astrophysical and cosmic-ray-induced neutrinos dates back to the independent suggestions by Reines, Greisen and Markov in 1960. The first neutrinos of extraterrestrial origin to be detected were of relatively low energy, from SN1987A and from the Sun. Studies of atmospheric and solar neutrinos with the densely instrumented detectors Super-Kamiokande and SNO led to the discovery of neutrino oscillations. The development of more sparsely instrumented detectors with target volumes large enough to detect TeV neutrinos from distant sources of high-energy cosmic rays proceeded in parallel, with DUMAND, Baikal, AMANDA, ANTARES and IceCube. The history of neutrino astronomy is nicely described in the paper by Christian Spiering (Eur. Phys. J. H37, 515–565, 2012).

This review volume is stimulated by the discovery of astrophysical neutrinos by IceCube. The first four chapters review the physics of several potential sources of high-energy neutrinos: gamma-ray bursts, active galactic nuclei, star-forming galaxies and sources in the Milky Way. A common theme is the relation between gamma rays and neutrinos: neutrinos from decay of charged pions imply a corresponding flux of photons from decay of neutral pions. Hence the observed gamma rays place constraints on the sources and spectra of the neutrinos.

The IceCube results are described in Chapter 5. The discovery was first made by selecting events that start inside the detector then confirmed by measurement of the spectrum of neutrino-induced muons, which extends above the steeply falling background of atmospheric neutrinos. Next, searches for point sources with IceCube are described. The fact that no point source has yet been identified places limits on the contribution to the observed signal from rare, strong sources. Results from the ANTARES and Baikal detectors are presented in Chapters 7 and 8.

The chapter on multi-messenger astronomy introduces the connections among gamma rays, neutrinos and cosmic rays and describes the procedures for sending alerts for follow-up of high-energy neutrino events by optical, X-ray and gamma-ray detectors. The next chapter discusses two aspects of supernovae. On one hand, the burst of low-energy neutrinos from a nearby collapse can be observed as a sudden increase and subsequent decline in the counting rates of the optical modules in IceCube. In addition, collapse of massive, extragalactic stars exploding and accelerating cosmic rays into their progenitor winds could be sources of multi-TeV neutrinos. The chapter on dark matter summarizes the indirect searches from all currently operating neutrino telescopes.

The discovery of astrophysical neutrinos raises a number of questions that motivate the construction of larger detectors to obtain more events and provide complementary views of the sky. There are plans to expand Baikal to GVD, and construction of KM3NeT is beginning. A next-generation IceCube-Gen2 would have a target volume nearly an order of magnitude greater than at present. Both KM3NeT and IceCube-Gen2 would also have densely instrumented components aimed at neutrino physics in the multi-GeV range, including $\nu_\tau$ appearance and mass hierarchy. At the other extreme, radio detection of neutrinos would provide the much larger target volumes needed to detect cosmogenic neutrinos from interactions of ultra-high-energy cosmic rays in the cosmic background radiation.

High energy gamma rays, neutrinos and cosmic rays are the crucial messengers that together offer the prospect to reveal the nature of the cosmic accelerators and to identify them. An intriguing feature of the cosmic neutrino flux detected by IceCube is the fact that its power between 100 TeV and 10 PeV is of the same order of magnitude as the power in the observed extragalactic photon flux in the energy range of 1 to 100 GeV and the flux of extragalactic cosmic rays at energies above several EeV. This value is close to the upper bound for such a flux proposed by Waxman & Bahcall, and it is important to determine whether this is a signal of an underlying connection or a coincidence. More data with new detectors and their combined analysis will be needed to close in on the origin of the high energy particle fluxes.

This volume is intended to serve as a reference of the observational status, a review of the theoretical implication and a forward looking perspective to a generation of detectors to be realized within the next decade and beyond. It is intended to be useful for experts in the field, the broader scientific audience of high-energy astrophysics as well as students interested to explore the new view of the Universe.

# Chapter 1

# Gamma-Ray Bursts as Neutrino Sources

P. Mészáros

*Center for Particle and Gravitational Astrophysics,*
*Dept. of Astronomy & Astrophysics and Dept. of Physics,*
*Pennsylvania State University, University Park, PA 16802, USA*
*nnp@psu.edu*

Gamma-ray burst sources appear to fulfill all the conditions for being efficient cosmic ray accelerators, and being extremely compact, are also expected to produce multi-GeV to PeV neutrinos. I review the basic model predictions for the expected neutrino fluxes in classical GRBs as well as in low luminosity and choked bursts, discussing the recent IceCube observational constraints and implications from the observed diffuse neutrino flux.

## 1. Introduction

Gamma-ray bursts (GRBs) have been postulated to be sources of very high energy (TeV to PeV) neutrinos since at least 1997.[1] This is based on the realization that these objects may be good sites for accelerating ultra-high energy cosmic rays (UHE-CRs[2-4]), while having an extremely high photon luminosity which provides ideal target photons for photohadronic interactions. The observed electromagnetic radiation is typically interpreted in terms of shock-accelerated relativistic electrons undergoing synchrotron and inverse Compton losses. The standard fireball shock model of GRBs (see e.g. Ref. 5) leads to estimates for the shock region size $R$, comoving magnetic field strength $B'$ and comoving photon density $n'_\gamma$, which provide the basis for arguing that GRBs should also be sources of both UHECRsultra-high energy cosmic rays (UHECR) and very high energy neutrinos. The prediction that GRBs could be strong neutrino sources has served as one of the science goals motivating the building of the IceCube, ANTARES and the planned KM3NeT Cherenkov neutrino detectors.

## 2. GRB model and variants

GRBs are thought to be caused by a cataclysmic event at the end of the life cycle of some massive stars, such as the collapse of the fast-rotating central core of stars more massive than $\sim 28 M_\odot$ (giving rise to so-called "long GRBs" of gamma-ray durations $\gtrsim$ few seconds); or the merger of two neutron stars or a neutron star and a stellar mass black hole, which themselves resulted from the previous core collapse of somewhat less massive stars (giving rise to so-called "short GRBs," gamma-ray durations $\lesssim$ few seconds).[6,7] This is the central engine which can provide the huge energy needed to power the GRB emission which, while lasting only seconds, equals roughly the total luminous output emitted by the Sun over $10^{10}$ years, or that emitted by the entire galaxy over a hundred years, and is detectable out to the farthest reaches of the Universe.

The collapse or merger results in the liberation of a gravitational energy of order $E_{grav} \sim GM^2/r \sim 10^{54}$ erg on a very short timescale, in a region whose dimensions $r_0$ are of order of tens of kilometers, leading to a fireball of photons, $e^\pm$ pairs, magnetic fields and baryons.[a] This fireball, which is initially extremely optically thick, expands most easily along the rotation axis, driven by the radiation pressure. The expansion is highly relativistic, characterized by bulk Lorentz factors of the fireball plasma of order $\Gamma \sim 10^2 - 10^3$, as inferred from the observation of multi-GeV photons.[5] This requires the fireball to have a small baryon load $M_0 c^2$ compared to the fireball energy $E_0$, i.e. a high dimensionless entropy $\eta = E_0/M_0 c^2 \sim 10^2 - 10^3$. The actual outflow is inferred, from observations of the light-curves[6] to be collimated into jets of opening angle $\theta_j \sim 0.1$, which reduces the fireball energy requirements to $\mathcal{O}(10^{51} \mathrm{erg})$.

In the standard fireball model, if the inertia is dominated by the baryon load, the Lorentz factor grows as $\Gamma \propto r$ by converting internal into bulk kinetic energy, up to $\Gamma_f \simeq \eta \simeq$ constant, at $r_s \gtrsim r_0 \eta$ after which the ejecta coasts.[8-10] Beyond a photospheric radius $r_{ph}$ where the ejecta becomes optically thin to Thompson scattering the fireball quasi-thermal gamma-rays can escape freely. For $r > r_s$ however, most of the fireball energy is kinetic, rather than in the form of photons, and the spectrum is quasi-thermal, contrary to most of the observed burst spectra.

The most widely accepted paradigm for producing the $\gamma$-ray spectrum seen in the majority of GRBs is referred to as the standard fireball shock model, which naturally produces a non-thermal spectrum, and also increases the efficiency by tapping the large reservoir of expansion kinetic energy.[11-13] Collisionless shocks are expected outside the photosphere, where the ejecta is optically thin, leading to Fermi acceleration of particles to a relativistic power law distribution, leading to broken power-law synchrotron and inverse Compton spectra. Two types of shocks

---

[a]Most of the liberated $E_{grav} \sim 10^{54}$ erg, however, escapes as a $\sim$10 s burst of $\sim$10−30 MeV thermal neutrinos, as in supernovae, and as gravitational waves.

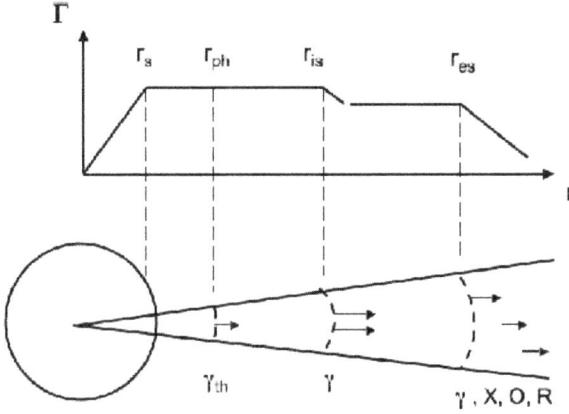

Fig. 1. Top: Schematic evolution of the bulk Lorentz factor for a GRB baryonic outflow. Bottom: The three main emission zones from which gamma-rays are detectable.

are expected: internal shocks at $r_{is} \gtrsim r_{ph}$ caused by variations in the ejecta's Lorentz factor,[13] which can give rise to the fast-varying prompt non-thermal $\gamma$-ray emission; and external shocks at $r_{es} \gtrsim r_{is} \gtrsim r_{ph}$,[11] giving rise to the longer-lasting X-ray, optical and radio afterglows.[14] The typical radii of the photosphere (for a baryon dominated ejecta) and the shocks are

$$r_{ph} \simeq (L_0 \sigma_T / 4\pi m_p c^3 \eta^3) \sim 4 \times 10^{12} L_{\gamma,52} \eta_{2.5}^{-3} \text{ cm}$$

$$r_{is} \simeq \Gamma^2 c t_v \sim 3 \times 10^{13} \eta_{2.5}^2 t_{v,-2} \text{ cm}$$

$$r_{es} \simeq (3E_0 / 4\pi n_{ext} m_p c^2 \eta^2)^{1/3} \tag{1}$$

$$\sim 2 \times 10^{17} (E_{53}/n_0)^{1/2} \eta_2^{2/3} \text{cm}.$$

Here we used $E_0, L_0, \eta \sim \Gamma, n_{ext}, t_v$ as the burst total energy, luminosity, initial dimensionless entropy, coasting bulk Lorentz factor, external density and intrinsic time variability, see Fig. 1. Figure 2 gives a schematic illustration of the different regions.[5,15]

The prompt $\gamma$-ray spectra are usually phenomenologically fitted with a "Band" broken power-law spectrum. In some bursts the fitted low energy spectral slopes appeared initially to be harder than $\alpha > -2/3$, which would violate the low energy asymptote of a synchrotron spectrum, see e.g. Ref. 5. To address this issue, a combination of a passive photospheric quasi-blackbody spectrum at low energies and a power-law shock synchrotron spectrum at high energies was considered,[16] and in fact, evidence for low energy quasi-thermal emission compatible with a photosphere has been detected in a number of bursts.[17,18] More recently, it has been confirmed[19,20] that considering the joint effects of a photosphere-like quasi-thermal spectrum together with a Band broken power-law non-thermal spectrum results in fitted slopes which are compatible with a synchrotron interpretation, which could arise in a shock outside the photosphere.

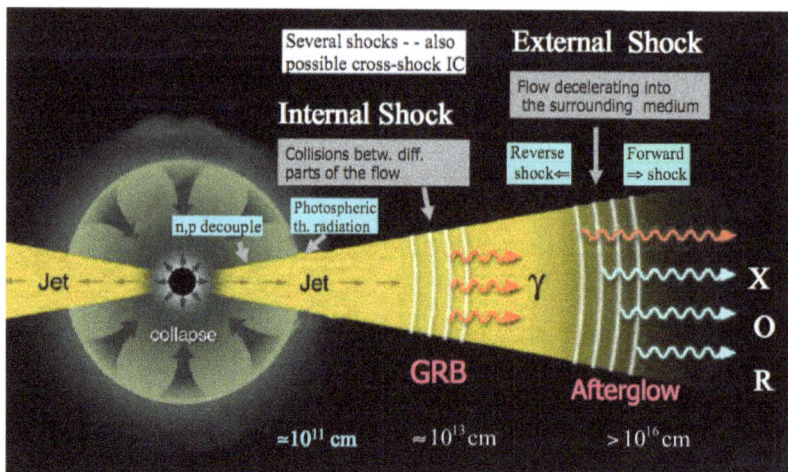

Fig. 2. Schematic GRB jet emission zones (either or both $\nu$ and $\gamma$), starting with sub-photospheric (innermost) to photosphere to internal shock (IS) to external shock (ES).

An alternative view of the origin of the prompt $\gamma$-ray emission, which similarly addresses the low energy slope issue and in addition also the efficiency issue, is that the entire prompt emission arises in a dissipative photosphere[21] (as opposed to a passive, adiabatic photosphere). The low energy slope is hard because it is self-absorbed, while the high energy slope is a power law due to comptonization, see e.g. Ref. 22. The dissipation could be due to internal shocks at or below the photosphere, or else it could be due to magnetic field reconnection[23] or it could be due to collisional effects following the decoupling of protons and neutrons below the photosphere.[24] In this case the high energy photon power-law slope extension is due to upscattering of the thermal photons by relativistic positrons from pion decay following *pn* collisions.

## 3. VHE neutrinos from GRBs

The co-acceleration of ions is natural in models where electrons are accelerated, as inferred from the observation of non-thermal gamma-rays. The latter is expected in a stellar core collapse or merger event, where the mass density is close to nuclear. The detailed model fits indicate that the baryon load is small but non-negligible, as inferred from termination bulk Lorentz factors $\Gamma_f \sim E_0/M_0c^2 \sim 10^2 - 10^3$. In such scenarios, VHE neutrino production in the shock or acceleration zones is expected from $p\gamma$ interactions. Other scenarios where the stress-energy is dominated by $e^\pm$ or magnetic fields (which could have much fewer or no baryons, as in pulsar models) are possible, and would imply negligible neutrino production. However, such models would naturally be associated with much larger bulk Lorentz factors than those inferred from observations. In such models the flow further out may decelerate due to pair drag[25] or it may pick up more baryons, but such scenarios involve more free parameters and are not widely considered in model fits. Models where the stress-energy is largely magnetic which do have small but appreciable baryon loads have been considered, see e.g. Refs. 26, 27, which lead to observationally acceptable final Lorentz factors, and having baryons, also fulfill a necessary condition for being potential neutrino sources.

The most straightforward prediction for VHE neutrino production is that associated with internal shocks.[1] This was initially based on a simplified internal shock, with given fixed shock radius parameterized by the total gamma-ray energy, Lorentz factor and outflow time variability, and approximating the photon spectrum as an average-slope Band broken power law, using the $\Delta$-resonance approximation for the photohadronic interaction. This simplified model was adopted[28] to make the first predictions of an expected diffuse VHE neutrino background, assuming a relativistic proton to electron luminosity ratio $f_p = f_e^{-1} = L_p/L_e$ and taking $L_\gamma \simeq L_e$, for a given set of electromagnetically observed bursts. Using 215 bursts with known $\gamma$-ray fluences, the first IceCube observations with 40 strings and later 56 strings were compared[29-31] to the diffuse flux predicted by this simplified internal shock (IS) model. They concluded that for a nominal ratio $f_p = f_e^{-1} = L_p/L_e = 10$ the model over-predicted the data by a factor 5, and a model-independent analysis comparing the observed diffuse neutrino flux to that expected if GRBs were the sources of the observed UHECR flux was also similarly off. This was a very important first result using a major Cherenkov neutrino facility to constrain astrophysical source models.

Subsequent analyses pointed out[32,33] that using the same fixed shock radius IS model but correcting for various approximations, and including the $\Delta$-resonance, multi-pion and Kaon channels, as well as interactions with the entire target photon spectrum, lower predicted fluxes are obtained which do not disagree with the 40+56 string data, and which indicates that 5 years of observations might be needed with the full 86 string array to rule out the simple IS model.

An issue with the fixed-radius IS models is that, even if one uses a distribution of radii $r_{is}$ in Eq. (2) based on observational distributions of variability times $t_v$ and bulk Lorentz factors $\Gamma$, this still denotes the shock initiation radius, and the baryon, photon and magnetic field densities decrease as the shocked mass shells expand beyond this radius, necessitating a time-dependent calculation in addition to a statistical averaging. Such time-dependent IS proton acceleration and full physics $p\gamma$ neutrino diffuse flux calculations result,[34,35] as expected, in a diminished predicted neutrino flux, as the $p\gamma$ opacity drops with distance away from the initial radii, giving a result which is well within the bounds of the 40+56 string upper limits, see Fig. 3.[31] It remains to redo such comparisons against the full array data.

Of course, an open question with GRBs is whether the basic internal shock model is correct for the prompt gamma-ray emission (and its related proton acceleration and $p\gamma$ production). The $\gamma$-ray spectral issues may no longer be a concern for IS models[19,20] but the mechanical radiative efficiency still remains a question. On the

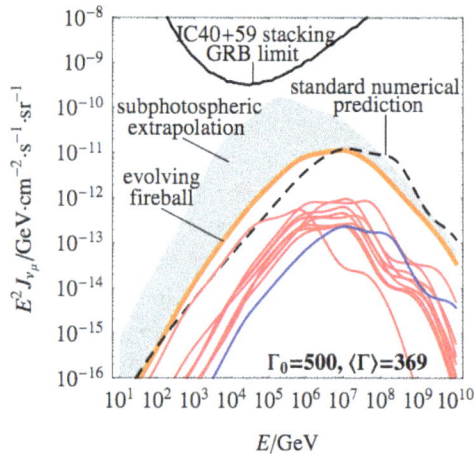

Fig. 3. Results from a time-dependent internal shock neutrino production calculation, including a sub-photospheric contribution, compared to IceCube 40+59. From Ref. 35.

other hand, dissipative photospheric models of the prompt $\gamma$-ray emission appear to address both the efficiency and spectra adequately, see e.g. Refs. 23, 24, 36. Early diffuse neutrino flux predictions from baryonic GRB photospheres were presented in Refs. 37 and 38. The results are different for magnetically dominated outflows, where the initial acceleration can be parameterized through $\Gamma(r) \propto r^\alpha$, where $1/3 < \alpha < 1$, as opposed, to $\Gamma(r) \propto r$ in the baryon-dominated case. Diffuse neutrino flux predictions from both baryon-dominated and magnetically dominated photospheres were calculated by indicating no violation of the 40+56 string IceCube constraints,[39–41] but approaching it. The main reason is that the photosphere radii are smaller than those of internal shocks, as in Eq. (2), hence photon and particle densities and $p\gamma$ optical depths are larger. Model-independent calculations confirm this.[42]

A more recent analysis of the four years of data from IceCube, including two years of the full array,[43] used the full physics of the $p\gamma$ interactions and the entire target photon spectrum to compare against the predictions of the standard IS model, a baryonic photosphere model and an ICMART[b] model, all three for fixed radius (steady state) emission zones, see Fig. 4. They concluded at 99% confidence level that less than 1% of the observed diffuse neutrino background can be contributed by the observed sample of 592 electromagnetically detected GRBs. This is a much larger burst sample with a more complete array, and while it would be important to redo this analysis with time-dependent expanding radius emission zones, this is likely to be a much stronger constraint. If the basic acceleration paradigm in these emission zones is correct, and the result continues to stand, it may indicate that the ratio $L_p/L_e = f_p \lesssim 1$. Other photospheric models with substantially different

---

[b]Internal collision magnetic reconnection and turbulence.

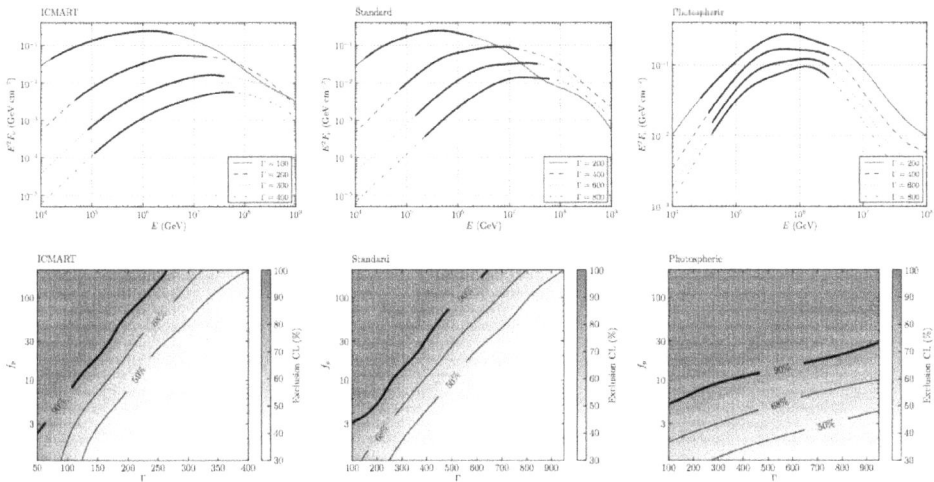

Fig. 4.   Top: Total normalized neutrino fluxes for ICMART, IS and (baryonic) photosphere models (left to right) for various $\Gamma$, scaling with $f_p$ (which is 10 here). Bottom: Allowed region for $f_p$ and $\Gamma$ for the different models. From Ref. 43.

neutrino production physics[44,45] have been investigated, but so far have been only qualitatively compared against the data.

## 4.  Other types of GRBs as possible neutrino sources

The classical GRBs discussed above are what may be called "overt" (electromagnetically detected) GRBs, the majority (70–90%) of which are long GRBs ascribed to core-collapse events located at redshifts $z \gtrsim 1$. The rest, about 10–30% of the classical GRBs, are short GRBs which appear to be compact mergers and which have a luminosity lower by about one order of magnitude, detected typically at lower redshifts $z \lesssim 1$–1.5. Their neutrino luminosity probably scales with their gamma-ray luminosity, and as such they are not expected to contribute much to the long GRB predicted diffuse neutrino fluxes.

There are, however, at least three other classes of GRBs in addition to the classical ones, which could contribute to the diffuse neutrino flux. These are the low-luminosity GRBs (LLGRBs), the choked GRBs, and the shock break-out GRBs (which may be an intermediate or transition class between the LLGRBs and the choked GRBs).

The low-luminosity GRBs, not surprisingly, have been discovered only at low redshifts, some as low as $z = 0.0085$ (GRB980425/SN1998bw); at the same time, because of the low redshift, in most of these an associated supernova of type Ic has been spectroscopically detected. While the total number of detected LLGRBs is only a handful, the inferred local rate appears to be about an order of magnitude higher (per unit volume and time) than that of classical GRBs, see e.g. Refs. 46–48. A simple scaling of the classical long GRB IS shock paradigm has led to the expectation that LLGRBs could contribute a significant, or perhaps even dominant fraction of the total GRB UHECR and VHE neutrino diffuse fluxes.[49–52]

Choked GRBs are core collapse objects similar in their dynamics to the observed classical long GRBs, where a relativistic jet has been launched from a central engine, the difference being that the jet did not make it out from the star, having stalled either because the accretion onto the central object did not last long enough, or because the stellar envelope is larger than in "successful" GRBs, where the jet has emerged.[53] Internal shocks can be expected in such jets while they are still below the stellar surface, which can also accelerate protons and undergo $p\gamma$ interactions leading to neutrinos that emerge.[53] In successful or overt GRBs (where the jet emerges and makes $\gamma$-rays) this neutrino burst from the sub-stellar phase of the jet acts as a precursor, while in truly choked jets, where the jet stalls and which are $\gamma$-dark, the neutrino burst reveals a failed GRB, which is a forerunner of a jet-boosted supernova. The spectrum should have a low energy (multi-GeV) $pp$ component as well as a higher energy $p\gamma$ TeV component, and these could be used to diagnose the stellar envelope extent and structure.[54–56] More detailed calculations[57–59] have explored possible signatures and their potential detectability by IceCube and Deep Core, preliminary results and limits having been presented.[60]

The shock which propagates through the envelope of core collapse supernovae (ccSNe) eventually breaks out of the envelope, and this may happen whether a jet was launched from the core or not, and whether such a jet eventually emerges or not from the envelope and/or the optically thick precursor stellar wind. When it does emerge, an X-ray flash is observed; in some cases this was observed in what appeared to be a LLGRB, see e.g. Ref. 61; in another case it was seen in a core-collapse supernova unassociated with a GRB.[62] It is tempting to identify the former with ccSNe where a jet was launched and emerged, and the latter with ccSNE where the jet did not emerge (or perhaps was not even launched, due to lack of enough angular momentum to feed a central accreting black hole or magnetar). The shock break-out occurs when the photons which previously were diffusively trapped become able to escape freely, which is thought to occur above the ejecting envelope, in the optically thick wind which precedes the SN explosion, see e.g. Refs. 63 and 64. If a jet was launched, an anisotropy of the envelope and the wind is expected, as indicated by the interpretation of GRB 060218.[63] This would be expected whether the jet emerged or not.[c] The ejecta of several of the SNe associated to LLGRB appear to have a semi-relativistic component,[48,61] which is interesting for the production of UHECR, see e.g. Ref. 65. The details of the shock break-out process, and whether the envelope becomes semi-relativistic (as appears to be the case in most of the SNe accompanying LLGRB) is of continued interest especially for its impact on their possible UHECR and VHE neutrino production, see e.g. Refs. 44, 66, 67.

---

[c]In fact, many SN remnants show at late stages optical polarization attributed to scattering by an anisotropic ejecta.

## 5. The TeV–PeV diffuse neutrino background

The observed diffuse flux of sub-PeV to PeV neutrinos detected by IceCube[68,69] and its extension down to the TeV range[70] is believed to be of astrophysical origin. So far, however, no significant spatial or temporal correlations have been found with classical (high-luminosity) GRBs detected electromagnetically by Swift or Fermi,[43] nor for that matter with any other type of known sources. Various possible types of candidate sources have been considered (a partial list is in Ref. 71). In particular starburst galaxies are a possibility (see Ref. 72 and other references cited in Ref. 71), within which the actual production sites are likely to be hypernovae and supernovae, see e.g. Ref. 71. Another possibility are low-luminosity GRBs (LLGRBs, see Sec. 4).[49–52] Their rate being higher than that of classical GRB, they could provide a significant neutrino background, and in $\gamma$-rays they are detectable only at low redshifts (a handful so far) but not at $z \gtrsim 0.5-1$. However, one would expect sooner or later a $\nu-\gamma$ coincidence at low redshifts. Another interesting possibility are the choked GRBs.[53,57–59,73] The shocks accelerating protons occur in the jet inside the star, hence they would be $\gamma$-dark, although the ejected envelope could lead to longer optical/IR longer transients or supernovae. Note that buried jets of arbitrary luminosity, see e.g. Ref. 73, are subject to the caveat[59] that high (classical) luminosities lead to radiation-dominated buried shocks, which prevents Fermi acceleration; collisionless shocks able to Fermi-accelerate protons are expected only for low-luminosity choked jets. Both LLGRBs and choked jets, being associated to ccSNe, are also likely to be predominantly located in starburst or starforming galaxies.

## Acknowledgments

Partial support from NASA NNX13AH50G as well as useful discussions with K. Murase and N. Senno are gratefully acknowledged.

## References

1. E. Waxman and J. Bahcall, High energy neutrinos from cosmological gamma-ray burst fireballs, *Phys. Rev. Lett.* **78** (1997) 2292–2295.
2. E. Waxman, Cosmological gamma-ray bursts and the highest energy cosmic rays, *Phys. Rev. Lett.* **75** (1995) 386–389.
3. M. Vietri, On the acceleration of ultrahigh-energy cosmic rays in gamma-ray bursts, *Astrophys. J.* **453** (1995) 883–889.
4. M. Milgrom and V. Usov, Gamma-ray bursters as sources of cosmic rays, *Astropart. Phys.* **4** (1996) 365–370.
5. P. Mészáros, Gamma-ray bursts, *Rept. Prog. Phys.* **69** (2006) 2259–2322.
6. N. Gehrels, E. Ramirez-Ruiz, and D.B. Fox, Gamma-ray bursts in the swift era, *Annu. Rev. Astron. Astrophys.* **47** (2009) 567–617. doi: 10.1146/annurev.astro.46.060407.145147.
7. S.E. Woosley and J.S. Bloom, The supernova gamma-ray burst connection, *Annu. Rev. Astron. Astrophys.* **44** (2006) 507–556. doi: 10.1146/annurev.astro.43.072103.150558.

8. B. Paczýnski, Gamma-ray bursters at cosmological distances, *Astrophys. J. Lett.* **308** (1986) L43–L46. doi: 10.1086/184740.

9. J. Goodman, Are gamma-ray bursts optically thick?, *Astrophys. J. Lett.* **308** (1986) L47–L50. doi: 10.1086/184741.

10. A. Shemi and T. Piran, The appearance of cosmic fireballs, *Astrophys. J. Lett.* **365** (1990) L55–L58. doi: 10.1086/185887.

11. M.J. Rees and P. Mészáros, Relativistic fireballs energy conversion and time-scales, *M.N.R.A.S.* **258** (1992) 41P–43P.

12. P. Mészáros and M.J. Rees, Relativistic fireballs and their impact on external matter — Models for cosmological gamma-ray bursts, *Astrophys. J.* **405** 278–284 (1993). doi: 10.1086/172360.

13. M.J. Rees and P. Mészáros, Unsteady outflow models for cosmological gamma-ray bursts, *Astrophys. J. Lett.* **430** (1994) L93–L96. doi: 10.1086/187446.

14. P. Mészáros and M.J. Rees, Optical and long-wavelength afterglow from gamma-ray bursts, *Astrophys. J.* **476** (1997) 232. doi: 10.1086/303625.

15. B. Zhang and P. Mészáros, Gamma-ray bursts: Progress, problems and prospects, *Int. J. Mod Phys A.* **19** (2004) 2385–2472. doi: 10.1142/S0217751X0401746X.

16. P. Mészáros and M.J. Rees, Steep slopes and preferred breaks in gamma-ray burst spectra: The role of photospheres and comptonization, *Astrophys. J.* **530** (2000) 292–298. doi: 10.1086/308371.

17. F. Ryde, C.-I. Björnsson, Y. Kaneko, P. Mészáros, R. Preece, and M. Battelino, Gamma-ray burst spectral correlations: Photospheric and injection effects, *Astrophys. J.* **652** (2006) 1400–1415. doi: 10.1086/508410.

18. F. Ryde, A. Pe'Er, T. Nymark, M. Axelsson, E. Moretti, C. Lundman, M. Battelino, E. Bissaldi, J. Chiang, M.S. Jackson, S. Larsson, F. Longo, S. McGlynn, and N. Omodei, Observational evidence of dissipative photospheres in gamma-ray bursts, *M.N.R.A.S* **415** (2011) 3693–3705. doi: 10.1111/j.1365-2966.2011.18985.x.

19. S. Guiriec, C. Kouveliotou, F. Daigne, B. Zhang, R. Hascoët, R.S. Nemmen, D.J. Thompson, P.N. Bhat, N. Gehrels, M.M. Gonzalez, Y. Kaneko, J. McEnery, R. Mochkovitch, J.L. Racusin, F. Ryde, J.R. Sacahui, and A.M. Ünsal, Toward a better understanding of the GRB phenomenon: A new model for GRB prompt emission and its effects on the new $L_{nTh} - E_{peak,nTh,rest}$ relation, *Astrophys. J.* **807** (2015) 148. doi: 10.1088/0004-637X/807/2/148.

20. S. Guiriec, R. Mochkovitch, T. Piran, F. Daigne, C. Kouveliotou, J. Racusin, N. Gehrels, and J. McEnery, GRB 131014A: A laboratory to study the thermal-like and non-thermal emissions in gamma-ray bursts, and the new $L_{nTh} - E_{peak,nTh,rest}$ relation, arXiv: 1507.06976 (2015).

21. M.J. Rees and P. Mészáros, Dissipative photosphere models of gamma-ray bursts and X-ray flashes, *Astrophys. J.* **628** (2005) 847–852. doi: 10.1086/430818.

22. A. Pe'er, P. Mészáros, and M.J. Rees, The observable effects of a photospheric component on GRB and XRF prompt emission spectrum, *Astrophys. J.* **642** (2006) 995–1003. doi: 10.1086/501424.

23. P. Mészáros and M.J. Rees, GeV emission from collisional magnetized gamma-ray bursts, *Astrophys. J. Lett.* **733** (2011) L40+. doi: 10.1088/2041-8205/733/2/L40.

24. A.M. Beloborodov, Collisional mechanism for gamma-ray burst emission, *M.N.R.A.S* **407** (2010) 1033–1047. doi: 10.1111/j.1365-2966.2010.16770.x.

25. P. Mészáros and M.J. Rees, Poynting jets from black holes and cosmological gamma-ray bursts, *Astrophys. J. Lett.* **482** (1997) L29+.

26. A. Tchekhovskoy, R. Narayan, and J.C. McKinney, Magnetohydrodynamic simulations of gamma-ray burst jets: Beyond the progenitor star, *New. Ast.* **15** (2010) 749–754. doi: 10.1016/j.newast.2010.03.001.

27. B.D. Metzger, D. Giannios, T.A. Thompson, N. Bucciantini, and E. Quataert, The protomagnetar model for gamma-ray bursts, *M.N.R.A.S* **413** (2011) 2031–2056. doi: 10.1111/j.1365-2966.2011.18280.x.

28. D. Guetta, D. Hooper, J. Alvarez-Muñiz, F. Halzen, and E. Reuveni, Neutrinos from individual gamma-ray bursts in the BATSE catalog, *Astroparticle Phys.* **20** (2004) 429–455. doi: 10.1016/S0927-6505(03)00211-1.

29. M. Ahlers, M.C. Gonzalez-Garcia, and F. Halzen, GRBs on probation: Testing the UHE CR paradigm with IceCube, *Astroparticle Phys.* **35** (2011) 87–94. doi: 10.1016/j.astropartphys.2011.05.008.

30. R. Abbasi, Y. Abdou, T. Abu-Zayyad, J. Adams, J.A. Aguilar, M. Ahlers, K. Andeen, J. Auffenberg, X. Bai, M. Baker *et al.*, Limits on neutrino emission from gamma-ray bursts with the 40 string IceCube detector, *Phys. Rev. Lett.* **106**(14) (2011) 141101. doi: 10.1103/PhysRevLett.106.141101.

31. R. Abbasi, Y. Abdou, T. Abu-Zayyad, M. Ackermann, J. Adams, J.A. Aguilar, M. Ahlers, D. Altmann, K. Andeen, J. Auffenberg *et al.*, An absence of neutrinos associated with cosmic-ray acceleration in $\gamma$-ray bursts *Nature* **484** (2012) 351–354. doi: 10.1038/nature11068.

32. Z. Li, Note on the normalization of predicted gamma-ray burst neutrino flux, *Phys. Rev. D.* **85**(2) (2012) 027301. doi: 10.1103/PhysRevD.85.027301.

33. S. Hümmer, P. Baerwald, and W. Winter, Neutrino emission from gamma-ray burst fireballs, revised, *Phys. Rev. Lett.* **108**(23) (2012) 231101. doi: 10.1103/PhysRevLett.108.231101.

34. K. Asano and P. Meszaros, Neutrino and cosmic-ray release from gamma-ray bursts: Time-dependent simulations, *Astrophys. J.* **785** (2014) 54. doi: 10.1088/0004-637X/785/1/54.

35. M. Bustamante, P. Baerwald, K. Murase, and W. Winter, Neutrino and cosmic-ray emission from multiple internal shocks in gamma-ray bursts, *Nature Communications* **6** (2015) 6783. doi: 10.1038/ncomms7783.

36. B. Zhang and H. Yan, The internal-collision-induced magnetic reconnection and turbulence (ICMART) model of gamma-ray bursts, *Astrophys. J.* **726** (2011) 90. doi: 10.1088/0004-637X/726/2/90.

37. K. Murase, Prompt high-energy neutrinos from gamma-ray bursts in photospheric and synchrotron self-Compton scenarios, *Phys. Rev. D.* **78**(10) (2008) 101302. doi: 10.1103/PhysRevD.78.101302.

38. X.-Y. Wang and Z.-G. Dai, Prompt TeV neutrinos from the dissipative photospheres of gamma-ray bursts, *Astrophys. J. Lett.* **691** (2009) L67–L71. doi: 10.1088/0004-637X/691/2/L67.

39. S. Gao, K. Asano, and P. Meszaros, High energy neutrinos from dissipative photospheric models of gamma ray bursts *J. Cosmol. Astropart.* **11** (2012) 58.

40. S. Gao and P. Mészáros, Multi-GeV neutrino emission from magnetized gamma-ray bursts, *Phys. Rev. D.* **85**(10) (2012) 103009. doi: 10.1103/PhysRevD.85.103009.

41. I. Bartos, A.M. Beloborodov, K. Hurley, and S. Márka, Detection prospects for GeV neutrinos from collisionally heated gamma-ray bursts with IceCube/DeepCore, *Phys. Rev. Lett.* **110**(24) (2013) 241101. doi: 10.1103/PhysRevLett.110.241101.

42. B. Zhang and P. Kumar, Model-dependent high-energy neutrino flux from gamma-ray bursts, *Phys. Rev. Lett.* **110**(12) (2013) 121101. doi: 10.1103/PhysRevLett.110.121101.

43. M.G. Aartsen, M. Ackermann, J. Adams, J.A. Aguilar, M. Ahlers, M. Ahrens, D. Altmann, T. Anderson, C. Arguelles, T.C. Arlen, *et al.*, Search for prompt neutrino emission from gamma-ray bursts with IceCube, *Astrophys. J. Lett.* **805** (2015) L5. doi: 10.1088/2041-8205/805/1/L5.

44. K. Murase, K. Kashiyama, and P. Mészáros, Subphotospheric neutrinos from gamma-ray bursts: The role of neutrons, *Phys. Rev. Lett.* **111**(13) (2013) 131102. doi: 10.1103/PhysRevLett.111.131102.

45. K. Kashiyama, K. Murase, and P. Mészáros, Neutron-proton-converter acceleration mechanism at subphotospheres of relativistic outflows, *Phys. Rev. Lett.* **111**(13) (2013) 131103. doi: 10.1103/PhysRevLett.111.131103.

46. E.J. Howell and D.M. Coward, A redshift-observation time relation for gamma-ray bursts: Evidence of a distinct subluminous population, *M.N.R.A.S.* **428** (2013) 167–181. doi: 10.1093/mnras/sts020.

47. F.J. Virgili, E.-W. Liang, and B. Zhang, Low-luminosity gamma-ray bursts as a distinct GRB population: A firmer case from multiple criteria constraints, *M.N.R.A.S.* **392** (2009) 91–103. doi: 10.1111/j.1365-2966.2008.14063.x.

48. A.M. Soderberg, S.R. Kulkarni, E. Nakar, E. Berger, P.B. Cameron, D.B. Fox, D. Frail, A. Gal-Yam, R. Sari, S.B. Cenko, M. Kasliwal, R.A. Chevalier, T. Piran, P.A. Price, B.P. Schmidt, G. Pooley, D.-S. Moon, B. E. Penprase, E. Ofek, A. Rau, N. Gehrels, J.A. Nousek, D.N. Burrows, S.E. Persson, and P.J. McCarthy, Relativistic ejecta from X-ray flash XRF 060218 and the rate of cosmic explosions, *Nature* **442** (2006) 1014–1017. doi: 10.1038/nature05087.

49. K. Murase, K. Ioka, S. Nagataki, and T. Nakamura, High-energy neutrinos and cosmic rays from low-luminosity gamma-ray bursts?, *Astrophys. J. Lett.* **651** (2006) L5–L8. doi: 10.1086/509323.

50. K. Murase, K. Ioka, S. Nagataki, and T. Nakamura, High–energy cosmic–ray nuclei from high and low luminosity gamma-ray bursts and implications for multimessenger astronomy, *Phys. Rev. D.* **78**(2) (2008) 023005. doi: 10.1103/PhysRevD.78.023005.

51. N. Gupta and B. Zhang, Neutrino spectra from low and high luminosity populations of gamma ray bursts, *Astroparticle Phys.* **27** (2007) 386–391. doi: 10.1016/j.astropartphys.2007.01.004.

52. R.-Y. Liu, X.-Y. Wang, and Z.-G. Dai, Nearby low–luminosity gamma–ray bursts as the sources of ultra-high-energy cosmic rays revisited, *M.N.R.A.S.* **418** (2011) 1382–1391. doi: 10.1111/j.1365-2966.2011.19590.x.

53. P. Mészáros and E. Waxman, TeV neutrinos from successful and choked gamma-ray bursts, *Phys. Rev. Lett.* **87**(17) (2001) 171102.

54. S. Razzaque, P. Meszaros, and E. Waxman, Tev neutrinos from core collapse supernovae and hypernovae, *Phys. Rev. Lett.* **93** (2004) 181101.

55. S. Razzaque, P. Mészáros, and E. Waxman, Neutrino tomography of gamma ray bursts and massive stellar collapses, *Phys. Rev.* **D68** (2003) 083001.

56. S. Ando and J.F. Beacom, Revealing the supernova gamma-ray burst connection with TeV neutrinos, *Phys. Rev. Lett.* **95**(6) (2005) 061103. doi: 10.1103/PhysRevLett.95.061103.

57. S. Horiuchi and S. Ando, High–energy neutrinos from reverse shocks in choked and successful relativistic jets, *Phys. Rev. D.* **77**(6) (2008) 063007. doi: 10.1103/PhysRevD.77.063007.

58. S. Horiuchi and S. Ando. Probing dark gamma-ray bursts with neutrinos. In eds. C. Balazs & F. Wang, *American Institute of Physics Conference Series*, Vol. 1178, pp. 97–103 (2009). doi: 10.1063/1.3264563.

59. K. Murase and K. Ioka, TeV-PeV neutrinos from low-power gamma-ray burst jets inside stars, *Phys. Rev. Lett.* **111**(12) (2013) 121102. doi: 10.1103/PhysRevLett.111. 121102.

60. I. Taboada. Multi-messenger observations of gamma-ray bursts. In eds. J. E. McEnery, J. L. Racusin, and N. Gehrels, *American Institute of Physics Conference Series*, Vol. 1358, pp. 365–370 (2011). doi: 10.1063/1.3621806.

61. S. Campana, V. Mangano, A.J. Blustin, P. Brown, D.N. Burrows, G. Chincarini, J.R. Cummings, G. Cusumano, M. Della Valle, D. Malesani, P. Mészáros, J.A. Nousek, M. Page, T. Sakamoto, E. Waxman, B. Zhang, Z.G. Dai, N. Gehrels, S. Immler, F.E. Marshall, K.O. Mason, A. Moretti, P.T. O'Brien, J. P. Osborne, K.L. Page, P. Romano, P.W.A. Roming, G. Tagliaferri, L.R. Cominsky, P. Giommi, O. Godet, J.A. Kennea, H. Krimm, L. Angelini, S.D. Barthelmy, P.T. Boyd, D.M. Palmer, A.A. Wells, and N.E. White, The association of GRB 060218 with a supernova and the evolution of the shock wave, *Nature* **442** (2006) 1008–1010. doi: 10.1038/nature04892.

62. A.M. Soderberg, E. Berger, K.L. Page, P. Schady, J. Parrent, D. Pooley, X.-Y. Wang, E.O. Ofek, A. Cucchiara, A. Rau, E. Waxman, J.D. Simon, D.C.-J. Bock, P.A. Milne, M.J. Page, J.C. Barentine, S.D. Barthelmy, A.P. Beardmore, M.F. Bietenholz, P. Brown, A. Burrows, D.N. Burrows, G. Byrngelson, S.B. Cenko, P. Chandra, J.R. Cummings, D.B. Fox, A. Gal-Yam, N. Gehrels, S. Immler, M. Kasliwal, A.K.H. Kong, H.A. Krimm, S.R. Kulkarni, T.J. Maccarone, P. Mészáros, E. Nakar, P.T. O'Brien, R.A. Overzier, M. de Pasquale, J. Racusin, N. Rea, and D.G. York, An extremely luminous X-ray outburst at the birth of a supernova, *Nature* **453** (2008) 469–474. doi: 10.1038/nature06997.

63. E. Waxman, P. Mészáros, and S. Campana, GRB 060218: A relativistic supernova shock breakout, *Astrophys. J.* **667** (2007) 351–357. doi: 10.1086/520715.

64. R.A. Chevalier and C. Fransson, Shock breakout emission from a type Ib/c supernova: XRT 080109/SN 2008D, *Astrophys. J. Lett.* **683** (2008) L135–L138. doi: 10.1086/591522.

65. X.-Y. Wang, S. Razzaque, P. Meszaros, and Z.-G. Dai, High-energy cosmic rays and neutrinos from semi-relativistic hypernovae, *Phys. Rev.* **D76** (2007) 083009. doi: 10. 1103/PhysRevD.76.083009.

66. B. Katz, N. Sapir, and E. Waxman, X-rays, gamma-rays and neutrinos from collisionless shocks in supernova wind breakouts, ArXiv e-prints (2011).

67. K. Kashiyama, K. Murase, S. Horiuchi, S. Gao, and P. Mészáros, High-energy neutrino and gamma-ray transients from trans-relativistic supernova shock breakouts, *Astrophys. J. Lett.* **769** (2013) L6. doi: 10.1088/2041-8205/769/1/L6.

68. M.G. Aartsen, R. Abbasi, Y. Abdou, M. Ackermann, J. Adams, J.A. Aguilar, M. Ahlers, D. Altmann, J. Auffenberg, X. Bai *et al.*, First observation of PeV-energy neutrinos with IceCube, *Phys. Rev. Lett.* **111**(2) (2013) 021103. doi: 10.1103/PhysRevLett.111.021103.

69. IceCube Collaboration, Evidence for high-energy extraterrestrial neutrinos at the IceCube detector, *Science* **342** (2013). doi: 10.1126/science.1242856.

70. M.G. Aartsen, M. Ackermann, J. Adams, J.A. Aguilar, M. Ahlers, M. Ahrens, D. Altmann, T. Anderson, C. Arguelles, T.C. Arlen *et al.*, Atmospheric and astrophysical neutrinos above 1 TeV interacting in IceCube, *Phys. Rev. D.* **91**(2) (2015) 022001. doi: 10.1103/PhysRevD.91.022001.

71. N. Senno, P. Mészáros, K. Murase, P. Baerwald, and M.J. Rees, Extragalactic star-forming galaxies with hypernovae and supernovae as high-energy neutrino and gamma-ray sources: The case of the 10 TeV neutrino data, *Astrophys. J.* **806** (2015) 24. doi: 10.1088/0004-637X/806/1/24.

72. A. Loeb and E. Waxman, The cumulative background of high energy neutrinos from starburst galaxies, *J. Cosmol. Astropart. Phys.* **5** (2006) 3. doi: 10.1088/1475-7516/2006/05/003.
73. N. Fraija, Could a multi-PeV neutrino event have as origin the internal shocks inside the GRB progenitor star? arXiv: 1508.03009 (2015).

# Chapter 2

# Active Galactic Nuclei as High-Energy Neutrino Sources

Kohta Murase

*Center for Particle and Gravitational Astrophysics; Department of Physics;*
*Department of Astronomy & Astrophysics, The Pennsylvania State University,*
*University Park, Pennsylvania 16802, USA*
*Institute for Advanced Study, Princeton, New Jersey 08540, USA*
*murase@psu.edu*

Active galactic nuclei (AGN) are believed to be promising candidates of extra-galactic cosmic-ray accelerators and sources, and associated high-energy neutrino and hadronic gamma-ray emission has been studied for many years. We review models of high-energy neutrino production in AGN and discuss their implications for the latest IceCube observation of the diffuse neutrino intensity.

## 1. Introduction

Active galactic nuclei (AGN), which are powered by accretion of mass onto supermassive black holes (SMBHs) at the center of their host galaxies and/or the rotational energy of SMBHs, are the most luminous persistent sources of electromagnetic radiation in the Universe. They have also been of interest as powerful high-energy cosmic-ray accelerators (CRs), including ultrahigh-energy cosmic rays (UHECRs). High-energy neutrinos from AGN have been discussed since the late 70s at least.[1–3] If protons are accelerated by, for example, the diffusive shock acceleration mechanism, because the optical and X-ray radiation density is rather high in the vicinity of a SMBH, the CRs may efficiently interact with the ambient photons. Early models that attempted to interpret X-ray emission with pair cascades[4] led to very large diffuse neutrino intensities.[5–7] The X-ray origin is now thought to be Comptonized disk emission[8] and the originally predicted fluxes have been largely constrained by neutrino observations themselves.[9] However, these early models motivated the searches for cosmic high-energy neutrinos with water- or ice-Cherenkov detectors.

Both observational and theoretical multi-wavelength efforts have helped us understand the physics of AGN, which also lead to different proposals for CR acceleration and associated neutrino production in AGN. In particular, radio-loud AGN and their on-axis objects "blazars" are most widely discussed as powerful non-thermal sources.[10] Our gamma-ray view of blazars has been drastically enriched by EGRET on the *Compton Gamma Ray Observatory*, *Fermi*, and various ground-based Cherenkov telescopes. They are the dominant sources in the extragalactic gamma-ray sky,[11] which has tempted us to speculate that the radio-loud AGN including blazars are powerful accelerators of protons as well as electrons.[12-17] The origin of gamma-ray emission is still under debate even in the *Fermi* era. The standard explanation is inverse-Compton emission by non-thermal electrons (leptonic scenario),[18-21] but lepto-hadronic scenarios have also been exploited to explain BL Lacertae objects (BL Lacs),[19,22,23] quasar-hosted galaxies (QHBs) including flat-spectrum radio quasars (FSRQs),[24-27] Fanaroff-Riley I (FR I) and Fanaroff-Riley II (FR II) radio galaxies.[28-30,32,33] A fraction of BL Lacs, so-called extreme BL Lacs, show very hard gamma-ray spectra, which could be explained by hadronic cascade emission induced by CRs propagating in intergalactic space.[34-38]

The IceCube's discovery of cosmic high-energy neutrinos[39-43] raises new questions about the non-thermal properties of AGN. Is the observed gamma-ray emission produced by high-energy CRs accelerated in blazars and radio galaxies? Do AGN make a dominant contribution to the observed diffuse neutrino intensity? Are they the sources of UHECRs? In this article, we discuss possibilities of neutrino production in AGN, with a focus on recent studies in light of the IceCube data. AGN, especially radio-loud AGN, have been excellent targets of multi-wavelength observations. We will see that they are also promising targets of the multi-messenger astronomy.

## 2. Models of AGN neutrino emission

It is convenient to divide AGN into radio-quiet and radio-loud AGN. In the radio-loud objects, the emission contribution from jets and bubbles or lobes is prominent especially at radio wavelengths. In the radio-quiet objects, the continuum emission comes from core regions within $\sim 1-100$ $(GM_{BH}/c^2)$, since jet and jet-related emission are weak. We first consider neutrino production in CR accelerators. The accelerators can be AGN jets of radio-loud AGN, or AGN cores of both radio-loud and radio-quiet AGN. Next, we consider the fate of CRs escaping from accelerators, and discuss neutrino production in CR reservoirs or during CR propagation.

The diffuse neutrino intensity from extragalactic AGN is given by[44]

$$\Phi_\nu = \frac{c}{4\pi H_0} \int^{z_{max}} dz \frac{1}{\sqrt{(1+z)^3 \Omega_m + \Omega_\Lambda}} \int dL_\nu \frac{dn_s}{dL_\nu}(z) \frac{L_{E'_\nu}}{E'_\nu}, \qquad (1)$$

where $dn_s/dL_\nu$ is the neutrino luminosity function of the sources (per comoving volume per luminosity) and $z_{max}$ is the maximum value of the redshift $z$ for a given source class. To make model predictions for the diffuse neutrino intensity, it is necessary to relate the calculated neutrino luminosity to the observed photon luminosity at some energy band. In addition, one needs to normalize the CR spectrum. It is ideal to calculate CR acceleration from first principles, but our present knowledge on particle acceleration is not sufficient. Phenomenologically, the neutrino flux can be normalized by the observed CR data or by the existing gamma-ray data. Alternatively, one can introduce a phenomenological parameter such as the CR loading factor ($\xi_{cr}$) to represent the efficiency of CR proton acceleration.

## 2.1. *AGN jets*

The most promising site of non-thermal AGN emission is the jet of radio-loud AGN (see Fig. 1). Their broadband emission has been studied at multi-wavelengths from radio to gamma rays. In particular, spectral energy distributions of blazars have been measured and modeled for many years. There are two main blazar sub-classes, namely BL Lacs and QHBs (mostly FSRQs). They differ mostly in their optical spectra, and FSRQs display strong broad emission lines, whereas BL Lacs are characterized by optical spectra showing at most weak emission lines or absorption features. Continuum radiation of both blazar classes (FSRQs and BL Lacs) typically consists of two humps (see the left panel of Fig. 2). The low-energy hump (peaking in the infrared to soft X-ray band) is well explained by synchrotron radiation from non-thermal electrons. The high-energy hump is conventionally attributed to inverse-Compton emission. The spectral energy distributions of high-energy-peaked BL Lac objects (HBLs) are interpreted as synchrotron and synchrotron self-Compton components. In contrast, those of low-energy-peaked BL Lac objects (LBLs) and FSRQs are generally well fit with synchrotron and external inverse-Compton components. External radiation fields are naturally provided by the accretion-disk radiation, its

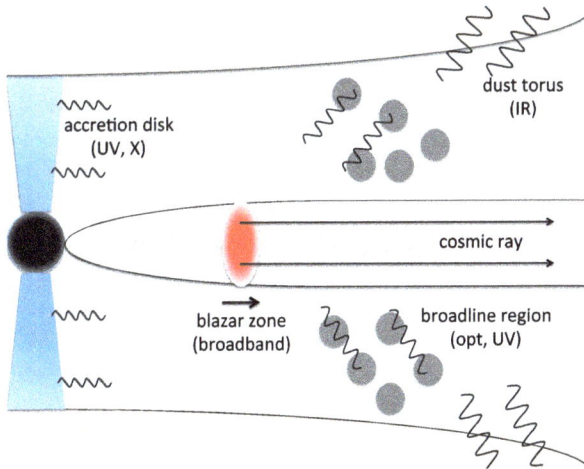

Fig. 1.   Schematic picture of photohadronic interactions by CRs in inner jets of radio-loud AGN.[44]

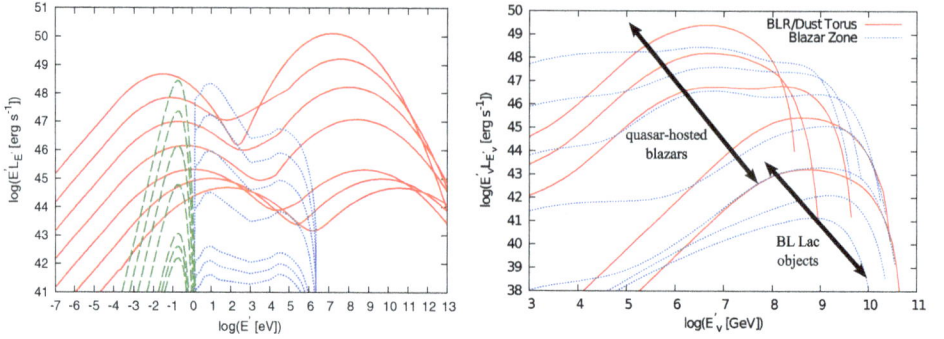

Fig. 2. *Left panel*: Differential continuum luminosity spectra of observed photons from blazars.[44] The solid, dotted, and dashed curves represent the non-thermal jet component, the accretion disk component, and the torus infrared component, respectively. The radio luminosity at 5 GHz varies as $\log(L_{5\mathrm{GHz}}) = 47, 46, 45, 44, 43, 42$, and $41$, in units of erg s$^{-1}$ from top to bottom. *Right panel*: Differential luminosity spectra of photohadronic neutrinos from blazars.[44] The muon neutrino spectrum is calculated for $s = 2.0$ and $\xi_{\mathrm{cr}} = 10$, with neutrino mixing. From top to bottom, the radio luminosity varies corresponding to the left panel.

scattered radiation from the broadline region (BLR), and infrared (IR) radiation from the dust torus surrounding a SMBH and the BLR (see Fig. 1). Typical quasars show such broad optical and UV emission lines from the BLR, and the dust torus plays a key role in the AGN unification scheme.[45]

High-energy protons may be accelerated by diffusive shock acceleration or stochastic acceleration in a jet. They interact with synchrotron photons provided by non-thermal electrons that are co-accelerated in jets.[17,46–50] The effective optical depth to photomeson production is estimated to be

$$f_{p\gamma}(E'_p) \approx \frac{2\kappa_\Delta \sigma_\Delta}{1+\beta} \frac{\Delta\bar{\varepsilon}_\Delta}{\bar{\varepsilon}_\Delta} \frac{L^s_{\mathrm{rad}}}{4\pi r \Gamma^2 c E'_s} \left(\frac{E'_p}{E'^s_p}\right)^{\beta-1}, \tag{2}$$

where $\sigma_\Delta \sim 5 \times 10^{-28}$ cm$^2$, $\kappa_\Delta \sim 0.2$, $\bar{\varepsilon}_\Delta \sim 0.3$ GeV, $\Delta\bar{\varepsilon}_\Delta \sim 0.2$ GeV, $L^s_{\mathrm{rad}}$ is the jet synchrotron luminosity at the synchrotron peak $E'_s$, $r$ is the emission radius, $\Gamma$ is the bulk Lorentz factor of jets, and $\beta$ is the photon index of target photons. Note that primes refer to quantities in the black-hole rest frame. The characteristic energy of protons that interact with target photons with $E'_s$ is given by

$$E'^b_p \approx 0.5\Gamma^2 m_p c^2 \bar{\varepsilon}_\Delta (E'_s)^{-1}. \tag{3}$$

For BL Lac objects with $L^s_{\mathrm{rad}} \sim 10^{45}$ erg/s and $E'_s \sim 10$ eV, we have

$$f_{p\gamma}(E'_p) \simeq 7.8 \times 10^{-4} L^s_{\mathrm{rad},45} \Gamma_1^{-4} \delta t'^{-1}_5 (E'_s/10 \text{ eV})^{-1} \begin{cases} (E'_\nu/E'^b_\nu)^{\beta_h-1} & (E'_p \lesssim E'^b_p) \\ (E'_\nu/E'^b_\nu)^{\beta_l-1} & (E'^b_p < E'_p) \end{cases} \tag{4}$$

where $\delta t'$ is the variability time in the black hole rest frame, and $\beta_l \sim 1.5$ and $\beta_h \sim 2.5$ are the low-energy and high-energy photon indices, respectively. When cooling of mesons and muons is negligible, the characteristic neutrino energy corresponding to $E'^b_p$ is

$$E'^b_\nu \approx 0.05 E'^b_p \simeq 80 \text{ PeV } \Gamma_1^2 (E'_s/10 \text{ eV})^{-1}. \tag{5}$$

We immediately see the following features. For a power-law CR spectrum such as $E_p^{-2}$, the resulting neutrino spectra should be hard since $f_{p\gamma}$ increases with energy. As an example, let us consider BL Lacs, where external radiation fields are not relevant. As shown in the right panel of Fig. 2, neutrino spectra of BL Lacs rise up to EeV energies, and the peak energy is much higher than $\sim 1$ PeV and the Glashow resonance energy at 6.3 PeV. Secondly, $f_{p\gamma}$ is quite sensitive to $\Gamma$. This is one of the reasons why blazar neutrino models have large uncertainties in their predictions for the normalization of the neutrino flux.

Next, we consider interactions with external photons provided by the BLR clouds and the IR dust torus. The importance of BLR photons and IR photons for the neutrino production has been studied by several authors.[24,25,44,49] For the calculation, one can use empirical relations between the BLR/torus size and accretion-disk luminosity $L_{\text{AD}}$.[51,52] Then, assuming an isotropic distribution in the black hole rest frame, the photomeson production efficiency in the BLR is estimated to be[44]

$$f_{p\gamma} \approx \hat{n}_{\text{BL}} \sigma_{p\gamma}^{\text{eff}} r_{\text{BLR}} \simeq 5.4 \times 10^{-2} f_{\text{cov},-1} L_{\text{AD},46.5}^{1/2}, \tag{6}$$

above the pion production threshold energy, where $f_{\text{cov}}$ is the covering factor and $\sigma_{p\gamma}^{\text{eff}}$ is the attenuation cross section of the photomeson production. Similarly, the photomeson production efficiency for CR protons propagating in IR radiation fields supplied by the dust torus is estimated to be[44]

$$f_{p\gamma} \simeq 0.89 \, L_{\text{AD},46.5}^{1/2} (T_{\text{IR}}/500 \text{ K})^{-1}, \tag{7}$$

above the pion production threshold energy. Importantly, $f_{p\gamma}$ does not depend on $\Gamma$ and $\delta t'$, which implies that the results on neutrino fluxes are much more insensitive to model parameters compared to the case of internal synchrotron target photon fields. The photomeson production with external radiation fields is important and should not be neglected for luminous blazars such as LBLs and QHBs, leading to spectral bumps in the PeV and EeV range (see the right panel of Fig. 2). Note that the accretion-disk emission is also important if $\tau_{\text{sc}} \gtrsim f_{\text{cov}}$.

Final results of the diffuse neutrino intensity depend on the neutrino luminosity function. It has been suggested that the spectral energy distributions of blazars evolve with luminosity, which is the so-called blazar sequence[51,53] (see the left panel of Fig. 2). In the simple one-zone leptonic model, since $f_{p\gamma}$ increases with the observed photon luminosity, photohadronic interactions with broadline and IR emission in LBLs and QHBs play an important role.[44] As a result, the neutrino spectrum is roughly expressed by

$$E'_\nu L_{E'_\nu} \approx \frac{3}{8} \min[1, f_{p\gamma}] (E'_p L_{E'_p}) \begin{cases} (E'_\nu/E'^b_\nu)^2 & \text{(for } E'_\nu \leqq E'^b_\nu) \\ (E'_\nu/E'^b_\nu)^{2-s} & \text{(for } E'^b_\nu < E'_\nu) \end{cases} \tag{8}$$

As shown in the right panel of Fig. 2, the resulting neutrino spectra are quite hard above PeV energies because of IR photons from the dust torus as well as internal synchrotron photons. One of the advantages of this simple model is that the results are not sensitive to details of the blazar sequence. This is because photohadronic interactions with external radiation fields are dominant, where $f_{p\gamma}$ is not sensitive to $\Gamma$ and $\delta t'$. Also, target photon fields have narrow distributions at UV and IR bands, so that predictions for the neutrino spectral shape are reasonably robust for a given CR spectrum. Note that Eq. (4) typically governs the neutrino spectral shape for HBLs, where external fields are not relevant. Even in such models, as long as the CR spectrum extends to sufficiently high energies, PeV–EeV neutrino detections are crucial to test the models.

In general, relating the neutrino luminosity to photon luminosity is model-dependent. For example, one can abandon the simple one-zone leptonic scenario for observed continuum spectra. Instead, one can adopt lepto-hadronic scenarios, where gamma rays are attributed to proton synchrotron radiation or $p\gamma$-induced cascade emission, although huge CR luminosities are usually required.[54,55] In the lepto-hadronic scenarios, the relationship between the neutrino and photon luminosities is different.[56,57] One of the appealing points is that the neutrino flux can be calibrated by the gamma-ray flux, and the diffuse gamma-ray intensity could be explained simultaneously with the diffuse neutrino intensity at ∼1 PeV.[57]

For comparison, predictions of various blazar neutrino models are shown in the left panel of Fig. 3. Except for an early HZ97 model, the models shown here lie in the range of the MID14 model with $\xi_{cr} = 3–50$. One sees that a hard neutrino spectrum is a generic trend of the blazar neutrino models as long as a flat CR spectrum is used. It is possible to invoke a specific case that explains only neutrino events around PeV energies,[57] but there remains a strong tension with the absence of Glashow resonance events at 6.3 PeV. (Note that electron anti-neutrinos come from $\pi^-$s produced via higher resonances and multi-pion production.) An obvious solution to reduce this tension is to introduce a spectral cutoff in the CR spectrum. This might be realized if CR acceleration is caused by stochastic acceleration rather than shock acceleration.[58] Explaining ∼100 TeV diffuse neutrinos is also possible by adopting a multi-zone emission model. One of the physically motivated models is the spine-sheath model for the AGN jet structure.[59,60]

The simple leptonic and lepto-hadronic scenarios of blazars have the problem that the predicted neutrino spectra are too hard to explain sub-PeV events and have tensions with IceCube upper limits above PeV energies. Also, models are being constrained by searches for extremely high-energy neutrinos above PeV energies.[61] In addition, since blazars are rare objects, the absence of auto- and cross-correlation leads to strong constraints,[62,63] implying that their contribution to the diffuse neutrino intensity is sub-dominant especially below PeV energies.

Note that, even if blazars are not responsible for observed diffuse neutrinos, it does not mean that they are excluded as the main sources of UHECRs.

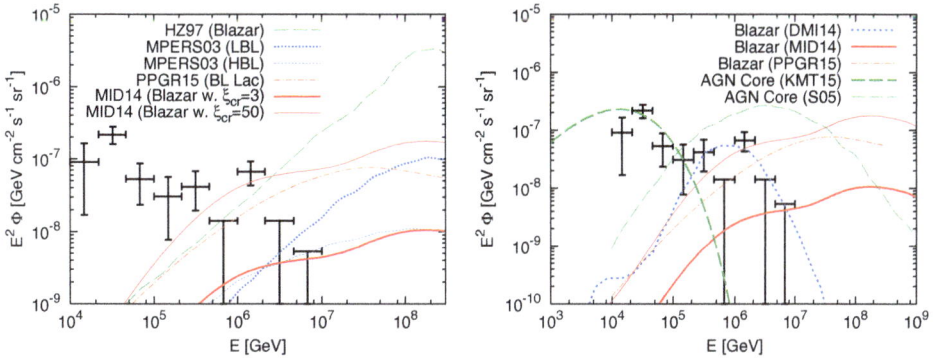

Fig. 3. *Left panel:* All-flavor diffuse neutrino intensity calculations of various AGN jet models: (HZ97) an early blazar model by Halzen and Zas,[48] (MPERS03-LBL) a LBL model by Mücke *et al.*,[17] (MPERS03-HBL) a HBL model by Mücke *et al.*,[17] (PPGR15) a BL Lacs model by Padovani *et al.*,[57] (MID14 with $\xi_{cr} = 50$) a blazar model by Murase *et al.*,[44] and (MID14 with $\xi_{cr} = 3$) a blazar model by Murase *et al.*[44] normalized with the observed UHECR luminosity density. The diffuse neutrino intensity data from the IceCube combined likelihood analysis[42] are also shown. *Right panel:* All-flavor diffuse neutrino intensity calculations for various AGN jet and core models: (DMI14) a FSRQ jet model by Dermer *et al.*[58] normalized to the IceCube data at PeV energies assuming an average redshift $\bar{z} = 2$, (MID14) a blazar jet model by Murase *et al.*[44] with $\xi_{cr} = 3$, $\xi_{cr} = 50$ based on the leptonic scenario, (PPGR15) a BL Lacs jet model by Padovani *et al.*[57] based on the lepto-hadronic scenario, (KMT15) a LL AGN core model by Kimura *et al.*[69] and (S05) a radio-quiet AGN core model by Stecker.[73,80] The diffuse neutrino intensity data from the IceCube combined likelihood analysis[42] are also shown.

Based on the leptonic scenario, Murase *et al.*[44] calculated the diffuse neutrino intensity based on the hypothesis that UHECRs are produced in inner jets of radio-loud AGN (where UHECRs can be largely isotropized in bubbles, cocoons, lobes and large scale structures[30]). The expected diffuse neutrino intensity reaches $E_\nu^2 \Phi_\nu \sim 10^{-8}$ GeV cm$^{-2}$ s$^{-1}$ sr$^{-1}$ at 100 PeV energies. Since values of $f_{p\gamma}$ for external radiation fields are more robust than those for internal radiation fields, the AGN-UHECR hypothesis can be tested in this model.

Recently, it was claimed that a major outburst of the blazar PKS B1424-418 occurred in temporal and spatial coincidence with the 2 PeV neutrino event observed in IceCube.[31] The probability for a chance coincidence is $\sim$5%, so this cannot be regarded as evidence for the blazar origin of IceCube neutrinos. Nevertheless, such the temporal and spatial coincidence can significantly reduce atmospheric backgrounds, and blazar flares are intriguing sources for high-energy neutrinos even if blazars are sub-dominant sources of the diffuse neutrino flux.

Neutrino and gamma-ray emissions from blazar jets are boosted by the relativistic beaming effect, and the corresponding high-energy emission from a single radio galaxy is expected to be weaker. Nevertheless, the diffuse flux could be comparable because the absence of the boost for off-axis observers can be compensated by the number density,[44] and nearby radio galaxies such as Centaurus A are observed at multi-wavelengths. Although the origin of TeV gamma rays is still under debate, spectral energy distributions of radio galaxies also consist of two spectral humps, suggesting an emission mechanism similar to that of blazars. Photohadronic production of high-energy neutrinos in jets of radio galaxies has been discussed in the literature.[64–66] However, as in the blazar case, predicted neutrino spectra are too hard to explain the IceCube data for a flat CR spectrum.

Note that $pp$ interactions are not important in inner jets, although they could be relevant for blazar flares in some specific setups (e.g. jet-star/cloud interactions) or AGN core emission.[67–69] Assuming high gas densities in the steady jets leads to serious energetics problems.[24] Radio-loud AGN have large scale jets at kpc scales. For such a large scale jet, observations of X-ray knots suggest column densities of $N_H \sim 10^{20} - 10^{22}$ cm$^2$, implying an effective $pp$ optical depth $f_{pp} \sim 10^{-4} N_{H,21}$ for a jet propagating with one-third the speed of light. While CRs in the jet are advected, sufficiently high-energy CRs can escape from the jet and can be confined in the ambient environment.[44,70] This possibility will be discussed later. In addition, high-energy CRs may also be accelerated at jet-cocoon boundaries, or hot spots, cocoon shocks, and radio lobes of FR II radio galaxies. Although they could even be relevant for UHECR production, the neutrino production there is usually inefficient.[71]

### 2.2. *AGN cores*

Both radio-loud AGN and radio-quiet AGN typically show X-ray emission. The cosmic X-ray background, which is much larger than the cosmic gamma-ray background, is known to be dominated by AGN, especially Seyfert galaxies. Seyferts and quasars (mostly radio-quiet AGN) show so-called blue bumps at the UV band, which are naturally explained as multi-color blackbody emission from geometrically-thin, optically-thick accretion disks.[72] On the other hand, hard X-ray emission is naturally explained as Comptonized emission[8] by hot thermal electrons with $T \sim 10^9$ K. It is believed that hot coronae are powered by the accretion disk via, for example, magnetic reconnection.

Although the hadronic interpretation of the observed X-ray emission is disfavored by, for example, the existence of a spectral cutoff in the spectrum, one may still consider possibilities of CR acceleration near the accretion disk. Assuming the standard disk photon spectrum, the photomeson production efficiency for CR protons interacting with accretion-disk photons around the maximum disk temperature $kT_{\max}$ is estimated to be[44,58]

$$f_{p\gamma} \sim 600 \, L_{\mathrm{AD},46.5}(kT_{\max}/20 \text{ eV})^{-1} r_{14.5}^{-1}. \tag{9}$$

Thus, if CRs can be accelerated, they are efficiently depleted and high-energy neutrinos should be produced. The typical accretion-disk temperature is $\sim$10 eV, so the neutrino spectrum resulting from $p\gamma$ interactions with the accretion-disk photons is expected to have a peak at PeV energies.[5] Note that $pp$ interactions are less important at high energies[5,7] since X-ray observations indicate a column density of $N_H \sim 10^{20} - 10^{24}$ cm$^2$. Since CRs are depleted in the AGN core models, the neutrino flux is often normalized by X-ray and gamma-ray observations,[73] and the earlier models have overestimated neutrino production by many orders of magnitude.[9,43]

The CR acceleration mechanism is not clear in the AGN core models. Shock dissipation may occur[3,7] but CR acceleration will be inefficient when the system is radiation-dominated. When the accretion rate is high enough, protons and electrons are thermalized via Coulomb scattering within the infall time. Stochastic acceleration is unlikely in the bulk of the accretion flow, although non-thermal proton acceleration in the corona may be possible.[74] Perhaps, electrostatic acceleration might operate, but the formation of a gap around a SMBH seems difficult except for sufficiently low-luminosity objects starved for plasma.[75] Another type of AGN core model was suggested by Kimura *et al.*[69] for low-luminosity AGN (LL AGN). Contrary to Seyferts and quasars, LLAGN do not have standard or slim disks, since their spectra show no blue bump.[76] Instead, their radiatively inefficient accretion flows (RIAFs)[77] are expected to be collisionless,[78] where particle acceleration may be possible via turbulence or magnetic reconnection.[79] If CRs are accelerated by either stochastic acceleration or magnetic reconnection or electrostatic acceleration, high-energy neutrinos can be produced via both $pp$ and $p\gamma$ interactions,[69,82] which may be responsible for the neutrino data in the 10–100 TeV range (see the right panel of Fig. 3).

AGN core models may explain the IceCube data,[69,80–82] and such hidden CR accelerators are suggested by the latest IceCube data.[83] However, all the models have large uncertainties. Both the flux normalization and maximum energy depend on model parameters and underlying assumptions. Although the luminosity functions, which are based on the observational data, are relatively well-known, the influence of the model uncertainties is stronger. The number densities of radio-quiet AGN ($n_s \sim 10^{-3}$ Mpc$^{-3}$) and low-luminosity AGN ($n_s \sim 10^{-2}$ Mpc$^{-3}$) are much higher than the number densities of blazars ($n_s \sim 10^{-7}$ Mpc$^{-3}$) and radio-loud AGN ($n_s \sim 10^{-4}$ Mpc$^{-3}$). Thus, the AGN core models are presently allowed by neutrino tomography constraints based on searches for neutrino event clustering.

## 2.3. *AGN in cosmic-ray reservoirs*

AGN may be important as CR accelerators, even if AGN themselves are not strong neutrino or gamma-ray emitters. Radio-loud AGN are the most popular CR accelerators, and high-energy CRs and possibly UHECRs may come from jets, hot spots, cocoon shocks and lobes. Radio-quiet AGN may also have weak jets, which can also supply high-energy CRs and possibly UHECRs.[84] In addition, disk-driven winds including ultra-fast outflows can serve as CR accelerators.[44,70] AGN are often located in galaxy clusters and groups, which have magnetic fields of $B \sim 0.1-1$ $\mu$G. Low-energy CRs escaping from AGN can be confined in the large scale structure containing galaxy assemblies for $\sim 1-10$ Gyr, and produce neutrinos and gamma rays.[85] In this scenario, the dominant process is $pp$ interactions with intracluster or intragroup gas. Using typical intracluster densities $\bar{n} \sim 10^{-4}$ cm$^{-3}$, with a possible enhancement factor $g \sim 1-3$, we obtain[70,86]

$$f_{pp} \simeq 1.1 \times 10^{-2} \, g\bar{n}_{-4}(t_{\text{int}}/3 \text{ Gyr}), \tag{10}$$

where $t_{\text{int}}$ is the duration of hadronic interactions. Murase *et al.*[86] suggested that CRs above the second knee may come from such galaxy assemblies, while CRs below the second knee are confined in the reservoirs and should produce neutrinos and gamma rays. The case of UHECR production by AGN in clusters and groups was also studied.[87] Interestingly, these models predicted a diffuse neutrino intensity $E_\nu^2 \Phi_\nu \sim 10^{-9}-10^{-8}$ GeV cm$^{-2}$ s$^{-1}$ sr$^{-1}$, which may explain the high-energy neutrino data, although the contribution to the diffuse gamma-ray intensity is sub-dominant.[86,88] Several authors[86,87] showed a model for a central point source, where the neutrino flux is somewhat enhanced because of the higher intracluster density in the cluster/group center. Interestingly, high-energy gamma-ray emission from the Virgo cluster center around the radio galaxy M87 can be explained by pionic gamma rays produced by interactions with the intracluster gas.[89] In reality, AGN have finite lifetimes of $\sim 1-10$ Myr, and they may not be located at the center. In the limit that CRs are injected over the whole reservoir, the neutrino spectrum is close to the injection spectrum up to the diffusion break energy, above which it becomes steeper. Note that low mass clusters and groups, which allow us to have positive redshift evolutions, are needed for the scenario to be consistent with other gamma-ray constraints,[a] including those from the gamma-ray background anisotropy and individual cluster observations.[70,90] Also, AGN are not the only CR accelerators in this scenario. Not only radio-loud AGN but also radio-quiet AGN, transients in galaxies (such as supernovae and gamma-ray bursts) can contribute to the resulting neutrino and gamma-ray intensities.

As noted above, before IceCube's discovery, Kotera *et al.*[87] obtained a required level of the diffuse neutrino intensity, assuming that radio-loud AGN are the UHECR accelerators. Recently, Giacinti *et al.*[92] attempted to explain the gamma-ray intensity as well as the observed UHECR intensity and diffuse neutrino intensities, assuming blazars without modeling the multi-messenger emission. However, as discussed above, main blazar emission itself is unlikely to be of $pp$ origin. Non-thermal emission from radio-loud AGN, including blazars, is typically variable, which is most naturally attributed to inverse-Compton or perhaps $p\gamma$-induced cascade or proton synchrotron radiation. Thus, in these scenarios, a promising possibility would be that neutrino and gamma-ray emission are mainly produced in CR reservoirs containing radio-loud AGN. In this model, gamma rays from

---

[a]Although it is argued that the early work predicted much larger diffuse fluxes,[90] actually, the early calculations for only massive clusters[86,88] also show $E_\gamma^2 \Phi_\gamma \sim 10^{-9}-10^{-8}$ GeV cm$^{-2}$ s$^{-1}$ sr$^{-1}$, i.e., the level of diffuse gamma-ray intensities is very similar.

galaxy clusters and groups contribute to the diffuse gamma-ray background significantly,[70] and they are expected to be detected soon or have their emissivity constrained.[90,91]

Host galaxies may also be regarded as CR reservoirs.[44] However, powerful jets will leave their host galaxy, whereas weak jets or disk-driven winds from an AGN lie in the galaxy. If CRs are accelerated by these outflows and escape from the AGN, they should interact with interstellar gas until they leave the galaxy. Although hadronuclear production of neutrinos in radio galaxies is expected to be typically inefficient,[69] it can be important when AGN co-exist with starburst galaxies.[44,70]

## 2.4. *AGN in intergalactic space*

CRs escaping from AGN further interact with the cosmic microwave background (CMB) and extragalactic background light (EBL). The most famous example is the cosmogenic neutrino production by UHECRs interacting with the CMB.[93] As shown in the left panel of Fig. 4,[94] cosmogenic neutrinos are expected in the EeV range, although the resulting neutrino intensity depends on the UHECR composition and redshift evolution of the sources. The matter density in intergalactic space is so small that $pp$ interactions in cosmic voids are negligible. Thus, for the production of PeV neutrinos, interactions with the EBL in the UV range are relevant, and the photomeson production efficiency is estimated to be

$$f_{p\gamma} \approx \hat{n}_{\mathrm{EBL}} \sigma^{\mathrm{eff}}_{p\gamma} d \simeq 1.9 \times 10^{-4} \, \hat{n}_{\mathrm{EBL},-4} d_{28.5}, \qquad (11)$$

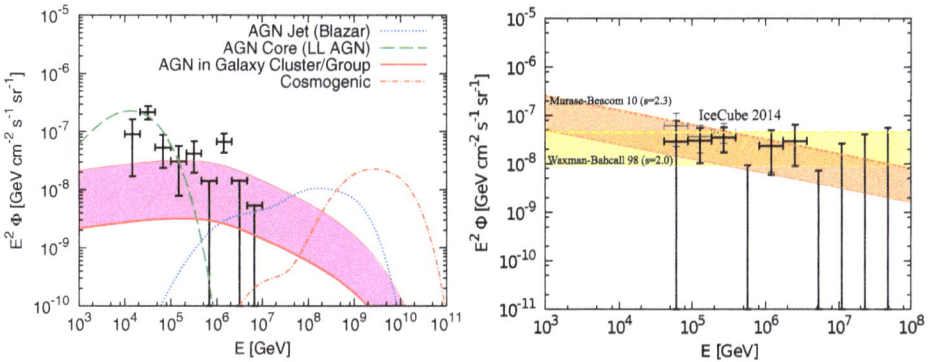

Fig. 4. *Left panel:* All-flavor diffuse neutrino intensity calculations for various AGN-related models: (AGN Jet) a blazar model by Murase *et al.*,[44] (AGN Core) a LLAGN core model by Kimura *et al.*,[69] (AGN in Galaxy Cluster/Group) a CR reservoir model by Murase *et al.*[86] but with optimistic and moderate normalization, (Cosmogenic) a cosmogenic neutrino model by Takami *et al.*[94] for the ankle-transition scenario. One should keep in mind that each model has large uncertainty in its prediction. The diffuse neutrino intensity data from the IceCube combined likelihood analysis[42] are also shown. *Right panel:* The nucleon-survival landmark by Waxman and Bahcall[102] and nucleus-survival landmark by Murase and Beacom.[105] The diffuse neutrino intensity data from the IceCube three-year high-energy starting event analysis[41] are also shown.

where $\hat{n}_{EBL} \sim 10^{-4}$ cm$^{-3}$ is the number of EBL photons and $d$ is the particle's traveling distance. Since the efficiency is tiny, for intergalactic neutrino production to be relevant, optimistic EBL models and large CR luminosity densities are required.[95] As in the blazar case, one has to optimize the CR maximum energy not to overproduce neutrino events above a few PeV energies, and the intergalactic origin of sub-PeV neutrinos is unlikely.[96,97]

## 3. Discussion and summary

Radio-loud AGN have powerful jets, which are promising CR accelerators. The spectral energy distributions and luminosity function of blazars have been measured reasonably well. For power-law CR injections, most blazar neutrino models predicted hard neutrino spectra and the peak energy is expected in the 10-100 PeV range. The absence of neutrino events at multi-PeV energies, and the lack of auto and cross correlations imply that the simple jet models of radio-loud AGN are already disfavored as the main origin of the observed diffuse neutrinos. More complicated scenarios may be necessary. Possibly, only neutrino events around ~1 PeV could be explained by blazars.[57,58] Or one can invoke multi-zone emission models and/or introduce non-power-law CR spectra such as a log-parabolic function motivated by stochastic acceleration.[58,59]

However, blazars do not have to be dominant sources of the observed neutrinos in IceCube. They may produce very-high-energy neutrinos without explaining the sub-PeV neutrinos because of their hard neutrino spectra. Indeed, some models[44] predict that, in addition to whatever produces the IceCube neutrinos, there might also be a low level of very-high-energy neutrinos from blazars that become prominent above a few PeV energies. In my personal view, searches for $10 - 100$ PeV (or higher-energy) neutrinos with IceCube, KM3Net, IceCube-Gen2,[98] ARA,[99] ARIANNA[100] and GRAND,[101] seem more interesting. Improving sensitivities in this very-high-energy energy range will allow us to constrain a significant part of the parameter space of various blazar neutrino models. In particular, their connection to UHECRs can be critically examined. Apparently, the observed diffuse neutrino intensity is compatible with the Waxman-Bahcall bound for a spectral index $s = 2.0$.[102] However, in the blazar neutrino models, a flat CR spectrum leads to a hard neutrino spectrum since $f_{p\gamma}$ increases with energy, so the simultaneous explanation of the observed UHECR and neutrino intensities is difficult.[44,103,104] Nevertheless, steeper CR spectra might help. Indeed, the observed diffuse neutrino intensity is also compatible with the nucleus-survival bound for a spectral index $s = 2.3$[105] (see the right panel of Fig. 4). Note that, for blazars to produce UHECRs, the composition is expected to be heavier at ultrahigh energies. Inner jets of FSRQs and FR II galaxies could accelerate protons to $\sim 10^{20}$ eV. However, for such luminous blazars, the IR dust-torus component can deplete UHECR protons and neutrons.[44] Since the photodisintegration cross section is higher, this is even more so the case for heavy nuclei. On the other hand, inner jets of BL Lacs and FR I galaxies are not powerful enough to accelerate protons to $\sim 10^{20}$ eV,[36] especially in the leptonic model for gamma rays. As a result, the AGN-UHECR hypothesis in the simple model predicts that $\sim 10-100$ PeV neutrinos large come from FSRQs while UHECRs come from BL Lacs and FR I galaxies.

Not only radio-loud AGN but also radio-quiet AGN can be the sources of diffuse neutrinos. For Seyferts and quasars, if CRs are accelerated in the vicinity of a SMBH, efficient $p\gamma$ interactions with UV and X-ray photons from the standard accretion disk and corona are expected. Alternatively, LL AGN associated with RIAFs have been considered as potential non-thermal particle emitters, and high-energy CRs may be accelerated by turbulence or magnetic reconnection. Although it is possible for the models to fit the IceCube data, model uncertainties are quite large at present and further theoretical and observational studies may be needed.

In contrast, CR reservoir scenarios have a strong predictive power. In this work, we discussed CR reservoir scenarios involving AGN, which are different from the starburst model.[70] Both radio-loud and radio-quiet AGN embedded in such reservoirs may contribute to the observed diffuse neutrino intensity, and a spectral break due to CR diffusion was expected before IceCube's discovery.[86,87] Although a part of the parameter space has been constrained by multi-messenger data, it is appealing that the connection between observed CRs above the second knee (that may include UHECRs) and PeV neutrinos is expected in this model. The contribution to the diffuse gamma-ray background would be sub-dominant.[70]

AGN are widely considered as promising CR accelerators and neutrino sources. However, many problems related to their non-thermal activities remain unresolved. High-energy neutrinos have provided us with a new probe of the physics of AGN, and detailed comparisons to various theoretical models have been made possible. I hope that further multi-messenger studies will help us solve some of the mysteries, especially the long-standing question whether AGN are the sources of UHECRs.

K.M. acknowledges Chuck Dermer, Tom Gaisser, Gwenael Giacinti, Shigeo Kimura, Peter Mészáros, Foteini Oikonomou, Maria Petropoulou, Shigeru Yoshida, and Fabio Zandanel for discussion and comments.

# References

1. V.S. Berezinsky, in *15th International Cosmic Ray Conference Proceedings*, Plovdiv, Bulgaria, August 13–26, 1977, pp. 84–107.
2. D. Eichler, *Astrophys. J.* **232** (1979) 106.
3. M.C. Begelman, B. Rudak, and M. Sikora, *Astrophys. J.* **362** (1990) 38.
4. D. Kazanas and D.C. Ellison, *Astrophys. J.* **304** (1986) 187.
5. F.W. Stecker, C. Done, M.H. Salamon, and P. Sommers, *Phys. Rev. Lett.* **66** (1991) 2697 [*Phys. Rev. Lett.* **69** (1992) 2738].
6. A.P. Szabo and R.J. Protheroe, *Astropart. Phys.* **2** (1994) 375, arXiv: astro-ph/9405020.

7. J. Alvarez-Muniz and P. Mészáros, *Phys. Rev. D* **70** (2004) 123001, arXiv: astro-ph/0409034.
8. F. Harrdt and L. Maraschi, *Astrophys. J.* **380** (1991) L51
9. A. Achterberg *et al.* [IceCube Collaboration], arXiv: astro-ph/0509330.
10. C.D. Dermer, arXiv:1202.2814 [astro-ph.HE].
11. Y. Inoue, arXiv:1412.3886 [astro-ph.HE].
12. P.L. Biermann and P.A. Strittmatter, *Astrophys. J.* **322** (1987) 643.
13. K. Mannheim, W.M. Krülls, and P.L. Biermann, *Astron. Astrophys.* **251** (1991) 723.
14. K. Mannheim, *Astron. Astrophys.* **269** (1993) 67.
15. F.A. Aharonian, *New Astron.* **5** (2000) 377, arXiv: astro-ph/0003159.
16. A. Mücke and R.J. Protheroe, *Astropart. Phys.* **15** (2001) 121, arXiv: astro-ph/0004052.
17. A. Mücke, R.J. Protheroe, R. Engel, J.P. Rachen, and T. Stanev, *Astropart. Phys.* **18** (2003) 593, arXiv: astro-ph/0206164.
18. G. Ghisellini, F. Tavecchio, L. Foschini, G. Ghirlanda, L. Maraschi, and A. Celotti, *Mon. Not. Roy. Astron. Soc.* **402** (2010) 497, arXiv:0909.0932 [astro-ph.CO].
19. M. Böttcher, A. Reimer, K. Sweeney, and A. Prakash, *Astrophys. J.* **768** (2013) 54, arXiv:1304.0605 [astro-ph.HE].
20. M. Cerruti, C.D. Dermer, B. Lott, C. Boisson, and A. Zech, *Astrophys. J.* **771** (2013) L4, arXiv:1305.4159 [astro-ph.HE].
21. C.D. Dermer, M. Cerruti, B. Lott, C. Boisson, and A. Zech, *Astrophys. J.* **782** (2) (2014) 82, arXiv:1304.6680 [astro-ph.HE].
22. M. Cerruti, A. Zech, C. Boisson, and S. Inoue, *Mon. Not. Roy. Astron. Soc.* **448** (2015) 910, arXiv:1411.5968 [astro-ph.HE].
23. D. Yan and L. Zhang, *Mon. Not. Roy. Astron. Soc.* **447** (2015) 2810, arXiv:1412.1544 [astro-ph.HE].
24. A.M. Atoyan and C.D. Dermer, *Astrophys. J.* **586** (2003) 79, arXiv: astro-ph/0209231.
25. C.D. Dermer, K. Murase, and H. Takami, *Astrophys. J.* **755** (2012) 147, arXiv:1203.6544 [astro-ph.HE].
26. C. Diltz, M. Böttcher, and G. Fossati, *Astrophys. J.* **802** (2) (2015) 133, arXiv:1502.03950 [astro-ph.HE].
27. M. Petropoulou and S. Dimitrakoudis, *Mon. Not. Roy. Astron. Soc.* **452** (2015) 1303, arXiv:1506.05723 [astro-ph.HE].
28. M. Kachelriess, S. Ostapchenko and R. Tomas, *New J. Phys.* **11** (2009) 065017, arXiv:0805.2608 [astro-ph].
29. A. Atoyan and C.D. Dermer, *Astrophys. J.* **687** (2008) L75, arXiv:0808.0161 [astro-ph].
30. C.D. Dermer, S. Razzaque, J.D. Finke and A. Atoyan, *New J. Phys.* **11** (2009) 065016, arXiv:0811.1160 [astro-ph].
31. M. Kadler *et al.*, arXiv:1602.02012 [astro-ph.HE].
32. S. Sahu, B. Zhang and N. Fraija, *Phys. Rev. D* **85** (2012) 043012, arXiv:1201.4191 [astro-ph.HE].
33. Y. Fujita, S.S. Kimura, and K. Murase, *Phys. Rev. D* **92** (2) (2015) 023001, arXiv:1506.05461 [astro-ph.HE].
34. W. Essey and A. Kusenko, *Astropart. Phys.* **33** (2010) 81, arXiv:0905.1162 [astro-ph.HE].
35. W. Essey, O. Kalashev, A. Kusenko, and J.F. Beacom, *Astrophys. J.* **731** (2011) 51, arXiv:1011.6340 [astro-ph.HE].

36. K. Murase, C.D. Dermer, H. Takami and G. Migliori, *Astrophys. J.* **749** (2012) 63, arXiv:1107.5576 [astro-ph.HE].
37. H. Takami, K. Murase and C.D. Dermer, *Astrophys. J.* **771** (2013) L32, arXiv:1305.2138 [astro-ph.HE].
38. D. Yan, O. Kalashev, L. Zhang, and S. Zhang, *Mon. Not. Roy. Astron. Soc.* **449** (1) (2015) 1018, arXiv:1412.4894 [astro-ph.HE].
39. M.G. Aartsen *et al.* [IceCube Collaboration], *Phys. Rev. Lett.* **111** (2013) 021103, arXiv:1304.5356 [astro-ph.HE].
40. M.G. Aartsen *et al.* [IceCube Collaboration], *Science* **342** (2013) 1242856, arXiv:1311.5238 [astro-ph.HE].
41. M.G. Aartsen *et al.* [IceCube Collaboration], *Phys. Rev. Lett.* **113** (2014) 101101, arXiv:1405.5303 [astro-ph.HE].
42. M.G. Aartsen *et al.* [IceCube Collaboration], *Astrophys. J.* **809** (1) (2015) 98, arXiv:1507.03991 [astro-ph.HE].
43. M.G. Aartsen *et al.* [IceCube Collaboration], *Phys. Rev. Lett.* **115** (8) (2015) 081102, arXiv:1507.04005 [astro-ph.HE].
44. K. Murase, Y. Inoue and C.D. Dermer, *Phys. Rev. D* **90** (2) (2014) 023007, arXiv:1403.4089 [astro-ph.HE].
45. R. Antonucci, *Ann. Rev. Astron. Astrophys.* **31** (1993) 473.
46. K. Mannheim, T. Stanev, and P.L. Biermann, *Astron. Astrophys.* **260** (1992) L1.
47. K. Mannheim, *Astropart. Phys.* **3** (1995) 295.
48. F. Halzen and E. Zas, *Astrophys. J.* **488** (1997) 669, arXiv: astro-ph/9702193.
49. A. Atoyan and C.D. Dermer, *Phys. Rev. Lett.* **87** (2001) 221102, arXiv: astro-ph/0108053.
50. S. Dimitrakoudis, M. Petropoulou, and A. Mastichiadis, *Astropart. Phys.* **54** (2014) 61, arXiv:1310.7923 [astro-ph.HE].
51. G. Ghisellini and F. Tavecchio, *Mon. Not. Roy. Astron. Soc.* **387** (2008) 1669, arXiv:0802.1918 [astro-ph].
52. M. Kishimoto, S.F. Hoenig, R. Antonucci, F. Millour, K.R.W. Tristram, and G. Weigelt, *Astron. Astrophys.* **536** (2011) A78, arXiv:1110.4290 [astro-ph.CO].
53. G. Fossati, L. Maraschi, A. Celotti, A. Comastri, and G. Ghisellini, *Mon. Not. Roy. Astron. Soc.* **299** (1998) 433, arXiv: astro-ph/9804103.
54. M. Sikora, L. Stawarz, R. Moderski, K. Nalewajko, and G. Madejski, *Astrophys. J.* **704** (2009) 38, arXiv:0904.1414 [astro-ph.CO].
55. A.A. Zdziarski and M. Böttcher, *Mon. Not. Roy. Astron. Soc.* **450** (1) (2015) L21. arXiv:1501.06124 [astro-ph.HE].
56. M. Petropoulou, S. Dimitrakoudis, P. Padovani, A. Mastichiadis, and E. Resconi, *Mon. Not. Roy. Astron. Soc.* **448** (3) (2015) 2412, arXiv:1501.07115 [astro-ph.HE].
57. P. Padovani, M. Petropoulou, P. Giommi, and E. Resconi, *Mon. Not. Roy. Astron. Soc.* **452** (2) (2015) 1877, arXiv:1506.09135 [astro-ph.HE].
58. C.D. Dermer, K. Murase and Y. Inoue, *JHEAp* **3–4** (2014) 29, arXiv:1406.2633 [astro-ph.HE].
59. F. Tavecchio, G. Ghisellini, and D. Guetta, *Astrophys. J.* **793** (2014) L18, arXiv:1407.0907 [astro-ph.HE].
60. F. Tavecchio and G. Ghisellini, *Mon. Not. Roy. Astron. Soc.* **451** (2015) 1502, arXiv:1411.2783 [astro-ph.HE].
61. M.G. Aartsen *et al.* [IceCube Collaboration], arXiv:1510.05223 [astro-ph.HE].
62. T. Glüsenkamp [IceCube Collaboration], arXiv:1502.03104 [astro-ph.HE].
63. B. Wang and Z. Li, *Sci. China Phys. Mech. Astron.* **59** (1) (2016) 619502, arXiv:1505.04418 [astro-ph.HE].

64. A. Cuoco and S. Hannestad, *Phys.Rev. D* **78** (2008) 023007, arXiv:0712.1830 [astro-ph].
65. H.B. J. Koers and P. Tinyakov, *Phys. Rev. D* **78** (2008) 083009, arXiv:0802.2403 [astro-ph].
66. I.B. Jacobsen, K. Wu, A.Y.L. On, and C.J. Saxton, *Mon. Not. Roy. Astron. Soc.* **451** (2015) 3649, arXiv:1506.05916 [astro-ph.HE].
67. L. Nellen, K. Mannheim, and P.L. Biermann, Phys. Rev. D **47** (1993) 5270, arXiv: hep-ph/9211257.
68. J. Becker Tjus, B. Eichmann, F. Halzen, A. Kheirandish and S.M. Saba, *Phys. Rev. D* **89** (12) (2014) 123005, arXiv:1406.0506 [astro-ph.HE].
69. S.S. Kimura, K. Murase and K. Toma, *Astrophys. J.* **806** (2015) 159, arXiv:1411.3588 [astro-ph.HE].
70. K. Murase, M. Ahlers and B.C. Lacki, *Phys. Rev. D* **88** (12) (2013) 121301, arXiv:1306.3417 [astro-ph.HE].
71. N. Fraija, *Astrophys. J.* **783** (2014) 44, arXiv:1312.6944 [astro-ph.HE].
72. N.I. Shakura and R.A. Sunyaev, *Astron. Astrophys.* **24** (1973) 337.
73. F.W. Stecker, *Phys. Rev. D* **72** (2005) 107301, arXiv:astro-ph/0510537.
74. C.D. Dermer, J.A. Miller and H. Li, *Astrophys. J.* **456** (1996) 106, arXiv: astro-ph/9508069.
75. J. Aleksic *et al.* [MAGIC Collaboration], *Science* **346** (2014) 1080, arXiv:1412.4936 [astro-ph.HE].
76. L.C. Ho, *Ann. Rev. Astron. Astrophys.* **46** (2008) 475, arXiv:0803.2268 [astro-ph].
77. R. Narayan and I.S. Yi, *Astrophys. J.* **428** (1994) L13, arXiv: astro-ph/9403052.
78. F. Takahara and M. Kusunose, *Prog. Theor. Phys.* **73** (1985) 1390.
79. F. Yuan, E. Quataert and R. Narayan, *Astrophys. J.* **598** (2003) 301, arXiv: astro-ph/0304125.
80. F.W. Stecker, *Phys. Rev. D* **88** (4) (2013) 047301, arXiv:1305.7404 [astro-ph.HE].
81. O. Kalashev, D. Semikoz and I. Tkachev, *J. Exp. Theor. Phys.* **120** (3) (2015) 541.
82. B. Khiali and E.M. de Gouveia Dal Pino, *Mon. Not. Roy. Astron. Soc.* **455** (1) (2016) 838, arXiv:1506.01063 [astro-ph.HE].
83. K. Murase, D. Guetta, and M. Ahlers, *Phys. Rev. Lett.* **116** (7) (2016) 071101, arXiv:1509.00805 [astro-ph.HE].
84. A. Pe'er, K. Murase, and P. Mészáros, *Phys. Rev. D* **80** (2009) 123018, arXiv:0911.1776 [astro-ph.HE].
85. V.S. Berezinsky, P. Blasi and V.S. Ptuskin, *Astrophys. J.* **487** (1997) 529, arXiv: astro-ph/9609048.
86. K. Murase, S. Inoue and S. Nagataki, *Astrophys. J.* **689** (2008) L105, arXiv:0805.0104 [astro-ph].
87. K. Kotera, D. Allard, K. Murase, J. Aoi, Y. Dubois, T. Pierog, and S. Nagataki, *Astrophys. J.* **707** (2009) 370, arXiv:0907.2433 [astro-ph.HE].
88. K. Murase, S. Inoue and K. Asano, *Int. J. Mod. Phys. D* **18** (2009) 1609.
89. C. Pfrommer, *Astrophys. J.* **779** (2013) 10, arXiv:1303.5443 [astro-ph.CO].
90. F. Zandanel, I. Tamborra, S. Gabici, and S. Ando, *Astron. Astrophys.* **578** (2015) A32, arXiv:1410.8697 [astro-ph.HE].
91. K. Murase and J.F. Beacom, JCAP **1302** (2013) 028, arXiv:1209.0225 [astro-ph.HE].
92. G. Giacinti, M. Kachelrie, O. Kalashev, A. Neronov, and D.V. Semikoz, *Phys. Rev. D* **92** (8) (2015) 083016, arXiv:1507.07534 [astro-ph.HE].
93. V.S. Berezinsky and G.T. Zatsepin, *Phys. Lett. B* **28** (1969) 423.
94. H. Takami, K. Murase, S. Nagataki, and K. Sato, *Astropart. Phys.* **31** (2009) 201, arXiv:0704.0979 [astro-ph].

95. O.E. Kalashev, A. Kusenko, and W. Essey, *Phys. Rev. Lett.* **111** (4) (2013) 041103, arXiv:1303.0300 [astro-ph.HE].

96. E. Roulet, G. Sigl, A. van Vliet, and S. Mollerach, *JCAP* **1301** (2013) 028, arXiv:1209.4033 [astro-ph.HE].

97. R. Laha, J.F. Beacom, B. Dasgupta, S. Horiuchi, and K. Murase, *Phys. Rev. D* **88** (2013) 043009, arXiv:1306.2309 [astro-ph.HE].

98. M.G. Aartsen *et al.*, [IceCube-Gen2 Collaboration], arXiv:1412.5106 [astro-ph.HE].

99. P. Allison *et al.*, *Astropart. Phys.* **35** (2012) 457, arXiv:1105.2854 [astro-ph.IM].

100. S.W. Barwick, *J. Phys. Conf. Ser.* **60** (2007) 276, arXiv:astro-ph/0610631.

101. O. Martineau-Huynh *et al.*, [GRAND Collaboration], *EPJ Web Conf.* **116** (2016) 03005, arXiv:1508.01919 [astro-ph.HE].

102. E. Waxman and J.N. Bahcall, *Phys. Rev. D* **59** (1999) 023002, arXiv:hep-ph/9807282.

103. S. Yoshida and H. Takami, *Phys. Rev. D* **90** (12) (2014) 123012, arXiv:1409.2950 [astro-ph.HE].

104. M.D. Kistler, T. Stanev, and H. Yüksel, *Phys. Rev. D* **90** (12) (2014) 123006, arXiv:1301.1703 [astro-ph.HE].

105. K. Murase and J.F. Beacom, *Phys. Rev. D* **81** (2010) 123001, arXiv:1003.4959 [astro-ph.HE].

# Chapter 3

# The Origin of IceCube's Neutrinos: Cosmic Ray Accelerators Embedded in Star Forming Calorimeters

E. Waxman

*Particle Physics and Astrophysics Department,*
*Weizmann Inst. of Science, Rehovot 76100, Israel*

The IceCube collaboration reports a detection of extra-terrestrial neutrinos. The isotropy and flavor content of the signal, and the coincidence, within current uncertainties, of the 50 TeV to 2 PeV flux and the spectrum with the Waxman-Bahcall bound, suggest a cosmological origin of the neutrinos, related to the sources of ultra-high energy, $>10^{10}$ GeV, cosmic-rays (UHECR). The most natural explanation of the UHECR and neutrino signals is that both are produced by the same population of cosmological sources, producing CRs (likely protons) at a similar rate, $E^2 d\dot{n}/dE \propto E^0$, over the [1 PeV, $10^{11}$ GeV] energy range, and residing in "calorimetric" environments, like galaxies with high star formation rates, in which $E/Z < 100$ PeV CRs lose much of their energy to pion production.

A tenfold increase in the effective mass of the detector at $\gtrsim$100 TeV is required in order to significantly improve the accuracy of current measurements, to enable the detection of a few bright nearby starburst "calorimeters", and to open the possibility of identifying the CR sources embedded within the calorimeters, by associating neutrinos with photons accompanying transient events responsible for their generation. Source identification and a large neutrino sample may enable one to use astrophysical neutrinos to constrain new physics models.

## 1. Introduction

The sources and acceleration mechanism of cosmic-rays of different energies have not been reliably identified despite many decades of research.[1,2] Particularly challenging to models are the observations of UHECRs, since most models cannot account for the highest observed particle energies.[2,3] One of the main goals of the construction of high energy neutrino telescopes is to resolve the open questions associated with these long standing puzzles.[3,4]

Assuming that UHECRs are charged nuclei accelerated electromagnetically to high energy in astrophysical objects, some fraction of their energy is expected to be converted to high energy neutrinos through the decay of charged pions produced by the interaction of cosmic-ray protons/nuclei with ambient gas and radiation.

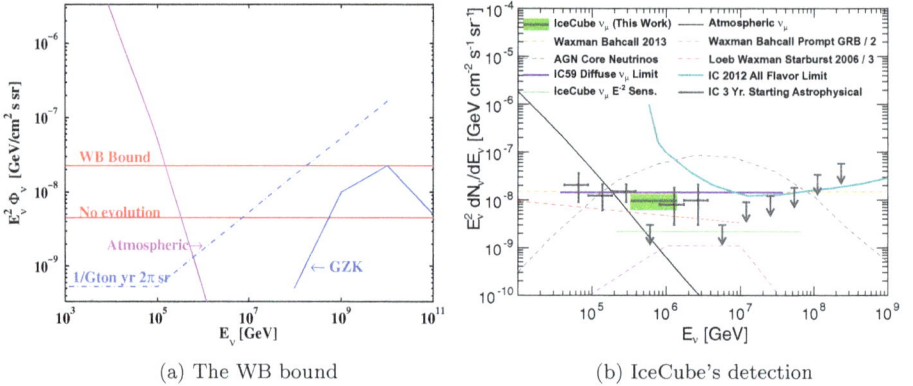

|                          |                          |
| :----------------------: | :----------------------: |
| (a) The WB bound         | (b) IceCube's detection  |

Fig. 1.   (a) The upper bound imposed by UHECR observations on the extra-galactic (all flavor) high energy neutrino intensity (lower-curve: no evolution of the energy production rate, upper curve: assuming evolution following the star formation rate), Eq. (2), compared with the atmospheric neutrino background. The curve labelled "GZK" shows the neutrino intensity expected from UHECR proton interactions with microwave background photons.[25] The dash-dotted blue line shows the muon neutrino intensity that would produce one neutrino event per year in a detector with an effective mass of 1 Gton. (b) The extra-terrestrial neutrino signal (single flavor, assuming a $\nu_e : \nu_\mu : \nu_\tau = 1 : 1 : 1$ flavor ratio) detected by IceCube in the energy range of $\sim 50$ TeV to $\sim 2$ PeV (points with error bars: "starting events" analysis:[7] green shaded area muon analysis;[8] adopted from Ref. 8). Within the current relatively large uncertainties, the detected extra-terrestrial signal coincides with the WB bound (Eq. (2)). Theoretical model predictions of the emission from particular astrophysical sources (starbursts,[26] GRBs,[27] AGNs[28]) are also shown.

A detection of this neutrino signal may enable the identification of the sources and will provide qualitative new constraints on accelerator models. The upper bound derived by Waxman & Bahcall (WB) on the neutrino intensity produced by the CR sources[5,6] implies that a gigaton neutrino telescope is required to detect the expected flux in the energy range of $\sim 1$ TeV to $\sim 1$ PeV, and that a much larger effective mass is required at higher energy (see Fig. 1(a)).

In this chapter we explain the reasoning leading to the conclusion, that the extra-terrestrial flux of neutrinos detected by the gigaton IceCube detector[7,8] is produced by UHECR sources embedded in "calorimetric" environments, characteristic of the conditions in galaxies with high star formation rate. The derivation of the WB bound is described in Sec. 2. The implications of IceCube's detection and the constraints it imposes on the sources of high energy neutrinos and CRs are described in Sec. 3. The main conclusions, the main open questions and the prospects for progress in the study of CR sources using high energy neutrinos of astrophysical origin are described in Sec. 4.

## 2.  The Waxman–Bahcall bound

### 2.1.  *UHECR composition, production rate and spectrum*

The composition of UHECRs is controversial, with air-shower data from the Fly's Eye, HiRes and Telescope Array observatories[9–11] suggesting a proton-dominated composition and data from the Pierre Auger Observatory[12] suggesting a transition to heavier elements above $10^{10}$ GeV. Due to this discrepancy, and due to the experimental and theoretical uncertainties in the relevant high energy particle interaction cross sections used for modeling the shape of the air showers, it is impossible to draw a definite conclusion regarding composition based on air-shower data at this time (the anisotropy signal provides an indication for a proton-dominated composition,[13] but is so far detected with only a $\sim 2\sigma$ confidence level[2]).

The observed flux and spectrum of $E > 10^{10.2}$ GeV CRs is consistent with a cosmological distribution of cosmic-ray sources, producing protons at a rate of[14–16]

$$Q_{\mathrm{UHE}} \equiv \left(E_p^2 d\dot{n}_p/dE_p\right)_{z=0} = 0.5 \pm 0.15 \times 10^{44} \text{ erg/Mpc}^3\text{yr}. \tag{1}$$

The energy density of CRs (at different energies) is determined by the (energy dependent) CR production rate and energy loss time. The energy loss of protons is dominated at high energy by the production of pions in interaction with cosmic microwave background (CMB) photons.[17,18] The rate estimate of Eq. (1) is based on the direct measurement of UHECRs and on the well understood physics of proton-CMB interaction. It is therefore accurate (to $\sim 30\%$) as long as the composition is dominated by protons. Observations determine only the local, $z = 0$, proton production rate and spectrum, since the energy loss time of protons is much smaller than the Hubble time (e.g., $\sim 3 \times 10^8$ yr corresponding to a propagation distance of $\sim 100$ Mpc at $10^{11}$ GeV).

The observed UHECR spectrum is consistent with a "flat" proton generation spectrum, $d \log \dot{n}/d \log E \approx -2$ with equal energy produced per logarithmic CR energy interval, modified by interaction with CMB photons. This supports a proton-dominated composition, since a flat generation spectrum is observed in a wide range of astrophysical environments (e.g. CR protons in the galaxy,[19,20] electrons in supernova remnants[19,20] and in $\gamma$-ray bursts[21]), and is a robust prediction of the best understood and most widely accepted model for particle acceleration in astrophysical objects — Fermi acceleration in collisionless shocks[19,22,23] (although a first principles understanding of the process is not yet available).

If the composition is dominated by heavier nuclei up to iron (e.g. O, Si, Fe), the inferred energy generation rate at $10^{10.5}$ GeV would change by a factor of only a few. This is due to the fact that the energy loss distance of protons is not very different from that of heavy nuclei (due to photo-disintegration interactions with the infra-red background).[24] However, the different energy dependence of the energy loss distances of heavy nuclei and protons, implies that for a heavy nuclei composition, a generation spectrum different from $E^2 d\dot{n}/dE \propto E^0$ or an ad-hoc energy dependent composition would be required to fit the observed spectrum.[24] We consider this as further evidence supporting the proton-dominated flux hypothesis, which we adopt in what follows. We return to the possibility of heavy nuclei domination in Sec. 3.2.

## 2.2. *The neutrino intensity limit*

The energy production rate, shown in Eq. (1), sets an upper bound to the neutrino intensity produced by sources, which are optically thin for high-energy nucleons in $p\gamma$ and $pp(n)$ interactions. For sources of this type, the energy generation rate of neutrinos can not exceed the energy generation rate implied by assuming that all the energy injected as high-energy protons (Eq. (1)) is converted to pions (via $p\gamma$ and $pp(n)$ interactions). The resulting all-flavor upper bound is[5,6]

$$E_\nu^2 \Phi_{\text{WB, all flavor}} = 3.4 \times 10^{-8} \frac{\xi_z}{3} \left[ \frac{(E_p^2 d\dot{n}_p/dE_p)_{z=0}}{0.5 \times 10^{44} \text{erg/Mpc}^3 \text{yr}} \right] \text{GeV/cm}^2 \text{s sr}, \quad (2)$$

where $\xi_z$ is (a dimensionless parameter) of order unity, which depends on the redshift evolution of $E_p^2 d\dot{n}_p/dE_p$. The value $\xi_z = 3$ is obtained for rapid redshift evolution, $\Phi(z) = (1+z)^3$ up to $z = 2$ and constant at higher $z$, corresponding approximately to that of the star-formation rate or AGN luminosity density evolution ($\xi_z = 0.6$ for no evolution). The numerical value (3.4) given in Eq. (1) is obtained for equal production of charged and neutral pions, as would be the case for $p\gamma$ interactions dominated by the $\Delta$ resonance.[5] For $p\gamma$ interactions at higher energy, or $pp(n)$ interactions, the charged to neutral pion ratio may be closer to 2:1, increasing the bound flux by $\sim30\%$.

Figure 1(a) illustrates that a gigaton neutrino telescope is required to detect the expected flux in the energy range of $\sim1$ TeV to $\sim1$ PeV, and that a much larger effective mass is required at higher energy. For a proton-dominated UHECR flux, the WB bound is expected to be saturated at $\sim10^{10}$ GeV, since most of the UHECR protons that have been produced over a Hubble time would have lost all their energy to pion production in interactions with the CMB photons.[25]

## 3. The origin of IceCube's neutrinos

### 3.1. *The extra-terrestrial component of the neutrino flux*

The IceCube collaboration reported[7] a detection of 37 neutrinos in the energy range of $\sim50$ TeV to $\sim2$ PeV, which constitutes a $5.7\sigma$ excess above the expected atmospheric neutrino and muon backgrounds (the uncertain contribution of atmospheric neutrinos from charmed meson decay is constrained by the angular distribution of the detected events). The excess neutrino spectrum is consistent with a "flat", $E_\nu^2 dn_\nu/dE_\nu \propto E_\nu^0$, spectrum, its angular distribution is consistent with isotropy, and its flavor content is consistent with $\nu_e : \nu_\mu : \nu_\tau = 1 : 1 : 1$.[29] Assuming a flat spectrum, the best fit normalization of the intensity is $E_\nu^2 \Phi_\nu = (2.85 \pm 0.9) \times 10^{-8}$ GeV/cm$^2$s sr. Assuming that this intensity extends to high energy, 3 events should have been detected on average above 2 PeV. The absence of such events suggests a suppression of the flux above 2 PeV, or a softer than $E_\nu^2 dn_\nu/dE_\nu \propto E_\nu^0$ spectrum. Fitting a power law excess, $dn_\nu/dE_\nu \propto E_\nu^{-\alpha}$ extending beyond 2 PeV, to the data yields $\alpha = 2.3 \pm 0.3$ (90% confidence).

The results of Ref. 7 were obtained by an analysis limited to events for which the neutrino interaction occurs within the detector ("starting events"). A recent analysis[8] searching for high energy neutrino induced muon events, not limited to neutrino interactions within the detector, revealed a muon-neutrino flux which constitutes a $3.7\sigma$ excess above the expected atmospheric neutrino and muon backgrounds in the energy range of $\sim$300 TeV to $\sim$1 PeV, with flux and spectrum consistent with those of the "starting events" analysis. The results of the "starting events" and muon analyses described above are shown graphically in Fig. 1(b).

Extending the analysis to lower energy, $\sim$10 TeV, where an astrophysical signal would be strongly dominated by the atmospheric flux, an excess of events is found at $\sim$30 TeV[30] above an extension of the flat, $E_\nu^2 dn/dE_\nu \propto E_\nu^0$, > 50 TeV spectrum. This may indicate a new low energy component or a steeper ($\alpha \approx 2.5$) spectrum across the entire observed range. Robust conclusions cannot yet be drawn, since the significance of the excess is not high, $2\sigma$,[31] and it is sensitive to the choice of energy bins (e.g. Fig. 12 of Ref. [30]). The possible low energy excess does not affect the analysis presented here, which is focused on the higher energy, >50 TeV, neutrinos.

A $\nu_e : \nu_\mu : \nu_\tau = 1 : 1 : 1$ flavor ratio is consistent with that expected for pion decay in cosmologically distant sources, for which oscillations modify the original $1 : 2 : 0$ ratio to a $1 : 1 : 1$ ratio.[32] However, with the current limited statistics, the data are consistent with any initial (i.e. at the source) flavor ratio.[29]

## 3.2.  *UHECR sources in "cosmic calorimeters"*

The extra-terrestrial neutrino flux is unlikely to be dominated by (yet unknown) galactic sources, which, unlike the observed signal, are expected to be strongly concentrated along the galactic disk. This, and the coincidence with the WB bound, suggests an extra-galactic origin. IceCube's neutrinos are therefore most likely emitted by the decay of charged pions produced in interactions of high energy CRs with ambient (low energy) photons or protons in extra-galactic sources.

Let us consider first the case where the parent CRs are protons. The fraction of the parent proton energy carried by each neutrino is approximately 1/20 (for production by interaction with either photons or protons). Since the neutrino flux coincides with the bound given in Eq. (2) over the $\sim$50 TeV to $\sim$2 PeV energy range, a lower limit of $E_p^2 d\dot{n}_p/dE_p \geq Q_{\mathrm{UHE}}$ is implied on the local, $z = 0$, proton production rate in the energy range of $\sim$1 PeV to $\sim$50 PeV. Two distinct scenarios may be considered. The observed neutrinos may be produced by sources accelerating protons at a rate $E_p^2 d\dot{n}_p/dE_p \sim Q_{\mathrm{UHE}}$, provided that protons of energy $E_p < 50$ PeV lose most of their energy to pion production either within the sources or at the sources' environment. Alternatively, the observed neutrinos may be produced by sources accelerating protons at a rate $(E_p^2 d\dot{n}_p/dE_p) \gg Q_{\mathrm{UHE}}$, provided that protons lose only a small fraction, $f(E) \ll 1$, of their energy to pion production. In the latter case, the small (and likely energy dependent) energy loss fraction $f(E)$ should compensate the large energy production rate, $(E_p^2 d\dot{n}_p/dE_p)/Q_{\text{UHE}} \gg 1$, to reproduce the observed flux and spectrum over two decades of $\nu$ energy, and the coincidence of the observed neutrino flux and spectrum with the WB bound would be an accident.

The simpler explanation, which we consider to be more likely, is that both the neutrinos and the UHECRs are produced by the same population of cosmological sources, producing protons with a flat spectrum, $d\log\dot{n}/d\log E \approx -2$ as observed in a wide range of astrophysics accelerators and as expected theoretically for electromagnetic acceleration in collisionless shocks (see Sec. 2.1), over the [1 PeV, $10^{11}$ GeV] energy range, and residing in "calorimetric" environments, in which protons of energy $E_p < 50$ PeV lose much of their energy to pion production.

Let us consider next the case where the parent CRs are heavy nuclei of atomic mass $A$. The fraction of the parent nucleus energy carried by each neutrino emitted by the decay of charged pions produced in inelastic collisions with ambient protons is approximately $1/20A$, while interactions with ambient photons may lead to photo-disintegration of the nuclei, roughly preserving $E/A$. In this case therefore, a lower limit of $E_A^2 d\dot{n}_A/dE_A \geq Q_{\text{UHE}}$ is implied on the local, $z = 0$, production rate of nuclei in the energy range of $\approx 1A$ PeV to $\approx 50A$ PeV. As mentioned in Sec. 2.1, if the UHECR flux is dominated by heavy nuclei, the energy production rate required to account for the observed UHECR flux would differ from that given by Eq. (1) by a factor of a few at $10^{10.5}$ GeV. Thus, the observed neutrino flux and spectrum, and the CR flux at $10^{10.5}$ eV could be explained by a single population of sources producing heavy nuclei with a flat spectrum, $d\log\dot{n}/d\log E \approx -2$ over the [1$A$ PeV, $10^{11}$ GeV] energy range, and residing in "calorimetric" environments, in which nuclei of energy $E/A \approx E/2Z < 50$ PeV lose much of their energy to pion production.

Unlike the case of protons, if the UHECR flux is dominated by heavy nuclei, the $>10^{10}$ GeV spectrum cannot be simply explained by an extension of a flat, $d\log\dot{n}/d\log E \approx -2$, spectrum to high energy (see Sec. 2.1). We consider this an additional evidence supporting a proton-dominated UHECR flux.

For all production channels, $p(A) - p(\gamma)$, similar flavor content (1 : 1 : 1) and particle/anti-particle content ($I_\nu = I_{\bar{\nu}}$) are expected, except for $p\gamma$ interactions dominated by the $\Delta$ resonance (which may be obtained in environments with soft, $d\log n_\gamma/d\log E_\gamma < -1$, photon spectra), where an excess of particles over anti-particles is expected.

Finally, we note that the absence of neutrino detection above a few PeV, which suggests a suppression of the neutrino flux above this energy, may be due to efficient escape of $E/Z > 100$ PeV CRs from the environments in which they produce the pions (as was predicted to be the case for sources residing in starburst galaxies[26]), and need not imply a cutoff in the CR production spectrum (see Sec. 3.3.1).

## 3.3. *Star forming galaxies*

### 3.3.1. *Starburst calorimeters*

Starburst galaxies have been predicted[26] to act as cosmic-calorimeters, producing an extra-galactic neutrino background comparable to the WB bound at energies $E < 0.5$ PeV, and possibly extending to higher energy (see Fig. 1(b)). Starbursts may be defined as galaxies with specific star formation rate (sSFR), i.e. SFR per galaxy stellar mass, which is much higher than the average sSFR of galaxies at a similar redshift.[33] Starbursts in the local universe are characterized by disks of typical radii $\ell$ of several hundred parsecs, with column densities $\Sigma_g > 0.1\,\mathrm{g/cm}^2$ [34–36] and magnetic fields $B \sim 1$ mG ($B \propto \Sigma_g$),[37] which are much larger than those of "normal" spiral galaxies ($\Sigma_g \approx 0.003\mathrm{g/cm}^2$, $B \sim 5\,\mu$G in the Milky Way). The large disk densities imply that the energy loss time of CR protons, due to inelastic *pp* collisions, is much shorter in starburst galaxies than in normal spirals, and the enhanced magnetic field implies that the confinement time of the protons is expected to be larger than in normal spirals. This in turn implies that, unlike normal spirals, starburst galaxies may act as proton calorimeters.

Starburst galaxies have long been argued[34] to act as calorimeters for few GeV protons, based on the FIR-radio correlation. The recent detection of GeV and TeV emission from the nearby starburst galaxies M82 and NGC253 indicate that these galaxies are calorimetric for protons of energy exceeding 10 TeV.[38] The theoretical arguments given in Ref. 26, and reproduced in a more general way below, suggest that calorimetry holds to energies exceeding 10 PeV.

Protons will lose all their energy to pion production provided that the energy loss time is shorter than both the starburst lifetime and the magnetic confinement time within the starburst gas. In the energy range of interest, the inelastic nuclear collision cross section is $\sigma_{\mathrm{pp}} \approx 50$ mb, with inelasticity of $\sim$0.5. The energy loss time, $\tau_{\mathrm{loss}} \approx (0.5n\sigma_{\mathrm{pp}}c)^{-1}$ where $n$ is the interstellar nucleon density, would be shorter than the starburst lifetime, which is at least the dynamical time given by the ratio of $\ell$ to the characteristic gas velocities $v$, $\sim(2\ell/v)$, as long as

$$\Sigma_{\mathrm{gas}} \gtrsim \Sigma_{\mathrm{crit}} \equiv \frac{m_p v}{\sigma_{\mathrm{pp}} c} = 0.03(v/300 \text{ km s}^{-1}) \text{ g cm}^{-2}. \qquad (3)$$

For characteristic gas velocities, $v =$ few hundred km/s, the critical surface density, $\Sigma_{\mathrm{crit}}$, is comparable to the minimum $\Sigma_{\mathrm{gas}}$ of known starburst galaxies.[34,37]

The ratio of confinement time and loss time is less straightforward to estimate, since magnetic confinement of CRs is not well understood. In the Milky Way, the total gas column density traversed by CRs of energy $E/Z \leq 1$ TeV before they escape the galaxy is $\Sigma_{\mathrm{conf,MW}} \approx 9(E/10Z\mathrm{GeV})^{-0.4}$ g cm$^{-2}$.[19,39] If the propagation of CRs in starburst galaxies is similar to that in our galaxy, we may expect $\Sigma_{\mathrm{conf,SB}}(E/Z) = (n_{\mathrm{SB}}/n_{\mathrm{MW}})\Sigma_{\mathrm{conf,MW}}(B_{\mathrm{MW}}E/B_{\mathrm{SB}}Z)$, since the propagation of the CRs in the magnetic field is determined by $E/ZB$. For $n_{\mathrm{SB}}/n_{\mathrm{MW}} = B_{\mathrm{SB}}/B_{\mathrm{MW}} = 100$, the fraction of proton energy lost to pion production before escape is $f_{\pi,\mathrm{SB}} \approx 1(E/1\mathrm{PeV})^{-0.4}$. Since the neutrino flux is expected to be dominated by starbursts at $z \gtrsim 1$, for which the typical surface density should be even higher than in local starbursts, it is reasonable to assume that most of the energy injected into starburst galaxies in $E \lesssim 10$ PeV protons is converted to pions.

### 3.3.2. The fraction of CR production occurring in calorimetric galaxies

The energy loss of $E \lesssim 10$ PeV protons in starburst galaxies would produce a neutrino flux and intensity similar to the WB bound at $E \lesssim 1$ PeV, provided that the sources of UHECRs produce most of their energy output in starburst environments. This would be the case if, as assumed in Ref. 26, (i) the rate of CR production is proportional to the SFR and (ii) most (or a significant fraction) of the stars in the universe have been produced in starburst episodes.

The leading UHECR accelerator candidates are $\gamma$-ray bursts (GRBs) and transient accretion events onto massive black holes residing at the centers of galaxies.[2,3] The assumption that the CR production rate is proportional to the SFR is natural for sources like GRBs, which are related to the deaths of massive stars, and may not be inconsistent with models based on activity around massive central black holes. On the other hand, the assumption that a significant fraction of the stars in the universe have been produced in starburst episodes (as suggested e.g. by Ref. 40) is not necessarily valid.

Recent observations indicate that for $z < 2$ the sSFR is narrowly distributed around a $z$-dependent average, which increases by a factor of $\sim$30 from $z = 0$ to $z = 2$.[33] Outliers with high sSFR, which may be categorized as starbursts, are found to contribute only $\sim$10% of the total SFR.[33] The interpretation of these results is still debated. In particular, starburst activity has been commonly argued to be triggered by galaxy mergers, and whether or not the narrow distribution of the sSFR rules out major mergers as the cause for the rapid increase of the average sSFR with redshift is still debated.[41,42] In other words, if starburst activity is driven by mergers and major mergers are responsible for the increase in the SFR, all galaxies at higher redshift may be classified as starbursts.

While the definition of a starburst is debateable and the fraction of star-formation occurring in starburst episodes is still uncertain, it should be noted that the typcial galaxies at high $z$, which are rapidly forming stars, may well be calorimetric. CO observations of six $z = 1.5$ galaxies with "normal" (i.e. close to the average) sSFR show that they are characterized by rapid SFR, $\sim$100$M_\odot$/yr, and high column density massive molecular disks, $\Sigma_g \sim 0.1$ g/cm$^2$.[41] In the local universe, galaxies with such column density and SFR are calorimetric (see Eq. (3) and Ref. 38). While the disk structure of high $z$ galaxies may be different, and the determination of the high $z$ molecular gas content is uncertain, these results support the hypothesis that a large fraction of the SFR occurred in calorimetric environments.

### 3.4. *A lower limit to the density of steady sources*

The non-detection by IceCube of point sources producing multiple neutrino events, combined with the measured "diffuse" neutrino intensity, sets a lower limit to the density of the sources producing the neutrinos, and an upper limit to their neutrino luminosity. We give below a simple order of magnitude estimate of these limits.

Consider a population of "standard candle" sources, with density $n_s(z)$ and 0.1 to 1 PeV muon neutrino luminosity $L_{\nu_\mu}$. We consider only neutrino-induced muon events, for which the arrival direction may be determined with good accuracy (better than 1 deg). The coincidence of the neutrino intensity with the WB bound implies, using Eq. (2) and a measured $1:1:1$ flavor ratio,

$$n_0 L_{\nu_\mu} \approx 2 \times 10^{43} \left( \frac{\xi_z}{3} \right)^{-1} \text{erg/Mpc}^3\text{yr}, \qquad (4)$$

where $n_0 \equiv n_s(z=0)$. The flux of neutrinos of energy $E_\nu$ required for the detection of more than one neutrino induced muon event is approximately given by $f_m = E_\nu / A T P_{\mu\nu}$, where $A$ is the detector's effective area, $T$ is the integration time, and the probability that a 1 TeV to 1 PeV neutrino with a propagation track crossing the detector would produce a muon going through the detector is $P_{\mu\nu} \approx 10^{-6}(E_\nu/1\text{TeV})$.[4] This yields $f_m \approx 2 \times 10^{-12}(AT/3\text{km}^2\text{yr})^{-1}$ erg/cm²s. The limit inferred below on $n_0$ implies that the distance $d_m$ below which the flux exceeds $f_m$, $d_m = (L_{\nu_\mu}/4\pi f_m)^{1/2}$, is $\ll c/H_0$. Thus, the number of sources producing multiple upward-moving (and hence neutrino-induced) muon events is approximately

$$N_m \approx \frac{2\pi}{3} n_0 d_m^3 \approx 1 \left( \frac{\xi_z}{3} \right)^{-3/2} \left( \frac{n_0}{10^{-7}\text{Mpc}^{-3}} \right)^{-1/2} \left( \frac{A}{1\text{km}^2} \frac{T}{3\text{yr}} \right)^{3/2}. \qquad (5)$$

The requirement $N_m < 1$ sets a lower limit to the density of steady sources, $n_0 > 10^{-7}$ Mpc$^{-3}$ (implying $L_{\nu_\mu} < 10^{43}$ erg/s and $d_m < 200$ Mpc), consistent with that of the detailed analysis of Ref. 43. It rules out rare candidate sources, like bright $L \sim 10^{47}$ erg/s AGN with $n_0 \sim 10^{-9}$ Mpc$^{-3}$, and is consistent with the density of starburst galaxies, $n_0 \sim 10^{-5}$ Mpc$^{-3}$.

## 4. Summary and discussion

### 4.1. *The calorimetric star-forming galaxies model and its uncertainties*

IceCube's measurements of extra-terrestrial neutrinos are consistent with a model in which both the neutrinos and the observed UHECRs are produced by the same population of cosmological sources, producing CR protons at a similar rate, $E^2 d\dot{n}/dE \propto E^0$, over the [1 PeV, $10^{11}$ GeV] energy range, and residing in "calorimetric" environments, in which $E < 50$ PeV protons lose much of their energy to pion production (Sec. 3.2). This model is a natural explanation of IceCube's results since it relies on a known (although not yet identified) population of sources — the

UHECR sources, and since it does not depend on ad-hoc choices or fine tuning of model parameters: the neutrino flux normalization is determined by the observed UHECR flux, and the neutrino spectrum is consistent with that implied by the measured UHECR spectrum (Sec. 2.1). Moreover, a "flat" $d \log \dot{n}/d \log E \approx -2$ generation spectrum is observed in a wide range of astrophysical environments[19–21] and is a robust prediction of the most widely accepted and best understood (although not yet from first principles) model for particle acceleration in astrophysical objects-Fermi acceleration in collisionless shocks.[19,22,23]

A wide variety of (non-calorimetric) models for the neutrino origin were proposed following IceCube's discovery (including dark matter decay, active galactic nuclei of various types, see Fig. 1(b), and galaxy clusters, see Refs. 44 and 45 for reviews). These models generally rely on ad-hoc choices of model parameters, which cannot be derived theoretically or determined observationally, in order to reproduce the observed flux and spectrum of the neutrinos. Their predictive power is thus limited.

Radio to TeV observations of local starburst galaxies imply that they are calorimetric for protons of energy exceeding 10 TeV,[34,38] and theoretical arguments suggest that calorimetry holds to energies exceeding 10 PeV (Sec. 3.3.1). Starburst galaxies have thus been predicted[26] (Fig. 1(b)) to produce an extra-galactic neutrino background comparable to the WB bound at energies $E < 0.5$ PeV, and possibly extending to higher energy, provided that (i) the rate of CR production is proportional to the SFR, and (ii) most (or a significant fraction) of the stars in the universe have been produced in starburst environments. The validity of the first assumption is natural if UHECRs are produced by GRBs (and may hold also for models based on activity around massive central black holes). It is further supported by the observation that at low CR energy, $\sim 10$ GeV, the ratio of CR production rate to SFR is similar, $\sim 10^{47}$ erg for 1 $M_\odot$ of star formation, in the galaxy and in starburst galaxies, while their SFRs differ by orders of magnitude.[46]

The production rate per unit volume of CR protons at $\sim 10$ GeV, $E^2 d\dot{n}/dE|_{10 \, \text{GeV}}$, is only $\sim 10$ times the rate at UHE, suggesting that the same sources are responsible for the production of CRs of all energies,[46] from 1 GeV to $10^{11}$ GeV (implying $d \log \dot{n}/d \log E \simeq -2.1$). Alternatively, there may exist a population of sources, associated with star formation and embedded in starbursts (like supernovae), producing CRs at a rate and spectrum similar to that of the UHE sources but reaching only lower energy ($\ll 10^{10}$ GeV), and thus making a contribution to the lower energy CR flux and possibly to the 1 PeV neutrino flux,[47–49] which is similar to that of the UHECR sources. In this scenario, the similarity of the energy production rates of the lower $E$ and UHE sources requires an explanation.[a]

---

[a]The ad hoc explanation given in the next to last parag. of Ref. 26 is incorrect. It follows erroneous statements that are inconsistent with the rest of that article (that the local 1.4 GHz luminosity is dominated by starbursts, and that UHECRs do not escape starbursts).

The main uncertainty remaining is the fraction of stars formed in calorimetric environments. The observed neutrino intensity is dominated by sources at redshift 1–2. The "average" galaxies at these redshifts produce stars at a much higher rate than local galaxies,[33] and observations indicate that they are characterized by high column density molecular disks, $\Sigma_g \sim 0.1$ g/cm$^2$.[41] While this provides an indication that a significant fraction of the star formation occurred in calorimetric environments (Sec. 3.3), the structure of galaxy disks at this redshift range is poorly constrained.

The production of neutrinos by pion decay is accompanied by the production of high energy photons, which initiate electromagnetic cascades via interaction with infra-red background photons, leading to a background of $\lesssim 0.1$ TeV photons. In calorimetric models with a power-law proton spectrum, pion production by $\sim 1$ TeV protons contributes directly to the $\lesssim 0.1$ TeV photon background. The resulting background intensity is expected to contribute a significant fraction of the observed 0.1 TeV background,[51] limiting the proton spectra to $d\log\dot{n}/d\log E > -2.2$.[52–54]

## 4.2. *Open questions and the way towards their resolution*

Due to the limited statistics and flavor discrimination power, the current uncertainties in the determination of the neutrino spectrum, angular distribution and flavor content are large (Sec. 3.1). A significant reduction of uncertainties requires a significant (order of magnitude) expansion of the effective mass of the detector at $\gtrsim 100$ TeV. Such expansion is necessary for example for a study of the hints for spectral breaks at low, $<30$ TeV, and high, $>2$ PeV energy, which are currently detected with low statistical significance. Reduced uncertainties will provide much more stringent constraints on predictive models, such as the calorimetric model.

An identification of the sources by angular correlation with a catalog of nearby sources is unlikely, since the fraction of events originating from $z < 0.1$ sources is $\approx 1/20$, implying that the fraction of well-localized neutrino-induced muon events from $z < 0.1$ sources over $2\pi$ sr is $\sim 1/200$.

A detection of neutrino emission from few nearby bright starburst galaxies would constitute a major evidence in support of the calorimetric star-forming galaxy model. Since the local density of starbursts is $n_0 \sim 10^{-5}$ Mpc$^{-3}$, a $\sim$10-fold increase in the effective area of the detector at $\sim 100$ TeV is required in order to enable the detection of a few nearby sources (see Eq. 5 and Refs. 43, 50).

An identification of the neutrino sources is likely to identify the calorimeters within which the CR accelerators reside, but not the accelerators themselves. The neutrino flux that is produced within the accelerators is expected to be significantly lower than the total neutrino flux produced in the calorimeters surrounding them. For example, if UHECRs are protons produced in GRBs, the neutrino flux expected to be produced within the GRB accelerators is $\sim 10\%$ ($\sim 1\%$) of the WB flux at 1 PeV (0.1 PeV)[27] (Fig. 1(b)). Such a low flux would imply that had the accelerators been steady sources, even a tenfold increase in detector mass would have been unlikely to enable their identification. Luckily, the UHECR accelerators must be transient: The absence of (steady) sources with power output exceeding the minimum required for proton acceleration to $10^{11}$ GeV, $L > 10^{46}$ erg/s, within the $\sim 100$ Mpc propagation distance of such high energy protons, implies that the sources must be extremely bright transients.[2,3] Their identification may be possible by an association of a neutrino with an electromagnetic signal accompanying the transient event. In order to open the possibility for such associations, a tenfold increase in detector mass and a wide field electromagnetic transient monitoring program are required.

A large sample of high energy neutrinos may enable one to study both neutrino properties and possibly deviations from the Standard Model of particle physics (see Ref. 45 for a recent summary of such possibilities). The identification of the sources is important for such studies, in order to discriminate between spectral/flavor features originating from astrophysical and particle physics model effects (e.g. Ref. 55).

Finally, while we have argued (in Secs. 2.1 and 3.2) that a proton-dominated composition is more likely, a heavy-nuclei dominated composition cannot be excluded. A detection of (or stringent upper limit on) the GZK neutrino flux predicted for a proton dominated composition (Sec. 2.2, Fig. 1(a)) will provide a clear confirmation of (or will rule out) a proton-dominated composition. The recent analysis of Auger[56] sets an upper limit which is close to the WB bound at $\sim 10^{9.5}$ GeV, and the expected flux may be detectable by future radio experiments.[57-59]

## References

1. E.A. Helder, *et al.*, *Space Sci. Rev.* **173** (2012) 369–431.
2. M. Lemoine, *Journal of Physics Conference Series* **409**(1) (2013) 012007.
3. E. Waxman, in *Astronomy at the Frontiers of Science*, ed. J.-P. Lasota, Springer (2011) (arXiv:1101.1155).
4. T.K. Gaisser, F. Halzen, and T. Stanev, *Phys. Rep.* **258** (1995) 173–236.
5. E. Waxman and J. Bahcall, *PRD.* **59**(2) (1999) 023002.
6. J. Bahcall and E. Waxman, *PRD.* **64**(2) (2001) 023002.
7. M.G. Aartsen, *et al.*, *PRL.* **113**(10) (2014) 101101.
8. IceCube Collaboration, M.G. Aartsen, *et al.*, arXiv:1507.04005 (2015).
9. D.J. Bird, *et al.*, *PRL.* **71** (1993) 3401–3404.
10. R.U. Abbasi, *et al.*, (HiRes Collaboration), *ApJ.* **622** (2005) 910–926.
11. R.U. Abbasi, *et al.*, *Astroparticle Physics.* **64** (2015) 49–62.
12. A. Aab, *et al.*, *PRD.* **90**(12) (2014) 122006.
13. M. Lemoine and E. Waxman, *JCAP.* **11** (2009) 9.
14. E. Waxman, *ApJ.* **452** (1995) L1+.
15. J.N. Bahcall and E. Waxman, *Phys. Lett. B* **556** (2003) 1–6.
16. B. Katz, R. Budnik, and E. Waxman, *JCAP* **3** (2009) 20.
17. K. Greisen, *PRL.* **16** (1966) 748–750.
18. G.T. Zatsepin and V.A. Kuz'min, *JETP. Lett.* **4** (1966) 78.
19. R. Blandford and D. Eichler, *Phys. Rep.* **154** (1987) 1–75.
20. W.I. Axford, *ApJ. Supp.* **90** (1994) 937–944.
21. E. Waxman, *Plasma Physics and Controlled Fusion* **48** (2006) B137–B151.

22. J. Bednarz and M. Ostrowski, *PRL.* **80** (1998) 3911–3914.
23. U. Keshet and E. Waxman, *PRL.* **94**(11) (2005) 111102.
24. D. Allard, *Astropart. Phys.* **39** (2012) 33–43.
25. V.S. Beresinsky and G.T. Zatsepin, *Phys. Lett. B* **28** (1969) 423–424.
26. A. Loeb and E. Waxman, *JCAP.* **5** (2006) 003.
27. E. Waxman and J. Bahcall, *PRL.* **78** (1997) 2292–2295.
28. F.W. Stecker, *PRD.* **72**(10) (2005) 107301.
29. M.G. Aartsen, *et al.*, *PRL.* **114**(17) (2015) 171102.
30. M.G. Aartsen, *et al.*, *PRD.* **91**(2) (2015) 022001.
31. M.G. Aartsen, *et al.*, *APJ.* **809** (2015) 98.
32. J.G. Learned and S. Pakvasa, *Astropart. Phys.* **3** (1995) 267–274.
33. M.T. Sargent, *et al.*, *ApJ.* **747** (2012) L31.
34. H.J. Voelk, *A&A.* **218** (1989) 67–70.
35. J.J. Condon, *et al.*, *ApJ.* **378** (1991) 65–76.
36. R.C. Kennicutt, Jr., *ApJ.* **498** (1998) 541–552.
37. T.A. Thompson, *et al.*, *ApJ.* **645** (2006) 186–198.
38. B.C. Lacki, *et al.*, *ApJ.* **734** (2011) 107.
39. K. Blum, B. Katz, and E. Waxman, *PRL.* **111**(11) (2013) 211101.
40. S. Juneau, *et al.*, *ApJ.* **619** (2005) L135–L138.
41. E. Daddi, *et al.*, *ApJ.* **713** (2010) 686–707.
42. M. Puech, *et al.*, *MNRAS.* **443** (2014) L49–L53.
43. M. Ahlers, F. Halzen *et al.*, *PRD.* **90**(4) (2014) 043005.
44. P. Mészáros, *Nucl. Phys. B Proc. Supp.* **256** (2014) 241–251.
45. K. Murase. *AIP Conference Series*, **1666** (2015) 040006.
46. B. Katz, E. Waxman, T. Thompson, and A. Loeb, arXiv:1311.0287 (2013).
47. H.-N. He, *et al.*, *PRD.* **87**(6) (2013) 063011.
48. R.-Y. Liu, *et al.*, *PRD.* **89**(8) (2014) 083004.
49. N. Senno, *et al.*, *ApJ.* **806** (2015) 24.
50. L.A. Anchordoqui, *et al.*, *PRD.* **89**(12) (2014) 127304.
51. T.A. Thompson, E. Quataert, and E. Waxman, *ApJ.* **654** (2007) 219–225.
52. K. Murase, M. Ahlers, and B.C. Lacki, *PRD.* **88**(12) (2013) 121301.
53. X.-C. Chang and X.-Y. Wang, *ApJ.* **793** (2014) 131.
54. I. Tamborra, S. Ando, and K. Murase, *JCAP.* **9** (2014) 043.
55. T. Kashti and E. Waxman, *JCAP* **5** (2008) 6.
56. A. Aab, *et al.*, *PRD.* **91**(9) (2015) 092008.
57. P.W. Gorham, *Nucl. Phys. B Proc. Supp.* **243** (2013) 231–238.
58. ARA Collaboration, P. Allison, *et al.*, arXiv:1404.5285 (2014).
59. S.W. Barwick, *et al.*, *Astroparticle Physics.* **70** (2015) 12–26.

# Chapter 4

## Galactic Neutrino Sources

Markus Ahlers

*Wisconsin IceCube Particle Astrophysics Center (WIPAC) and
Department of Physics, University of Wisconsin-Madison,
Madison, WI 53706, USA*

There is general consensus that the flux of cosmic rays (CRs) below a few PeV originates in galactic sources. Around 3 PeV the CR spectrum shows a significant break, the so-called CR *knee*, that could originate from the fact that galactic CR sources reach their maximal acceleration energy, or that CR escape from the Milky Way becomes more efficient. The sources are unknown, but it has long been speculated that galactic core-collapse supernovae could be responsible.[1] These events produce ejecta with kinetic energy of the order of $\mathcal{E}_{\rm ej} \simeq 10^{51}\mathcal{E}_{\rm ej,51}$ erg per supernova (SN) explosion at a rate $R_{\rm SN} = 0.03 R_{\rm SN,-1/5}$ yr$^{-1}$. Diffuse shock acceleration taking place in remnant shocks could convert a significant fraction $\epsilon_{\rm CR} \simeq 0.1\epsilon_{\rm CR,-1}$ of this kinetic energy into a non-thermal population of cosmic rays. The CR spectrum per sources is expected to follow a power-law normalized as

$$E_{\rm CR}^2 N_{\rm CR} \simeq \frac{\epsilon_{\rm CR}\mathcal{E}_{\rm ej}}{\mathcal{R}_0}\left(\frac{E}{\rm GeV}\right)^{2-\Gamma}, \tag{1}$$

with bolometric correction factor $\mathcal{R}_0 \simeq (1 - (E_{\rm max}/{\rm GeV})^{2-\Gamma})/(\Gamma - 2)$ (or $\mathcal{R}_0 \simeq \ln(E_{\rm max}/{\rm GeV})$ for $\Gamma = 2$).

For a given source power density (not necessarily related to SNRs), the level of the observed galactic CR spectrum can then be estimated in the following way. The survival time of CRs in the galactic environment is limited by diffusion out of the plane. The galactic diffusion environment can be approximated as a cylinder with radius $R_{\rm MW} \simeq 20 R_{\rm MW,20}$ kpc and half height $H \simeq 1 H_0$ kpc. The diffusion constant is expected to scale with rigidity as $D \simeq (R/3\,{\rm GV})^{\delta} D_{28} 10^{28}$ cm$^2$/s where the power $\delta$ depends on the type of magnetic turbulence. The escape time $\tau_{\mathrm{esc}}(E) \simeq H^2/D$ then becomes energy dependent and leads to a softening of the emission spectra. Finally, the observed flux of CRs can be approximated as $E^2\phi_{\mathrm{CR}} \simeq (c/4\pi)(\tau_{\mathrm{esc}}(E)/V_{\mathrm{diff}})R_{\mathrm{SN}}E_{\mathrm{CR}}^2 N_{\mathrm{CR}}$ with diffusion volume $V_{\mathrm{diff}} \simeq 2H\pi R_{\mathrm{MW}}^2$, see for example Ref. 2 For Kolmogorov-type turbulence $\delta \simeq 1/3$ and power spectrum with $\Gamma \simeq 2.3$ and $E_{\mathrm{max}} \simeq 1$ PeV we can then estimate the diffuse flux of CRs as

$$E^2\phi_{\mathrm{CR}} \simeq 0.8\frac{\epsilon_{\mathrm{CR},-1}\mathcal{E}_{\mathrm{ej},51}H_0R_{\mathrm{SN},-1/5}}{R_{\mathrm{MW},20}^2 D_{28}}\left(\frac{E}{\mathrm{GeV}}\right)^{-0.63}\frac{\mathrm{GeV}}{\mathrm{cm^2\,s\,sr}}, \tag{2}$$

which is very close to observation.[3] The galactic power density of SNR would hence be sufficient to supply enough CRs to maintain the level of the observed CR spectrum for the given benchmark parameters.

On the other hand, galaxies similar to our own will also contain sources capable of accelerating CRs up to PeV energies and there are other extragalactic source candidates that could be more powerful emitters. However, there contribution to the low-energy CR spectrum is limited by the effect of magnetic horizons[4]: The diffusion time-scale of CRs in the extragalactic environment is typically much longer than in comparison to that of energy losses or the age of the Universe and they won't be able reach us. However, as we approach the CR *knee* there is the possibility that close-by extragalactic sources can contribute to the spectrum. For instance, Ref. 4 estimated the maximal distance of sources that can contribute to 10 PeV CRs for various intergalactic diffusion environments to be smaller than 100 Mpc.

Galactic sources can produce neutrinos at various stages of CR acceleration, emission and propagation. Point-source neutrino emission is expected from direct interaction of CRs in the sources or with interaction with close-by molecular clouds. The neutrino and CR nucleon ($N$) emission spectra are related via

$$\frac{1}{3}\sum_\alpha E_\nu^2 N_{\nu_\alpha}(E_\nu) \simeq \frac{1}{4}\frac{f_\pi K_\pi}{1+K_\pi}\left[E_N^2 N_N(E_N)\right]_{E_N=20E_\nu}, \tag{3}$$

where $f_\pi < 1$ is the pion production efficiency and $K_\pi$ the ratio of charged to neutral pions. In the case of hadronic interactions of CRs with gas (so-called *pp* interactions) we have $K_\pi \simeq 2$ and the pion production efficiency does only weakly depend on energy. In this case, the secondary neutrino and gamma-ray emission will follow the local CR spectrum.

Cosmic ray gas interactions in the vicinity of the sources are hence expected to produce hard emission spectra following the power law of freshly accelerated CRs. Once the CRs have diffused into the galactic medium the CR spectra and their secondary neutrino emission are expected to become softer due to the above described mechanism of rigidity dependent CR escape. The corresponding neutrino emission also becomes more and more diffuse and might reach angular diameters exceeding the point spread function of the neutrino observatory. However, extended neutrino emission can also be expected from other mechanisms. For instance the sum of neutrino sources below the detection threshold can be visible as quasi-diffuse flux of the Galactic Plane. There also exist extended gamma-ray emission regions above and below the Galactic Plane, so-called *Fermi Bubbles* that could be related to hadronic emission scenarios. Exotic scenarios like dark matter (DM) decay in the galactic DM halo would also be visible as extended emission.

The origin of the flux of TeV–PeV neutrinos recently discovered by the IceCube observatory is unknown. Considering hadronic production via $p\gamma$ or $pp$ interactions, this emission corresponds to an underlying CR population with energies reaching above the CR *knee*, but not necessarily above the *ankle* around 3 EeV. This is the transition region of galactic and extragalactic CRs and it is hence feasible that galactic sources might be responsible for the emission. However, the absence of a strong anisotropy in the arrival direction of events and upper neutrino limits on the individual point source emission limits the contribution of individual galactic point sources. A significant contribution to the observed diffuse flux is only possible for extended galactic diffuse emission.

Various sources might reach the required large maximal energies necessary for the IceCube observation, including hypernovae,[5] newly born pulsars or magnetars,[6,7] microquasars[8,9] or the unidentified sources of galactic TeV gamma-rays emission.[10–15] Possible contributions to super-TeV neutrinos are the diffuse neutrino emission of galactic CRs,[16–18] the joint emission of galactic PeV sources[19,20] or microquasars,[21] and extended galactic structures like the *Fermi Bubbles*[16,22,23] or the galactic halo.[24] A possible association with the sub-TeV diffuse galactic gamma-ray emission[25] and constraints from the non-observation of diffuse galactic PeV gamma-rays,[16,26] have also been investigated. More exotic scenarios have suggested a contribution of neutrino emission from decaying heavy dark matter.[27–30]

In the following sections we will summarize the main contenders for the sources of galactic CRs and related neutrino production scenarios and limits.

## 1. Discrete galactic sources

Many sources in our galaxy show high-energy GeV–TeV gamma-ray emission which has been studied in detail with the satellite missions EGRET and *Fermi*, by imaging atmospheric Cherenkov telescopes H.E.S.S., MAGIC and VERTIAS, and by water Cherenkov detectors Milagro and HAWC. These gamma-ray observations all indicate the presence of a population of non-relativistic particles in these sources that emit gamma-rays via leptonic or hadronic processes. In the latter case, we would also expect to see neutrino emission from these sources due to the production and decay of charged pions. The production rates of hadronic neutrinos and gamma rays are related as

$$\frac{1}{3}\sum_{\alpha} E_\nu^2 N_{\nu_\alpha}(E_\nu) \simeq \frac{K_\pi}{4}\left[E_\gamma^2 N_\gamma(E_\gamma)\right]_{E_\gamma=2E_\nu}, \qquad (4)$$

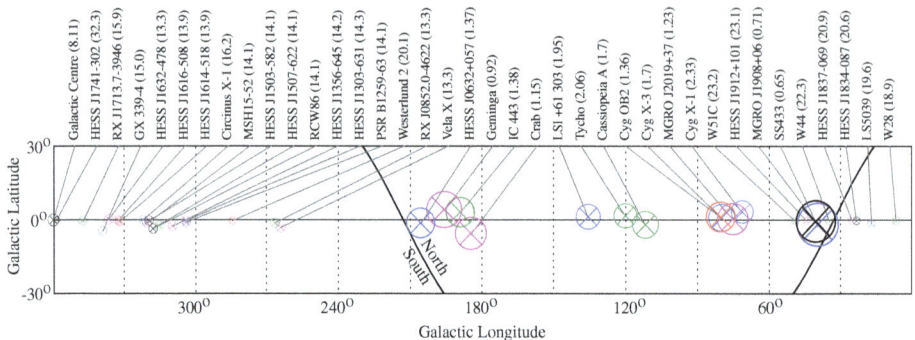

Fig. 1. Overview of galactic sources tested by IceCube and ANTARES. The different colors indicate SNR (green), binary systems (blue), pulsar/PWN (magenta), star clusters (red) and other sources (black). The solid line shows the celestial equator. The area of the symbols indicate the strength of the $E^{-2}$ neutrino limit, also indicate after each source name (in units of $E^2\phi_{\nu_\mu+\bar\nu_\mu}/(10^{12}\,\mathrm{TeV\,cm^{-2}s^{-1}}))$.[31,32]

with $E_{\nu_\alpha} = 0.5 E_\gamma$. Galactic TeV gamma-ray emission has been observed in association with SNRs and their close-by molecular clouds, pulsars and their nebula, binary systems, and massive star clusters. All these sources have also been speculated to be continuous or transient neutrino emitters, see e.g. Refs. 12, 33.

Figure 1 shows the location of galactic sources that are investigated by ANTARES[32] and IceCube[31] for continuous neutrino emission. The different colors indicate SNR (green), binary systems (blue), pulsar/PWN (magenta), star clusters (red) and other sources (black). The area of the symbols indicate the strength of the $E^{-2}$ neutrino limit with stronger limits having larger symbols. The corresponding value (in units of $E^2\phi_{\nu_\mu+\bar\nu_\mu}/(10^{12}\,\mathrm{TeV\,cm^{-2}s^{-1}})$) is also indicated after each source name. IceCube contributes with the strongest point source limits for sources located in the Northern hemisphere. However, many galactic sources including the Galactic Center are located in the Southern hemisphere, which is better studied with ongoing muon tracks from the location of ANTARES.

### 1.1. *Supernova remnants*

Supernova remnants are the prime candidates for the sources of galsactic CRs. If CRs are interacting inside the remnant or with close-by molecular clouds we expect that production of neutrinos from hadronic interactions. In fact, recently the *Fermi* satellite observed features in the gamma-ray spectra of the shell-type supernova remnants IC 443 and W44[34] that can be related to the mass threshold in pion-decay spectra. The maximal neutrino energy from hadronic interactions is expected to reach about 5% of the maximal CR energy in the production and decay of charged pions.

The maximum CR energy obtained in shock acceleration is expected to be reached by the end of the Sedov phase, when the SN remnant ejecta with radius starts to decelerate. The Sedov radius is related to the ejecta energy $\mathcal{E}_{ej}$, velocity $V_{ej}$ and ambient gas density $n$ as $R_{Sed} \simeq (3M_{ej}/4\pi n m_p)^{1/3}$ for a swept-up mass $M_{ej} \simeq 2\mathcal{E}_{ej}/V_{ej}^2$. The maximum energy can then be esimated as $E_{p,max} \simeq (3/20)eBR_{Sed}(V_{ej}/c)$, assuming Bohm diffusion in parallel shocks with magnetic field $B$.[35] For velocities $V_{ej} = V_{ej,-1}0.1c$, densities $n = n_0 \, cm^{-3}$, magnetic fields $B = 0.3B_{-3.5} \, mG$ and ejecta energy $\mathcal{E}_{ej} \simeq \mathcal{E}_{ej,51}10^{51}$ erg, the maximal proton energy can then be estimated to as $E_{p,max} \simeq 4.5B_{-3.5}(\mathcal{E}_{ej,51}V_{ej,-1}/n_0)^{1/3}$ PeV in the vicinity of the CR knee.

Neutrino emission from CR acceleration and interaction in supernova remnants have been studied in Refs. 36–39. Under the assumption that the TeV gamma-ray observation of these sources has a hadronic origin one can use Eq. (4) to normalize the neutrino emission. Promising neutrino candidate sources are the close-by shell-type SNR RX 0852.0-4622 and RX J1713.7-3946, which are part of the combined list of candidate point-source at IceCube and ANTARES (see Fig. 1).

## 1.2. *Pulsar and pulsar wind nebula*

The central compact object emerging from a core-collapse supernova can be neutron stars with initial small rotation period $P_i \simeq 10^{-3} \simeq P_{i,-3}$ s and high surface magnetic field strength $B \simeq 10^{13}B_{13}$ Gauss. The neutron star's moment of inertia is of the order of $I \simeq 10^{45}I_{45} \, g \, cm^2$ for one solar mass and radius of the order of 10 km. That means that the neutron star is born with a very high rotational energy of the order of $\mathcal{E}_{rot} \simeq 10^{51}I_{45}P_{-3}^{-2}$ erg.

The dipole field configuration of the rapidly rotating star can only co-rotate up to the distance of the light cylinder, $r_{lc} \simeq cP/2\pi$, and has to assume a quasi-azimuthal symmetry beyond this point with outward-spiraling magnetic field lines. The relativistic magnetic rotator can generate voltage drops across magnetic field lines. This generates a high-energy wind of electrons and positrons that energizes the surrounding nebula and gradual slows down the neutron star's rotation. The mis-alignment of the magnetic dipole axis and the rotation axis leads to an apparent pulsation of the gamma ray emission created in the vicinity of the light cylinder, hence the name *pulsar*.

Pulsars have also been considered as potential sites for CR acceleration.[40–44] However, many details of the mechanisms leading to extraction of rotational energy and acceleration of charged particles are vague. In the model of Ref. 41 the maximal energy of CRs was estimated as a fraction $\eta = 0.1\eta_{-1}$ that can be converted from the pulsar wind Poynting flux, which gives $E_{max} \simeq 7 \times 10^{18}\eta_{-1}ZB_{13}P_{i,-3}^{-2}$ eV. Neutrinos could be produced by CR interactions with thermal photons of the neutron star's surface,[40] with thermal radiation in the SN envelope,[42] or with hadrons in the supernova ejecta.[45–48] As in the case of SNRs, the relation between TeV gamma-ray and neutrino emission can also be exploited for the case of pulsar wind nebula.[38,49,50]

Pulsars born in massive star clusters like Cygnus OB2 or in the Galactic Center region would inject their CRs into the near environment with enhanced gas densities and magnetic field strength compared to the galactic average. This could enhance the neutrino signal from the local diffuse emission of CRs in hadronic interactions.[51,52]

### 1.3. *X-Ray binaries and microquasars*

X-ray binary system consists of a compact object (neutron star or black hole) that accretes mass from a companion star. The accretion of matter from the companion star into an accretion disk or onto the compact object creates the X-ray emission that can show periodicity corresponding to the orbital motion, but also intermittent emission or X-ray bursts. In some cases, so-called *microquasars*, the X-ray binary system develops a pair of collimated plasma jets perpendicular to the accretion disk similar to the morphology and mechanisms of blazar sources.[8,9,53]

X-ray binary and microquasars share not only a similar morphology to active galactic nuclei and blazars, respectively, but also similar physical processes of particle acceleration and emission that scale with the mass and dimension of the system. For instance, the typical luminosity of a microquasar, $L \simeq 10^{37}$ erg/s, corresponds to a mass inflow of about $\dot{M} \sim 10^{-9} M_\odot \, \text{yr}^{-1}$. For comparison, quasars need to accrete $\dot{M} \sim 10 M_\odot \, \text{yr}^{-1}$ to maintain their luminosity of $10^{47}$ erg/s. The temperature in the accretion disk at the last stable orbit scales with the mass of the central object as $T \sim 5 \times 10^7 (M/M_\odot)^{-1/4}$.[54] Hence, whereas the accretion disk of quasars is heated to a temperature of $T \sim 10^5$ K, corresponding to UV and optical radiation, microquasars radiate with a temperature of $T \sim 10^7$ K in X-rays.

X-ray binary system and microquasars are potential sources of CRs. Possible sites for CR acceleration are accretion shocks from the infall of matter onto the compact object[55,56] or internal shocks of the jets.[8] In the model of Ref. 8, the maximal proton energy is limited by the dynamical time and the energy loss due to photo-meson interactions. For typical jet parameters this is of the order of a few PeV. Hadronic interactions in microquasars are a possible source of galactic neutrino emission.[8,9,53] An important aspect of the flux calculation are relativistic beaming effects depending on the relative motion of the observed with respect to the interaction region. If the compact object of the X-ray binary is a pulsar, the external radiation and matter can provide additional targets for CRs accelerated by the pulsar.[57]

## 2. Diffuse galactic emission

In the introduction to this chapter we have already indicated that the diffusion of CRs in the galactic environment can reproduce the observed CR spectrum.

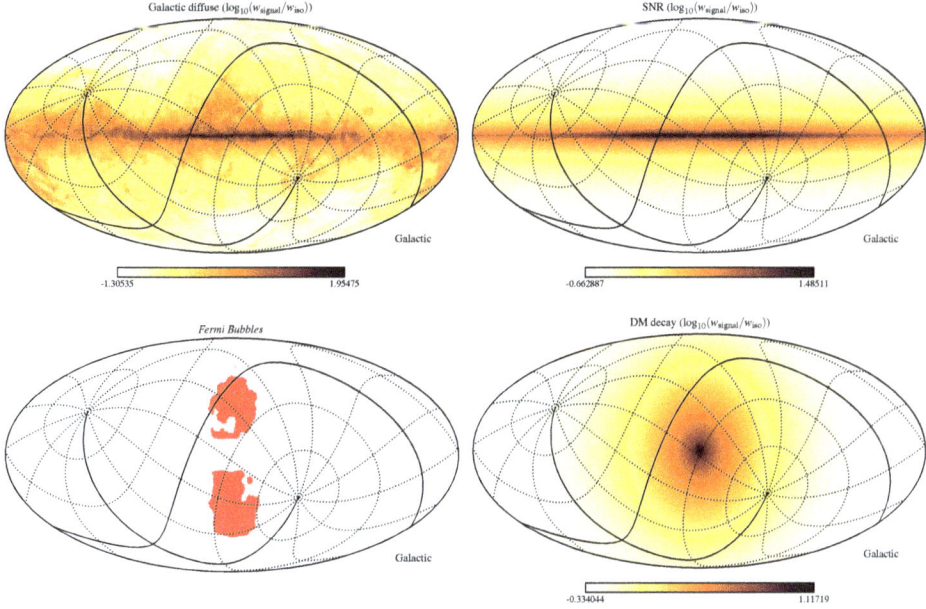

Fig. 2. Mollweide projections of four scenarios of extended galactic neutrino emission centered at the Galactic Center (from Ref. 58). From left to right and top to bottom, they are galactic diffuse ($E_\nu = 10$ TeV), supernovae remnants (Case), *Fermi Bubbles*, and decaying dark matter. The mesh indicates the equatorial coordinate system with right ascension $\alpha = 0°$ and declination $\delta = 0°$ indicated as solid lines.

In particular, the softness of the spectrum is a result of a hard emission spectrum with index $\Gamma$ and softening due to the rigidity dependence of the diffusion coefficient with index $\delta$. This mechanism also implies that the CR density in our Galaxy is smooth and we can thus approximate the CR distribution by the local CR density, $n_{CR} \simeq 4\pi\phi_{CR}/c$. This guarantees a diffuse galactic emission of neutrino and gamma-ray emission from CR interactions with gas in the vicinity of the Galactic Plane, see Fig. 2.[59–64]

The local emission rate of neutrinos (per flavor) can then be approximated as

$$E_\nu^2 Q_\nu(E_\nu) = \frac{1}{6} cn\kappa\sigma_{pp} \left[E_N^2 n_N(E_N)\right]_{E_N=20E_\nu}, \tag{5}$$

where $\sigma_{pp}$ is the inelastic proton-proton cross section with inelasticity $\kappa$ and $n$ is the average local gas density. In the following we use the approximation $\sigma_{pp} \simeq 34 + 1.88L + 0.25L^2$ mb for $L = \ln(E_p/\text{TeV})$[65,66] with $\kappa \simeq 0.5$. The average gas density is at the level of $n \simeq 1n_0$ cm$^{-3}$ in the Galactic Plane and drops exponentially in orthogonal direction $z$ as $\exp(-|z|/h)$ with $h \simeq 10.1 - h$ kpc. The overall factor of $1/6$ in Eq. (5) is a rough estimation of the total neutrino energy over the total proton energy accounting for the per flavor emission ($1/3$), the total neutrino energy fraction in the charge pion decay ($3/4$) and for the charged pion fraction in $pp$ collisions ($2/3$). Furthermore, for the individual neutrino energy, one can use the approximate relation $E_\nu \simeq E_p/20$ with a factor of $1/4$ for the neutrino (each flavor) energy in the pion decay and a factor of $1/5$ for the average energy of the pion produced in $pp$ interactions.

The diffuse flux of neutrinos from a fixed angular area $\Delta\Omega$ can then be expressed as the integrated emission of neutrinos along the line of sight. For this purpose it is convenient to express the neutrino emission rate density as $Q_\nu = \widehat{Q}_\nu \widehat{\rho}(r, z)$ where $\widehat{Q}_\nu$ is the total galactic emission rate (over the effective volume $2h\pi R_{\mathrm{MW}}^2$) and $\widehat{\rho}(r, z)$ is the normalized emission density in terms of cylindrical galacto-centric coordinates with $2\pi \int \mathrm{d}r \mathrm{d}z r \widehat{\rho} = 1$. The density of neutrinos from diffuse galactic CRs is approximately $\widehat{\rho} \simeq e^{-|z|/h}/(2h\pi R_{\mathrm{MW}}^2)$ for $r \leq R_{\mathrm{MW}}$ and $|z| < H$. The galacto-centric coordinates are related to spherical helio-centric coordinates $(s, \ell, b)$ as $r^2 = s^2 \cos^2 b + R_\odot^2 - 2sR_\odot \cos\ell \cos b$ and $z = s \sin b$. The distance from the Sun to the Galactic Center is $R_\odot \simeq 8.5$ kpc.

The galactic diffuse neutrino flux from an angular area $\Delta\Omega$ can then be expressed as

$$E_\nu^2 \phi_\nu(E_\nu) = \frac{1}{\Delta\Omega} \int \mathrm{d}\Omega J(b, \ell) E_\nu^2 \widehat{Q}_\nu(E_\nu), \tag{6}$$

where we define the normalized $J$-factor as the line of sight integral

$$J(b, \ell) = \frac{1}{4\pi} \int \mathrm{d}s \widehat{\rho}\left(r(s, \ell, b), z(s, \ell, b)\right). \tag{7}$$

Due to the (approximate) exponential distribution of target gas off the Galactic Plane the diffuse emission from Eq. (6) depends on the angular size $\Delta\Omega$ of the Galactic Plane. For a latitude range $|b| < 2°$ the per flavor diffuse neutrino flux can be numerically evaluated to

$$E_\nu^2 \phi_\nu^{\mathrm{diff}} \simeq 6.1 \times 10^{-5} n_0 h_{-1} R_{\mathrm{MW},20} \left[E_N^2 \phi_N(E_N)\right]_{E_N = 20 E_\nu}. \tag{8}$$

We approximate the locally observed CR nucleon flux as $E_N \phi_N \simeq 1.03/(\mathrm{cm}^2\ \mathrm{s}\ \mathrm{sr})(E_N/\mathrm{GeV})^{-\gamma}(1 + (E_N/E^*)^3)^{-\delta/3}$ with $\gamma = 1.64$, $\delta = 0.67$ and $E^* = 0.9\,\mathrm{PeV}$.[67] The result is shown as a red solid line in Fig. 3. This estimate agrees well with a more elaborate study using numerical CR propagation code to evaluate the CR density across the galaxy and using non-azimuthal target gas maps.[58] The corresponding distribution is shown in the top left plot of Fig. 2. We also show the diffuse flux of cosmic neutrinos observed by IceCube[68] which has the same level in the Galactic Plane at 10 TeV.

The galactic diffuse flux (8) drops below the level of the isotropic flux observed in IceCube for $E_\nu \gg 10$ TeV. It is thus unlikely that this galactic contribution has a strong impact on the interpretation of the IceCube data.[16–18,58] However, the above estimate relies on the assumption that the local CR flux is a good approximation for the average galactic CR density. This is not necessarily the case in more complex (and more realistic) diffusion scenarios and/or strongly inhomogeneous source and target distributions in the galaxy. For instance, diffusion along the regular component of the Galactic magnetic field is stronger compared to orthogonal directions[69]

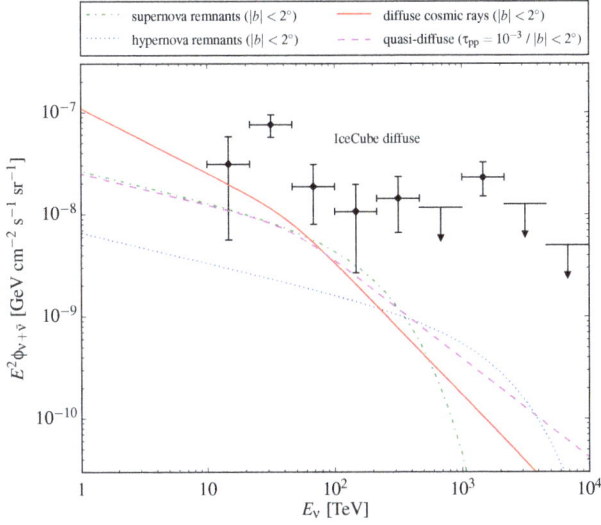

Fig. 3. Diffuse emission from the Galactic Plane ($|b| \leq 2°$) in comparison to the isotropic diffuse neutrino flux (per flavor) observed by IceCube.[68] We show the contribution of the Galactic diffuse flux of Eq. (8), the quasi-diffuse flux from Galactic sub-threshold sources with optical thickness $\tau_{pp} \simeq 10^{-3}$ and diffuse index $\delta = 1/3$ of in Eq. (13), and the galactic contribution of supernovae and hypernovae with $\Gamma = 2.3$ in Eq. (11).

and this effect can lead to local variations of the CR flux compared to the galactic average.[70] A similar effect can be be expected for non-azimuthally symmetric source distributions.[71,72] Also non-homogenous galactic diffusion models can enhance the hadronic gamma-ray and neutrino emissions in the multi-TeV region[73] as well as increased target gas densities, for instance in the Galactic Center region.[74,75] It was also argued that a time-dependent local CR injection episode could be responsible for local CR spectra that are softer than the galactic average,[25] again leading to an increase of the overall galactic diffuse emission.

However, galactic diffuse emission in the IceCube data is expected to be visible via a correlation with the Galactic Plane. The recent analysis[58] based on the 3-year high-energy starting event (HESE) data[76] showed that even with the poor angular resolution of cascade events the anisotropy produced by a strong galactic diffuse flux should be visible in data. The contribution to the high-energy data with deposited energy above 60 TeV is limited to about 50%. On the other hand, the recent analysis Ref. 77 claims that the 4-year HESE update shows evidence of galactic emission within latitudes $|b| \leq 10°$ above 100 TeV. However, the angular distribution of the muon neutrino data from the recent analysis[78] seem to challenge this claim.

## 3. Quasi-diffuse galactic emission

Another source of extended galactic emission is the cumulative quasi-diffuse flux of galactic sources.[79] This emission consists of sources that are below the detection threshold of the neutrino point source analysis, but could be identified as extended emission concentrated along the Galactic Plane. If $N_N(E_N)$ is the (time-integrated) CR nucleon spectrum of Eq. (1) we can define the galactic neutrino emission from interactions of CR with ambient gas as

$$E_\nu^2 \widehat{Q}_\nu(E_\nu) \simeq \frac{1}{6} c n \kappa \sigma_{pp} N_{\text{act}} \left[ E_N^2 N_N(E_N) \right]_{E_N=20E_\nu}, \qquad (9)$$

where $n = 1 n_0 \, \text{cm}^{-3}$ is the ambient gas density and $N_{\text{act}}$ are the number of *active* sources in the galaxy.

In the following, we consider the case of neutrino emission from SNRs in our Milky Way which can be described by the distribution[80]

$$\rho_{\text{SNR}}(r, z) = \rho_\odot \left( \frac{r}{R_\odot} \right)^\alpha \exp\left( -\beta \frac{r - R_\odot}{R_\odot} \right) \exp\left( -\frac{|z|}{h} \right), \qquad (10)$$

with $\alpha = 2$, $\beta = 3.53$ and $h = 0.181$ kpc. Analogous to the case of galactic diffuse emission we can define express the emission in terms of the total integrated neutrino emission rate of source, $\widehat{Q}_\nu$ and a $J$-factor defined in Eq. (7) with the distribution $\widehat{\rho} = \rho_{\text{SNR}}/N_s$ from Eq. (10) normalized by the total number of sources, $N_s = 2\pi \int dr dz r \rho_{\text{SNR}}(r, z)$. The observable diffuse neutrino flux is then given by Eqs. (6) and (7). The distribution of events is again highly concentrated along the Galactic Plane as shown in the top right plot of Fig. 2.

For SNRs the number of active sources can be estimated by the SN rate of $R_{\text{SN}} \simeq 0.03 R_{\text{SN},-1.5} \, \text{yr}^{-1}$ and the beginning of the snow-plow phase $t_{\text{sp}} \simeq 4 \times 10^4 \mathcal{E}_{\text{ej},51}^{4/17} n_0^{-9/17}$ yr which marks the end of the Sedov phase.[81] This gives $N_{\text{act}} \simeq R_{\text{SN}} t_{\text{sp}} \simeq 1000 N_{\text{act},3}$. From this we can estimate the (time-integrated) contribution of quasi-diffuse emission of SNR following the distribution (10) for $|b| < 2°$ as

$$E_\nu^2 \phi_\nu \simeq 1.4 \times 10^{-6} \frac{\text{GeV}}{\text{cm}^2 \, \text{s} \, \text{sr}}$$

$$\times n_0 \epsilon_{\text{CR},-1} \mathcal{E}_{\text{rot},51} N_{\text{act},3} \left[ \left( \frac{E_N}{\text{GeV}} \right)^{2-\Gamma} \frac{1}{\mathcal{R}_0} e^{-\frac{E_N}{E_{\text{max,p}}}} \right]_{E_N=20E_\nu}, \qquad (11)$$

with $E_{\text{max,p}} \simeq 4.6 B_{-3.5} (\mathcal{E}_{\text{ej},51} V_{\text{ej},-1}/n_0)^{1/3} \text{PeV}$. Energetic supenovae, so-called *hypernovae*, with ejecta energy $\mathcal{E}_{\text{ej}} \simeq 10^{52}$ erg and corresponding higher ejecta velocities may reach neutrino energies that are 10 times larger, but they are less frequent than normal supernovae with only 1%–2% of the supernovae rate.[16,19]

In Fig. 3 we show the estimated flux of supernova remnants (green dashed-dotted line) and hypernova remnants (blue dotted line) from our estimate (11) using $\Gamma \simeq 2.3$. Since the source emission spectrum is much harder than the diffuse CR spectrum the flux is expected to become more important at higher energies, corresponding to neutrino production of CRs close to the knee region.

Note that the previous estimate does not depend on the question if SNR are the main sources of galactic CRs. From the discussion in the introduction we know that any source population that reaches the total galactic CR power $P \simeq \epsilon_{\text{CR}} \mathcal{E}_{\text{ej}} R_{\text{SN}}$ can explain the observed diffuse flux in Eq. (2) above $E_{CR} \gtrsim 1\,\mathrm{GeV}$. We can re-write the total CR nucleon emission rate in terms of the local CR density $n_{CR} = (4\pi/c)\phi_{CR}$ as $E_N^2 \widehat{Q}_N(E_N) \simeq V_{\mathrm{diff}}/\tau_{\mathrm{esc}}(E) E_N^2 n_N(E_N)$. The total (per flavor) neutrino emission can then be related to the CR emission via an effective optical thickness $\tau_{pp}$ as

$$E_\nu^2 \widehat{Q}_\nu(E_\nu) \simeq \frac{1}{6}\kappa_{\mathrm{ine}}\tau_{pp}\left[E_N^2 \widehat{Q}_N(E_N)\right]_{E_N=20E_\nu}. \tag{12}$$

For a source emitting during a time-scale $t_{\mathrm{act}}$ and average gas density $n_{\mathrm{gas}}$ the optical thickness is given as $\tau_{pp} \simeq ct_{\mathrm{act}}n_{\mathrm{gas}}\sigma_{pp} \simeq 3 \times 10^{-4}n_0 t_{\mathrm{act},4}$. For instance, in the case of SNR we can estimate $t_{\mathrm{act}}$ by dynamical time-scale as $10^4$ yr, the beginning of the snowplow phase.[81] Using again the approximation of the local CR nucleon spectrum from Ref. 67, we arrive at the neutrino flux

$$E_\nu^2 \phi_\nu \simeq 7.3 \times 10^{-7}\tau_{pp,-3}R_{\mathrm{MW},20}^2 D_{28}H_0^{-1}\left[\left(\frac{E_N}{3\,\mathrm{GeV}}\right)^\delta E_N^2\phi_N(E_N)\right]_{E_N=20E_\nu}, \tag{13}$$

This flux is shown in Fig. 3 as a magenta dashed line assuming $\tau_{pp} \simeq 10^{-3}$ and diffusion index $\delta = 1/3$. Consistent with our previous estimates of supernova and hypernova remnants the flux envelopes the galactic contribution of these sources.

Similar to the case of the diffuse galactic emission, the contribution of weak galactic sources are constrained by the absence of anisotropies. In Ref. 58 it was shown that candidate galactic sources for the IceCube emission following the galactic distribution of supernova remnants[16,82] or pulsars[47,83] can only maximally contribute at a level of 65% to be consistent with the HESE three-year data.[58] The list of unidentified TeV gamma-ray sources considered in Refs. 19 and 20 is constrained to a level of about 25%.[58]

## 4. Extended galactic emission

In the previous sections we discussed neutrino emission of galactic sources, either directly from individual sources or indirectly via CR interactions in the Galactic Plane. A common feature of this emission is its strong anisotropy towards the Galactic Plane as can be seen in the top plots of Fig. 2. However, there are other emission scenarios that can also predict emission far off the Galactic Plane.

The *Fermi Bubbles* (FBs)[84,85] are hard and uniform emission regions of 1–100 GeV gamma rays detected by the *Fermi* satellite. Their morphology is shown in the lower left plot of Fig. 2. Assuming that the FB originate from the Galactic Center, their extension above and below the Galactic Plane reaches a distance of about 10kpc. The exact mechanism of this gamma-ray emission is unknown. It is possible that the emission is produced via inverse-Compton scattering of electrons that are accelerated throughout the bubble by repeated plasma shocks[86] or via stochastic acceleration.[87] Alternatively, this emission could be due to hadronuclear interactions of CRs, possibly accelerated by star-burst driven winds and convected

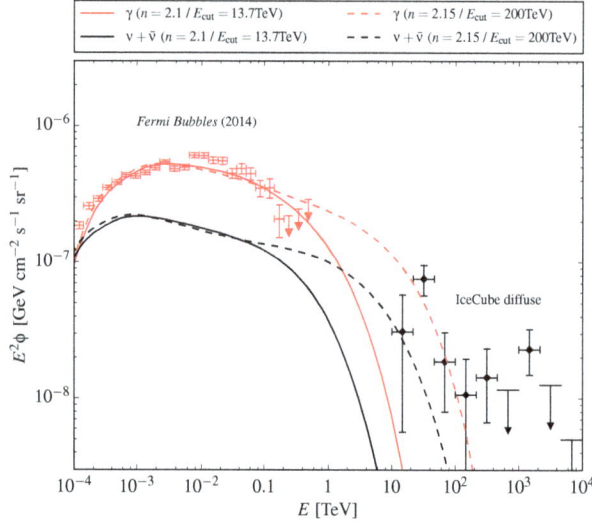

Fig. 4.   The diffuse flux from the *Fermi Bubbles*[85] compared to the diffuse per-flavor neutrino flux observed by IceCube.[68] We show hadronic models of $\gamma$-ray (red lines) and per-flavor neutrino (black lines) emission. The lines show power-law emission of CR protons following the model in Ref. 85 (Eq. (16)) with $n = 2.1$ and $E_{cut} = 13.7$ TeV (solid) or $n = 2.15$ and $E_{cut} = 200$ TeV (dashed), respectively. In the case of a large cutoff the neutrino emission extends into the energy region studied in Ref. 100.

from the galactic center region over time scales of the order of several billion years.[88] This mechanism would also predict the simultaneous emission of gamma rays and neutrinos in the FBs due to the interaction with gas.[16,23,88–90] In Fig. 4 we show two possible hadronic emission scenarios of the FB in comparison to the IceCube diffuse flux.

The HESE 2-year data[91] showed a statistically weak cluster of cascade events in an extended Galactic Center region. This could indicate extended neutrino emission from Galactic Center region, possibly by the FBs.[16,23] However, this weak excess has not grown statistically in two more years of HESE data, which would be consistent with the its interpretation as a background fluctuation. The maximum contribution of the *Fermi Bubble* region to the HESE three-year data has been shown to be limited to about 25%.[58] Future observations of multi-TeV gamma-rays with CTA or HAWC can help to identify the *Fermi Bubble* spectrum and to limit the possible contribution of neutrino emission.[16,92]

It was also speculated that galactic CRs diffusing from the galactic Plane into an extended galactic halo can be visible via hadronic emission.[93] The gas density of this halo is much lower than typical densities in the Galactic Plane but extends to a radius of a few hundred kilo-parsec. Provided that magnetic fields in this environment are sufficiently large, diffusive escape of CRs can take longer than $pp$ interactions and the halo then acts as a calorimeter with a hadronic emission spectrum following that of galactic CR sources. A possible association with the

IceCube observation has been discussed in Refs. 16 and 24. The anisotropy of this emission is expected to be very low and consistent with the neutrino data. A crucial test of this model could be the observation of PeV gamma rays, that are otherwise unobservable from extragalactic distances.[94]

More speculative scenarios assume that neutrinos can also be produced via the decay and/or annihilation of dark matter (DM) in the galactic DM halo, see e.g., Ref. 94 and references therein. Whereas the DM annihilation signal would be highly concentrated at the Galactic Center, the DM decay signal would be more extended. The corresponding morphology is shown in the lower right plot in Fig. 2 using the *Einasto* profile.[95] About 50% of the galactic signal will be within 60° around the Galactic Center, which makes the galactic anisotropy rather weak. As in the case of the *Fermi Bubbles*, the weak anisotropy at the Galactic Center observed in the HESE two-year data has led to speculations that decaying heavy dark matter might be responsible for the signal.[27-30,96-99]

While the initial motivation might have lost its validity with more years of HESE data, the DM decay scenario is still an interesting explanation of the IceCube flux. Since the anisotropy from galactic DM decay is rather weak and the HESE three-year data is consistent with the emission morphology.[58] In addition, the neutrino emission of extragalactic dark matter decay will be at a similar flux level as the galactic contribution as we will discuss in the next section. Hence, the high-energy neutrino events far off the Galactic Center can also be accounted for in this scenario without fine-tuning. Moreover, the galactic DM decay spectra can show various spectral features (line features and/or mass threshold effects) that can help to identify this exotic contribution. The corresponding neutrino emission from galaxies and galaxy clusters could also be identified as (extended) point-source emission in future IceCube searches.[98] It can be expected that the secondary emission from decaying dark matter scenarios will also include PeV gamma-rays that would be a *smoking gun* for galactic contributions.[94]

## 5. Comparison of galactic and extragalactic emission

Any mechanism that provides a strong anisotropic galactic neutrino emission is also expected to contribute via an isotropic emission from similar, but distant galaxies. If $\widehat{Q}_\nu(E)$ is the total spectral emission rate of our galaxy then the contribution of extragalactic sources can be written as the red-shift integral[101]

$$\phi_\nu^{\text{iso}}(E_\nu) = \frac{c}{4\pi} \int \frac{dz}{H(z)} \rho(z) \, \widehat{Q}_\nu((1+z)E_\nu), \tag{14}$$

where $H(z)/H_0 = \sqrt{\Omega_\Lambda + \Omega_{\text{m}}(1+z)^3}$ is the red-shift dependent Hubble constant in the matter-dominated era with $H_0 = 67 \text{ km s}^{-1} \text{Mpc}^{-1}$ and $\Omega_{\text{m}} = 1 - \Omega_\Lambda \simeq 0.31$.[102] The local density of normal galaxies is of the order of $\rho(0) \equiv \rho_0 \simeq (10^{-3}-10^{-2}) \text{ Mpc}^{-3}$ with evolution closely following that of the star formation rate (SFR).[103,104]

On the other hand, the galactic contribution (averaged over the whole sky) can be written as $\phi^{\text{Gal}} \simeq \widehat{Q}_\nu(E_\nu)\widehat{J}/4\pi$, with $J$-factor integrated over $4\pi$. For instance, for the quasi two-dimensional SNR distribution (10) we find $\widehat{J} \simeq (16\,\text{kpc})^{-2}$. Approximating the neutrino flux as power-law with index $\Gamma$, we have the relation

$$\frac{\phi_\nu^{\text{iso}}}{\phi_\nu^{\text{Gal}}} \simeq 10^{-3}\xi_z \left(\frac{\rho_0}{10^{-3}\,\text{Mpc}^{-3}}\right)\left(\frac{(16\,\text{kpc})^{-2}}{\widehat{J}}\right), \tag{15}$$

with source evolution factor $\xi_z = \mathcal{O}(1)$ ($\xi_z \simeq 0.5$ for no evolution and $\Gamma \simeq 2$ and $\xi_z \simeq 2.4$ for evolution following the SFR). The local galactic emission is then typically much stronger than the corresponding extragalactic contribution.

On the other hand, the three-dimensional dark matter distribution has a smaller integrated $J$-factor of $\widehat{J} \simeq (120\,\text{kpc})^{-2}$. In the case of dark matter decay the local density in Eq. (15) is given by $\rho_0 \simeq \Omega_{\text{cdm}}\rho_{\text{cr}}/M_{\text{DM}}^{\text{total}} \simeq 10^{-2}\,\text{Mpc}^{-3}$.[102] Therefore, up to uncertainties of the galaxy density and the dark matter distribution, the isotropic emission from extragalactic dark matter is expected to be at a similar level as the galactic emission.[28,98]

## Acknowledgments

I would like to thank Yang Bai, Vernon Barger, Ran Lu and Kohta Murase for collaborating on projects that contributed to this work. I acknowledge support by the U.S. National Science Foundation (NSF) under grants OPP-0236449 and PHY-0236449.

## References

1. W. Baade and F. Zwicky, Cosmic rays from supernovae, *Proceedings of the National Academy of Science.* **20** (1934) 259–263. doi: 10.1073/pnas.20.5.259.
2. P. Blasi and E. Amato, Diffusive propagation of cosmic rays from supernova remnants in the galaxy. I: spectrum and chemical composition, *JCAP.* **1201** (2012) 010. doi: 10.1088/1475-7516/2012/01/010.
3. K. Olive, *et al.*, Review of particle physics, *Chin. Phys.* **C38** (2014) 090001. doi: 10.1088/1674-1137/38/9/090001.
4. R. Aloisio and V.S. Berezinsky, Anti-GZK effect in UHECR diffusive propagation, *Astrophys. J.* **625** (2005) 249–255. doi: 10.1086/429615.
5. K. Ioka and P. Meszaros, Hypernova and gamma-ray burst remnants as TeV unidentified sources, *Astrophys. J.* **709** (2010) 1337–1342. doi: 10.1088/0004-637X/709/2/1337.
6. J. Gunn and J. Ostriker, Acceleration of high-energy cosmic rays by pulsars, *Phys. Rev. Lett.* **22** (1969) 728–731. doi: 10.1103/PhysRevLett.22.728.
7. K. Murase, P. Meszaros, and B. Zhang, Probing the birth of fast rotating magnetars through high-energy neutrinos, *Phys. Rev.* **D79** (2009) 103001. doi: 10.1103/PhysRevD.79.103001.
8. A. Levinson and E. Waxman, Probing microquasars with TeV neutrinos, *Phys. Rev. Lett.* **87** (2001) 171101. doi: 10.1103/PhysRevLett.87.171101.

9. C. Distefano, D. Guetta, E. Waxman, and A. Levinson, Neutrino flux predictions for known galactic microquasars, *Astrophys. J.* **575** (2002) 378–383. doi: 10.1086/341144.

10. L. Anchordoqui, F. Halzen, T. Montaruli, and A. O'Murchadha, Neutrino flux from cosmic ray accelerators in the cygnus spiral arm of the galaxy, *Phys. Rev.* **D76** (2007) 067301. doi: 10.1103/PhysRevD.76.067301,10.1103/PhysRevD.77.069906.

11. A. Kappes, J. Hinton, C. Stegmann, and F.A. Aharonian, Potential neutrino signals from galactic gamma-ray sources, *Astrophys. J.* **656** (2007) 870–896. doi: 10.1086/508936,10.1086/518161.

12. M.D. Kistler and J.F. Beacom, Guaranteed and prospective galactic TeV neutrino sources, *Phys. Rev.* **D74** (2006) 063007. doi: 10.1103/PhysRevD.74.063007.

13. J.F. Beacom and M.D. Kistler, Dissecting the cygnus region with TeV gamma rays and neutrinos, *Phys. Rev.* **D75** (2007) 083001. doi: 10.1103/PhysRevD.75.083001.

14. F. Halzen and A. O Murchadha, Neutrinos from cosmic ray accelerators in the cygnus region of the galaxy, *Phys. Rev.* **D76** (2007) 123003. doi: 10.1103/PhysRevD.76.123003.

15. M. Gonzalez-Garcia, F. Halzen, and S. Mohapatra, Identifying galactic PeVatrons with neutrinos, *Astropart. Phys.* **31** (2009) 437–444. doi: 10.1016/j.astropartphys.2009.05.002.

16. M. Ahlers and K. Murase, Probing the galactic origin of the IceCube excess with gamma-rays, *Phys. Rev.* **D90** (2014) 023010. doi: 10.1103/PhysRevD.90.023010.

17. J.C. Joshi, W. Winter, and N. Gupta, How many of the observed neutrino events can be described by cosmic ray interactions in the milky way?, *Mon. Not. Roy. Astron. Soc.* **439**(4) (2014) 3414–3419. doi: 10.1093/mnras/stu189,10.1093/mnras/stu2132. [Erratum: *ibid.* **446**(1) (2014) 892].

18. M. Kachelriess and S. Ostapchenko, Neutrino yield from galactic cosmic rays, *Phys. Rev.* **D90** (2014) 083002. doi: 10.1103/PhysRevD.90.083002.

19. D. Fox, K. Kashiyama, and P. Meszaros, Sub-PeV neutrinos from TeV unidentified sources in the galaxy, *Astrophys. J.* **774** (2013) 74. doi: 10.1088/0004-637X/774/1/74.

20. M. Gonzalez-Garcia, F. Halzen, and V. Niro, Reevaluation of the prospect of observing neutrinos from galactic sources in the light of recent results in gamma ray and neutrino astronomy, *Astropart. Phys.* **57–58** (2014) 39–48. doi: 10.1016/j.astropartphys.2014.04.001.

21. L.A. Anchordoqui, H. Goldberg, T.C. Paul, L.H.M. da Silva, and B.J. Vlcek, Estimating the contribution of galactic sources to the diffuse neutrino flux, *Phys. Rev.* **D90** (12) (2014) 123010. doi: 10.1103/PhysRevD.90.123010.

22. S. Razzaque, The galactic center origin of a subset of IceCube neutrino events, *Phys. Rev.* **D88** (2013) 081302. doi: 10.1103/PhysRevD.88.081302.

23. C. Lunardini, S. Razzaque, K.T. Theodoseau, and L. Yang, Neutrino events at IceCube and the fermi bubbles, *Phys. Rev.* **D90** (2014) 023016. doi: 10.1103/PhysRevD.90.023016.

24. A.M. Taylor, S. Gabici, and F. Aharonian, A galactic halo origin of the neutrinos detected by IceCube, *Phys. Rev.* **D89** (2014) 103003. doi: 10.1103/PhysRevD.89.103003.

25. A. Neronov, D. Semikoz, and C. Tchernin, PeV neutrinos from interactions of cosmic rays with the interstellar medium in the Galaxy, *Phys. Rev.* **D89** (2014) 103002. doi: 10.1103/PhysRevD.89.103002.

26. N. Gupta, Galactic PeV neutrinos, *Astropart. Phys.* **48** (2013) 75–77. doi: 10.1016/j.astropartphys.2013.07.003.

27. B. Feldstein, A. Kusenko, S. Matsumoto, and T.T. Yanagida, Neutrinos at IceCube from heavy decaying dark matter, *Phys. Rev.* **D88**(1) (2013) 015004. doi: 10.1103/PhysRevD.88.015004.

28. A. Esmaili and P.D. Serpico, Are IceCube neutrinos unveiling PeV-scale decaying dark matter?, *JCAP.* **1311** (2013) 054. doi: 10.1088/1475-7516/2013/11/054.

29. Y. Bai, R. Lu, and J. Salvado, Geometric compatibility of IceCube TeV-PeV neutrino excess and its galactic dark matter origin, *JHEP.* **01** (2016) 161. doi: 10.1007/JHEP01(2016)161.

30. J.F. Cherry, A. Friedland, and I.M. Shoemaker, Neutrino portal dark matter: From dwarf galaxies to IceCube. (2014).

31. M.G. Aartsen, *et al.*, Searches for extended and point-like neutrino sources with four years of IceCube data, *Astrophys. J.* **796**(2) (2014) 109. doi: 10.1088/0004-637X/796/2/109.

32. S. Adrian-Martinez, *et al.*, Searches for point-like and extended neutrino sources close to the galactic centre using the ANTARES neutrino telescope, *Astrophys. J.* **786** (2014) L5. doi: 10.1088/2041-8205/786/1/L5.

33. W. Bednarek, G.F. Burgio, and T. Montaruli, Galactic discrete sources of high energy neutrinos, *New Astron. Rev.* **49** (2005) 1. doi: 10.1016/j.newar.2004.11.001.

34. M. Ackermann, *et al.*, Detection of the characteristic pion-decay signature in supernova remnants, *Science.* **339** (2013) 807. doi: 10.1126/science.1231160.

35. T.K. Gaisser, *Cosmic Rays and Particle Physics.* 1990.

36. L.O. Drury, F.A. Aharonian, and H.J. Volk, The gamma-ray visibility of supernova remnants: A test of cosmic ray origin, *Astron. Astrophys.* **287** (1994) 959–971.

37. T.K. Gaisser, R.J. Protheroe, and T. Stanev, Gamma-ray production in supernova remnants, *Astrophys. J.* **492** (1998) 219. doi: 10.1086/305011.

38. J. Alvarez-Muniz and F. Halzen, Possible high-energy neutrinos from the cosmic accelerator RX J1713.7-3946, *Astrophys. J.* **576** (2002) L33–L36. doi: 10.1086/342978.

39. M.L. Costantini and F. Vissani, Expected neutrino signal from supernova remnant RX J1713.7-3946 and flavor oscillations, *Astropart. Phys.* **23** (2005) 477–485. doi: 10.1016/j.astropartphys.2005.03.003.

40. R.J. Protheroe, W. Bednarek, and Q. Luo, Gamma-rays and neutrinos from very young supernova remnants, *Astropart. Phys.* **9** (1998) 1–14. doi: 10.1016/S0927-6505(98)00014-0.

41. P. Blasi, R.I. Epstein, and A.V. Olinto, Ultrahigh-energy cosmic rays from young neutron star winds, *Astrophys. J.* **533** (2000) L123. doi: 10.1086/312626.

42. J.H. Beall and W. Bednarek, Neutrinos from early phase, pulsar driven supernovae, *Astrophys. J.* **569** (2002) 343–348. doi: 10.1086/339276.

43. K. Fang, K. Kotera, and A.V. Olinto, Newly-born pulsars as sources of ultrahigh energy cosmic rays, *Astrophys. J.* **750** (2012) 118. doi: 10.1088/0004-637X/750/2/118.

44. K. Fang, K. Kotera, and A.V. Olinto, Ultrahigh energy cosmic ray nuclei from extragalactic pulsars and the effect of their galactic counterparts, *JCAP.* **1303** (2013) 010. doi: 10.1088/1475-7516/2013/03/010.

45. W. Bednarek and R.J. Protheroe, Gamma-rays and neutrinos from the crab nebula produced by pulsar accelerated nuclei, *Phys. Rev. Lett.* **79** (1997) 2616–2619. doi: 10.1103/PhysRevLett.79.2616.

46. S. Nagataki, High-energy neutrinos produced by interactions of relativistic protons in shocked pulsar winds, *Astrophys. J.* **600** (2004) 883–904. doi: 10.1086/380095.

47. K. Fang, K. Kotera, K. Murase, and A.V. Olinto, Testing the newborn pulsar origin of ultrahigh energy cosmic rays with EeV neutrinos, *Phys. Rev.* **D90**(10) (2014) 103005. doi: 10.1103/PhysRevD.90.103005.

48. K. Fang, T. Fujii, T. Linden, and A.V. Olinto, Is the ultra-high energy cosmic-ray excess observed by the telescope array correlated with IceCube neutrinos?, *Astrophys. J.* **794**(2) (2014) 126. doi: 10.1088/0004-637X/794/2/126.

49. D. Guetta and E. Amato, Neutrino flux predictions for galactic plerions, *Astropart. Phys.* **19** (2003) 403–407. doi: 10.1016/S0927-6505(02)00221-9.

50. E. Amato, D. Guetta, and P. Blasi, Signatures of high energy protons in pulsar winds, *Astron. Astrophys.* **402** (2003) 827–836. doi: 10.1051/0004-6361:20030279.

51. W. Bednarek, Production of neutrons, neutrinos and gamma-rays by a very fast pulsar in the galactic center region, *Mon. Not. Roy. Astron. Soc.* **331** (2002) 483. doi: 10.1046/j.1365-8711.2002.05207.x.

52. W. Bednarek, Gamma-rays and cosmic-rays from a pulsar in Cygnus ob2, *Mon. Not. Roy. Astron. Soc.* **345** (2003) 847. doi: 10.1046/j.1365-8711.2003.06997.x.

53. I.F. Mirabel and L.F. Rodriguez, Sources of relativistic jets in the galaxy, *Ann. Rev. Astron. Astrophys.* **37** (1999) 409–443. doi: 10.1146/annurev.astro.37.1.409.

54. M.J. Rees, Black hole models for active galactic nuclei, *Ann. Rev. Astron. Astrophys.* **22** (1984) 471–506. doi: 10.1146/annurev.aa.22.090184.002351.

55. T.K. Gaisser and T. Stanev, Calculation of neutrino flux from Cygnus X-3, *Phys. Rev. Lett.* **54** (1985) 2265. doi: 10.1103/PhysRevLett.54.2265.

56. E.W. Kolb, M.S. Turner, and T.P. Walker, The production and detection of high-energy neutrinos from Cygnus X-3, *Phys. Rev.* **D32** 1145 (1985). doi: 10.1103/PhysRevD.33.859.2,10.1103/PhysRevD.32.1145. [Erratum: *idib.* **D33** (1986) 859].

57. M. Bartosik, W. Bednarek, and A. Sierpowska, Neutrinos produced by nuclei injected by young pulsars inside compact massive binaries, *Nucl. Phys. B: Proc. Sup.* **143** (2005) 531–531. doi: 10.1016/j.nuclphysbps.2005.01.196.

58. M. Ahlers, Y. Bai, V. Barger, and R. Lu, Galactic neutrinos in the TeV to PeV range, *Phys. Rev.* **D93**(1) (2016) 013009. doi: 10.1103/PhysRevD.93.013009.

59. F. Stecker, Diffuse fluxes of cosmic high-energy neutrinos, *Astrophys. J.* **228** (1979) 919–927. doi: 10.1086/156919.

60. G. Domokos, B. Elliott, and S. Kovesi-Domokos, Cosmic neutrino production in the Milky Way, *J. Phys.* **G19** (1993) 899–912. doi: 10.1088/0954-3899/19/6/010.

61. V. Berezinsky, T. Gaisser, F. Halzen, and T. Stanev, Diffuse radiation from cosmic ray interactions in the galaxy, *Astropart. Phys.* **1** (1993) 281–288. doi: 10.1016/0927-6505(93)90014-5.

62. D.L. Bertsch, T.M. Dame, C.E. Fichtel, S.D. Hunter, P. Sreekumar, J.G. Stacy, and P. Thaddeus, Diffuse gamma-ray emission in the galactic plane from cosmic-ray, matter, and photon interactions, *Astrophys.* **416** (1993) 587. doi: 10.1086/173261.

63. G. Ingelman and M. Thunman, Particle production in the interstellar medium. (1996).

64. C. Evoli, D. Grasso, and L. Maccione, Diffuse neutrino and gamma-ray emissions of the galaxy above the TeV, *JCAP.* **0706** (2007) 003. doi: 10.1088/1475-7516/2007/06/003.

65. S.R. Kelner, F.A. Aharonian, and V.V. Bugayov, Energy spectra of gamma-rays, electrons and neutrinos produced at proton-proton interactions in the very high energy regime, *Phys. Rev.* **D74** (2006) 034018. doi: 10.1103/PhysRevD.74.034018,10.1103/PhysRevD.79.039901. [Erratum: *ibid.* **D79** (2009) 039901.]

66. M.M. Block and F. Halzen, Experimental confirmation that the proton is asymptotically a black disk, *Phys. Rev. Lett.* **107** (2011) 212002. doi: 10.1103/PhysRevLett.107.212002.

67. T.K. Gaisser, Atmospheric leptons, *EPJ Web Conf.* **52** (2013) 09004. doi: 10.1051/epjconf/20125209004.

68. M.G. Aartsen, *et al.*, A combined maximum-likelihood analysis of the high-energy astrophysical neutrino flux measured with IceCube, *Astrophys. J.* **809**(1) (2015) 98. doi: 10.1088/0004-637X/809/1/98.

69. F. Casse, M. Lemoine, and G. Pelletier, Transport of cosmic rays in chaotic magnetic fields, *Phys. Rev.* **D65** (2002) 023002. doi: 10.1103/PhysRevD.65.023002.

70. F. Effenberger, H. Fichtner, K. Scherer, and I. Busching, Anisotropic diffusion of galactic cosmic ray protons and their steady-state azimuthal distribution, *Astron. Astrophys.* **547** (2012) A120. doi: 10.1051/0004-6361/201220203.

71. D. Gaggero, L. Maccione, G. Di Bernardo, C. Evoli, and D. Grasso, Three-dimensional model of cosmic-ray lepton propagation reproduces data from the alpha magnetic spectrometer on the international space station, *Phys. Rev. Lett.* **111** (2013) 021102. doi: 10.1103/PhysRevLett.111.021102.

72. M. Werner, R. Kissmann, A.W. Strong, and O. Reimer, Spiral arms as cosmic ray source distributions, *Astropart. Phys.* **64** (2014) 18–33. doi: 10.1016/j.astropartphys.2014.10.005.

73. D. Gaggero, D. Grasso, A. Marinelli, A. Urbano, and M. Valli, The gamma-ray and neutrino sky: A consistent picture of Fermi-LAT, Milagro, and IceCube results, *Astrophys. J.* **815** (2) (2015) L25. doi: 10.1088/2041-8205/815/2/L25.

74. R.M. Crocker, F. Melia, and R.R. Volkas, Neutrinos from the galactic center in the light of HESS, *Astrophys. J.* **622** (2005) L37–L40. doi: 10.1086/429539.

75. J. Candia, Detectable neutrino fluxes due to enhanced cosmic ray densities in the galactic center region, *JCAP.* **0511** (2005) 002. doi: 10.1088/1475-7516/2005/11/002.

76. M. Aartsen, *et al.*, Observation of high-energy astrophysical neutrinos in three years of IceCube data, *Phys. Rev. Lett.* **113** (2014) 101101. doi: 10.1103/PhysRevLett.113.101101.

77. A. Neronov and D.V. Semikoz, Evidence the galactic contribution to the IceCube astrophysical neutrino flux, *Astropart. Phys.* **75** (2016) 60–63. doi: 10.1016/j.astropartphys.2015.11.002.

78. M.G. Aartsen, *et al.*, Evidence for astrophysical muon neutrinos from the northern sky with IceCube, *Phys. Rev. Lett.* **115** (8) (2015) 081102. doi: 10.1103/PhysRevLett.115.081102.

79. S. Casanova and B.L. Dingus, Constraints on the TeV source population and its contribution to the galactic diffuse TeV emission, *Astropart. Phys.* **29** (2008) 63–69. doi: 10.1016/j.astropartphys.2007.11.008.

80. G.L. Case and D. Bhattacharya, A new sigma-d relation and its application to the galactic supernova remnant distribution, *Astrophys. J.* **504** (1998) 761. doi: 10.1086/306089.

81. J.M. Blondin, E.B. Wright, K.J. Borkowski, and S.P. Reynolds, Transition to the radiative phase in supernova remnants, *Astrophys. J.* **500** (1998) 342–354. doi: 10.1086/305708.

82. M. Mandelartz and J. Becker Tjus, Prediction of the diffuse neutrino flux from cosmic ray interactions near supernova remnants, *Astropart. Phys.* **65** (2014) 80–100. doi: 10.1016/j.astropartphys.2014.12.002.

83. P. Padovani and E. Resconi, Are both BL Lacs and pulsar wind nebulae the astro-physical counterparts of IceCube neutrino events?, *Mon. Not. Roy. Astron. Soc.* **443** (2014) 474–484. doi: 10.1093/mnras/stu1166.

84. M. Su, T.R. Slatyer, and D.P. Finkbeiner, Giant gamma-ray bubbles from Fermi-LAT: AGN activity or bipolar galactic wind?, *Astrophys. J.* **724** (2010) 1044–1082. doi: 10.1088/0004-637X/724/2/1044.

85. M. Ackermann, *et al.*, The spectrum and morphology of the Fermi Bubbles, *Astro-phys. J.* (2014). doi: 10.1088/0004-637X/793/1/64.

86. K.-S. Cheng, D.O. Chernyshov, V.A. Dogiel, C.-M. Ko, and W.-H. Ip, Origin of the Fermi Bubble, *Astrophys. J.* **731** (2011) L17. doi: 10.1088/2041-8205/731/1/L17.

87. P. Mertsch and S. Sarkar, Fermi gamma-ray 'bubbles' from stochastic acceleration of electrons, *Phys. Rev. Lett.* **107** (2011) 091101. doi: 10.1103/PhysRevLett.107.091101.

88. R.M. Crocker and F. Aharonian, The Fermi Bubbles: Giant, multi-billion-year-old reservoirs of galactic center cosmic rays, *Phys. Rev. Lett.* **106** (2011) 101102. doi: 10.1103/PhysRevLett.106.101102.

89. C. Lunardini and S. Razzaque, High energy neutrinos from the Fermi Bubbles, *Phys. Rev. Lett.* **108** (2012) 221102. doi: 10.1103/PhysRevLett.108.221102.

90. I. Cholis, Searching for the high-energy neutrino counterpart signals: The case of the Fermi bubbles signal and of dark matter annihilation in the inner Galaxy, *Phys. Rev.* **D88**(6) (2013) 063524. doi: 10.1103/PhysRevD.88.063524.

91. M.G. Aartsen, *et al.*, Evidence for high-energy extraterrestrial neutrinos at the Ice-Cube detector, *Science.* **342** (2013) 1242856. doi: 10.1126/science.1242856.

92. C. Lunardini, S. Razzaque, and L. Yang, Multimessenger study of the Fermi bubbles: Very high energy gamma rays and neutrinos, *Phys. Rev.* **D92**(2) (2015) 021301. doi: 10.1103/PhysRevD.92.021301.

93. R. Feldmann, D. Hooper, and N.Y. Gnedin, Circum-galactic gas and the isotropic gamma ray background, *Astrophys. J.* **763** (2013) 21. doi: 10.1088/0004-637X/763/1/21.

94. K. Murase and J.F. Beacom, Constraining very heavy dark matter using diffuse backgrounds of neutrinos and cascaded gamma rays, *JCAP.* **1210** (2012) 043. doi: 10.1088/1475-7516/2012/10/043.

95. A.W. Graham, D. Merritt, B. Moore, J. Diemand, and B. Terzic, Empirical models for dark matter halos. II. Inner profile slopes, dynamical profiles, and $\rho/\sigma^3$, *Astron.J.* **132** (2006) 2701–2710. doi: 10.1086/508990.

96. A. Bhattacharya, M.H. Reno, and I. Sarcevic, Reconciling neutrino flux from heavy dark matter decay and recent events at IceCube, *JHEP.* **1406** (2014) 110. doi: 10.1007/JHEP06(2014)110.

97. A. Esmaili, S.K. Kang, and P.D. Serpico, IceCube events and decaying dark matter: hints and constraints, *JCAP.* **1412** (12) (2014) 054. doi: 10.1088/1475-7516/2014/12/054.

98. K. Murase, R. Laha, S. Ando, and M. Ahlers, Testing the dark matter scenario for PeV neutrinos observed in IceCube, *Phys. Rev. Lett.* **115** (7) (2015) 071301. doi: 10.1103/PhysRevLett.115.071301.

99. A. Esmaili and P.D. Serpico, Gamma-ray bounds from EAS detectors and heavy decaying dark matter constraints, *JCAP.* **1510**(10) (2015) 014. doi: 10.1088/1475-7516/2015/10/014.

100. M.G. Aartsen, *et al.*, Atmospheric and astrophysical neutrinos above 1 TeV interact-ing in IceCube, *Phys. Rev.* **D91**(2) (2015) 022001. doi: 10.1103/PhysRevD.91.022001.

101. E. Waxman and J.N. Bahcall, High-energy neutrinos from astrophysical sources: An upper bound, *Phys. Rev.* **D59** (1999) 023002. doi: 10.1103/PhysRevD.59.023002.
102. P.A.R. Ade, *et al.*, Planck 2015 results. XIII. Cosmological parameters. (2015).
103. A.M. Hopkins and J.F. Beacom, On the normalisation of the cosmic star formation history, *Astrophys. J.* **651** (2006) 142–154. doi: 10.1086/506610.
104. H. Yuksel, M.D. Kistler, J.F. Beacom, and A.M. Hopkins, Revealing the high-redshift star formation rate with gamma-ray bursts, *Astrophys. J.* **683** (2008) L5–L8. doi: 10.1086/591449.

# Chapter 5

# Observations of Diffuse Fluxes of Cosmic Neutrinos

Christopher H. Wiebusch

*RWTH Aachen University, III. Physikalisches Institut,*
*Otto Blumenthal Strasse, 52074 Aachen, Germany,*
*wiebusch@physik.rwth-aachen.de*

In this contribution, current observations results of a diffuse flux of high-energy astrophysical neutrinos are reviewed. In order to understand the scientific implications, these measurements in different detection channels are discussed and results are compared. The discussion focuses on the energy spectrum, the flavor ratio, and large-scale anisotropy.

## 1. Introduction

For a long time, the detection of high-energy cosmic neutrinos to serve as cosmic messengers has been an outstanding goal of astroparticle physics. Their observation was already suggested by Markov[1] in the 1960. The proposed method was the detection of up going muons as the signature of a charged-current (CC) muon-neutrino interaction below the detector. Based on this signature, atmospheric neutrinos were discovered in deep underground detectors.[2,3] Soon it was realized that the expected astrophysical fluxes would be small and cubic-kilometer-sized detectors would be needed to accomplish the goal.[4] A key concept became the instrumentation of optically transparent natural media with photo sensors to construct large Cherenkov detectors. A major step was achieved by the BAIKAL collaboration,[5] which first succeeded in installing and operating a large volume Cherenkov detector using the deep water of Lake Baikal. This effort was rewarded by the first observation of atmospheric neutrino events in an open natural environment[6] (see contribution by V. Aynutdinov). Shortly after, the AMANDA neutrino telescope successfully demonstrated the feasibility of the construction and operation of a large Cherenkov detector in glacier ice.[7] It was the first neutrino telescope to observe high-energy atmospheric neutrinos in larger quantity[8] and to exclude optimistic astrophysical models.[9] In parallel with these efforts neutrino telescopes in deep oceans were also brought into operation. The ANTARES neutrino telescope[10] (see also contribution by P. Coyle) increased the effective area with respect to AMANDA and has been operational since 2006. In all these experiments, no indications of cosmic neutrinos had been found.[11]

Based on the success of AMANDA, the IceCube detector was designed.[12] In total, 5160 large area optical sensors have been deployed in the Antarctic ice at the geographical South Pole. They detect the Cherenkov light produced by secondary leptons and hadrons as a result of charged current (CC) and neutral current (NC) neutrino-nucleon interactions inside and outside the instrumented volume. The instrumented depth ranges from 1450 m to 2450 m below the surface of the ice. The sensors are attached to 86 vertical cable strings, with 60 sensors each, and have a horizontal spacing of about 125 m between strings on a hexagonal grid. Along the strings, the spacing between sensors is about 17 m, resulting in about 1 km$^3$ of instrumented volume. IceCube was completed in December 2010 and fully commissioned in its final configuration in May 2011. Already in its earlier configurations, with the partly-installed detector, a substantial exposure was accumulated and first indications for an astrophysical signal were obtained.[13,14]

## 2. Summary of detection signatures

The detection of neutrinos from cosmic accelerators (see contributions in this volume by Mészáros, K. Murase, E. Waxman, P. Lipari and M. Ahlers) is mainly based on their hard energy spectrum, which is expected to follow the spectrum of accelerated primary cosmic rays and thus to follow a power law with a hard spectral index $\phi \simeq \phi_0 \cdot E^{-2}$. The largest signal is expected from close below the horizon and above, as the Earth becomes almost opaque to neutrinos above $\sim$100 TeV–1 PeV. Background signals are from cosmic-ray-induced atmospheric muons and neutrinos. Those, however, exhibit a substantially softer spectrum at high energies. Other powerful methods to detect astrophysical neutrinos above this background rely on direction and time correlations of the measured events. However, this requires strong individual sources. Alternatively, the cumulative flux of all cosmic sources is expected to exceed the atmospheric backgrounds at high energies, typically above 100 TeV. In this paper, we focus on the detection of diffuse fluxes. Depending on the luminosity density and strength of the sources, this approach is very promising for the detection of a population of abundant but individually weak extra galactic sources. Detailed discussions can be found, e.g. in Refs. 15–17.

Searching for diffuse neutrino fluxes at high energies requires a rigorous rejection of the overwhelming atmospheric muon background and a precise modelling of the partly irreducible atmospheric neutrino backgrounds. The atmospheric muon background that penetrates from the surface to the depth of the detector can be rejected by focusing on up going events and/or events that interact within the instrumented volume. Atmospheric neutrinos can only be rejected if they arrive at the detector from above and the corresponding air shower is either tagged by a surface detector[18] or by observing correlated atmospheric muons.[19] For both types of background, the signal-to-background ratio increases with the energy threshold of the event selection.

Based on these signatures, neutrino telescopes can be sensitive to all neutrino flavors, in particular when combining different detection channels and strategies. The basic signature of a muon neutrino is a high-energy muon track from deep inelastic CC neutrino-nucleon interactions. Electron neutrinos produce an electromagnetic cascade superimposed by a hadronic cascade at the interaction vertex. The length scale of the cascades is small compared to the spacing of detector sensors. These events are called cascade-like events. Tau neutrinos mostly produce a cascade signature very similar to electron neutrinos with two exceptions. First, at high energies above a PeV, the tau travels typically $50\,\mathrm{m} \cdot E_\tau/\mathrm{PeV}$ before it decays. This results in a characteristic signature of two spatially separated cascades, called double-bang.[20] At all energies, the tau may decay leptonically into a muon with a branching ratio of $\sim 17\%$ contributing to the track-like signature of muon neutrinos. All flavors contribute equally to cascade-like events via NC interactions.

## 3. Observational status of different detection channels

### 3.1. *High-energy starting events*

Remarkably, the discovery of an astrophysical neutrino signal by IceCube was achieved not in the muon channel that has been the intuitively assumed baseline channel for decades but with a new type of analysis: the high-energy starting event analysis (HESE).[21,22]

The HESE analysis searches for neutrinos interacting inside the detector and is as such sensitive to all neutrino flavors and the full sky. The selection splits the detector into an outer region, which is used to tag and veto muon backgrounds from outside and an inner fiducial mass of about 0.4 Gt. Accepted events are required to deposit a visible energy of more than $20-30\,\mathrm{TeV}$ inside the detector. Furthermore, the earliest observed Cherenkov photons need to have been recorded within the fiducial volume, while in the veto region no early signal in excess of the noise level is allowed. The analysis is based on the combination of essentially four innovative new methods that were not available during the operation of AMANDA and were not available during the design and construction phases of IceCube.

(i) The selection of starting events allows for a very simple all-sky search of all flavors, with a high significance because a large fraction of the energy is deposited at the interaction vertex. This has greatly improved the sensitivity, e.g. with respect to the single flavor up going muon channel.

(ii) The precise analysis of measured photomultiplier waveforms enabled the possibility to reasonably reconstruct the direction and energy of cascade-like events including a good estimate of the uncertainty. This allowed the quantifying of the significance of each event as the backgrounds strongly depend on the observed zenith angle.

(iii) The usage of the outer layers of IceCube as veto allowed modelling and quantifying the remaining atmospheric muon background with good precision based on experimental data. It has been shown that the background related to the inefficiency of the veto falls off rapidly with energy and becomes insignificant above ~60 TeV.

(iv) It was realized that the method of an atmospheric neutrino veto[19] could be successfully applied, greatly reducing the atmospheric neutrino background in the down going region. The corresponding angular distribution is particularly important to unambiguously reject the hypothesis of a purely atmospheric origin, in particular as the high-energy atmospheric neutrino flux from prompt decays of heavy quarks is largely uncertain.[23]

Combining all four methods in interpreting the observation made the detection of a cosmic neutrino signal evident and allowed rejecting the hypothesis of atmospheric origin with high confidence. As a side remark, this underlines the importance of a not-too-specific optimization of large-scale instruments that aim to explore unknown physics. Multi-purpose instruments that deliver higher data and information quality than minimally required allow for a large flexibility in methods and foster unforeseeable innovations which evolve only during the operation of the instrument.

Currently, data from three years of operation have been published.[22] A pure atmospheric origin of the observed signal is rejected with a confidence level of $5.7\sigma$. Most recently, a fourth year of data has been preliminarily released[24] that further increases the significance of the observation to $6.5\sigma$. The observed distributions of energy and zenith, see Fig. 1, agree well with the expectation of a hard astrophysical spectrum and are clearly not compatible with the expectation from atmospheric backgrounds. The per flavor astrophysical flux measured with three years of data

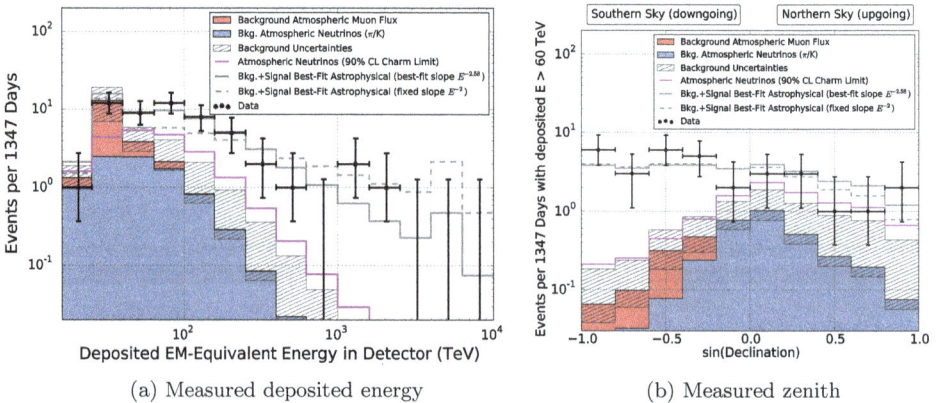

(a) Measured deposited energy                    (b) Measured zenith

Fig. 1.   Results for the high-energy starting event analysis of IceCube data.

is $E^2\phi(E) = 0.84 \pm 0.3 \times 10^{-8}\,\mathrm{GeVcm^{-2}s^{-1}sr^{-1}}$, assuming a spectral index $\gamma = 2$ in the energy range between 60 TeV and 2 PeV. A significant clustering of event directions has not been observed.

A particularly interesting extension of this analysis is to lower the energy threshold for this search. This is achieved by gradually increasing the veto thickness. The loss in fiducial volume is compensated by larger fluxes at lower energy. The first analysis of this type[25] using two years of data has been able to lower the energy threshold to about 1 TeV and extract the astrophysical signal down to about 10 TeV.

### 3.2. *Cascade channel*

Closely related to the analysis of starting events are searches dedicated to cascade-type events. Here, based on the event reconstruction, track-like event topologies are specifically rejected and a reasonably pure cascade-like sample is obtained. Backgrounds are atmospheric muons with catastrophic energy losses that outshine the muon and thus mimic a cascade-like signature. These can be suppressed by requiring containment of the interaction vertex similar to the HESE analysis. The flux from conventional atmospheric electron neutrinos is considerably smaller (about a factor of 20) than that of muon neutrinos, and hence poses a relatively small background. The largest background uncertainty arises from prompt atmospheric neutrinos. Advantages of this analysis are a good energy resolution for these contained cascades on the order of the energy scale uncertainty[26] of ~10 % and a lower energy threshold compared to the muon channel. Cascade analyses are usually sensitive to electron and tau neutrinos by CC interactions, with a small contribution of NC interactions by all flavors.

A pioneering analysis based on the configuration with only 40 installed detector strings had already found an excess of events above the atmospheric expectation.[13] Recently, the first year of data of the completed detector has been analyzed. The energy spectrum of atmospheric electron neutrinos was measured to be consistent with the theoretical expectation. No indication of a prompt signal was found, and the astrophysical component at high energies[27] was confirmed. The most recent results for the measurement of the astrophysical flux, based on two years of data, are reported in Ref. 28. This analysis substantially increases the number of observed high-energy events by also including partially contained events. It largely confirms the findings of the HESE analysis and observes a cosmic signal with significance of more than $4\sigma$.

### 3.3. *Muon channel*

The classical detection channel of neutrino telescopes is up going muon tracks from CC neutrino-nucleon interactions in and below the detector. The selection of up going tracks efficiently eliminates the background of down going cosmic-ray-induced atmospheric muons. The Earth's absorption increases with energy and results in a zenith-dependent expectation even for an isotropic diffuse cosmic signal. As the interaction can happen far outside the detector, the effective detection volume is much larger than the geometrical volume, resulting in larger event rates than for contained events. However, muons from neutrino interactions far away have lost a considerable and unknown fraction of their initial energy and carry little information to distinguish astrophysical from atmospheric neutrinos. In addition, muons that pass through the detector deposit only a small fraction of their energy inside the detector. Therefore, the muon energy has to be estimated by the observed energy loss, resulting in a resolution of about ∼50−70% for the muon energy as compared to ∼10% in the case of contained cascades. Diffuse searches using this channel contain potentially the largest number of cosmic neutrinos but require larger data-sets to observe the same significance as channels with good energy resolution, e.g. contained cascades. An important advantage of the muon channel compared to cascades is the good angular resolution, approaching about 0.1° at high energies. Though angular information is not of primary concern in a diffuse search, it is helpful for an efficient rejection of the atmospheric muon background and the selection of high purity data-sets.

The analysis is done as a two-dimensional likelihood fit of the measured energy and zenith angle. Fitting the full data from a few hundred GeV to high energies around a PeV makes it possible to strongly constrain systematic uncertainties.

The first indications of an astrophysical signal in the muon channels were found in the IceCube data of the 59-string configuration.[14] Though the final significance of a cosmic signal was only 1.8$\sigma$ with respect to the conventional atmospheric background-only hypothesis, this observation has been important not only in the context of promising indications of a cosmic neutrino signal but also in its power to constrain the conventional and prompt atmospheric neutrino backgrounds for the analysis of starting events. Particularly spectacular has been the highest energy muon, whose energy in the detector has been estimated at 400 TeV.

Since then, the evidence in this channel has been steadily increasing. A combined analysis using 35,000 events from the 79-string and first year of 86-string configurations (2010–2012) has found evidence for an astrophysical flux above 300 TeV, consistent with the HESE result at the 3.7$\sigma$ level.[29] Most recently, the full data from six years of IceCube operation including the 59-string and 79-string configurations as well as four years of IceCube with 86 strings (2009–2015) has been analyzed.[30] Here, the event selection efficiency has been optimized resulting in a total of about 340,000 muon neutrino events with an estimated purity of better than 99.9%. The significance of an astrophysical flux with the full six years is at the level of 6$\sigma$ (rejecting a pure atmospheric origin).

Also notable is the observation of an ultra-high-energy neutrino event,[31] see Fig. 2. It is a through going track that deposits an energy of 2.6 ± 0.3 PeV within

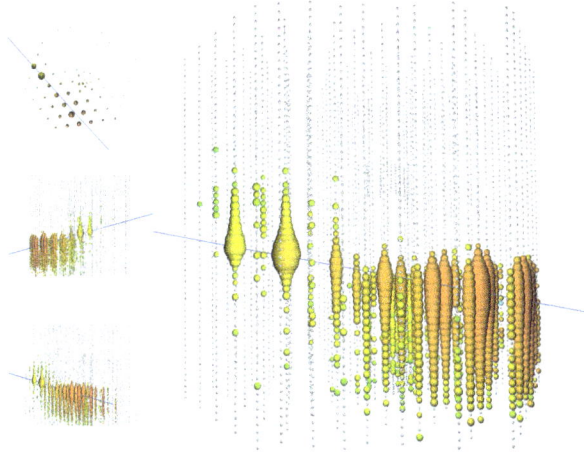

Fig. 2. Event view of the multi-PeV up going muon event detected with IceCube. Colored spheres indicated optical sensors that registered a signal, where the size encodes the logarithm of detected charge and the color the arrival time (red early and green late). On the left are three projected views. The reconstructed track is indicated as a line.

the instrumented volume of IceCube. Based on simulations of events with similar topology, such an energy loss would be expected from a muon of more than 4 PeV, and thus implies an even higher neutrino energy. This makes this event the highest energy neutrino detected to date. Based on the huge energy for this event alone the hypothesis of an atmospheric origin can be rejected[30] by about $4\sigma$. Though not relevant for the subject of this report, it is worthwhile to note that the directional uncertainty is less than 0.3°, but attempts to identify an astrophysical source have not been successful yet. The closest known source of GeV photon emission[32] is about 3° away and 11° for known TeV sources.[33] The direction is 11° off the Galactic Plane.

### 3.4. *Extremely high-energy neutrino searches*

Searches for extremely high-energy neutrinos are made by a channel targeting very large energy depositions related to events typically having $10^{17}$ eV energy. For example, such events are expected for neutrinos from the GZK effect. Because of the decreasing background of cosmic-ray-induced atmospheric muons, the energy threshold can be gradually reduced towards the horizon. For straight down going events, the surface detector IceTop is added as a veto. As a consequence of absorption within Earth, the region around and above the horizon is particularly important.

It was this type of analysis that initially observed the first neutrino events with PeV energy[34] close to its energy threshold. The analysis was based on an exposure of one year in the 79-string configuration and the first year of full IceCube operation.

An update to this analysis[35] covering six years of IceCube operation has further increased the sensitivity, particularly towards high energy. No further events with energies substantially above a few PeV have been observed, which results in the currently most constraining exclusion limits for ultra-high-energy neutrinos such as expected from the GZK effect.[36] As no further events of higher energy have been observed with recent data and thus no improved spectral information on the diffuse flux can be deduced, this channel is ignored in the following discussions.

### 3.5. *Tau channel*

The tau neutrino is interesting because, given neutrino flavor oscillations, about one third of astrophysical neutrinos are expected to be of tau flavor. Furthermore, due to "regeneration" in tau decays, the Earth is not fully opaque to tau neutrinos.[37] The atmospheric background of high-energy tau neutrinos is substantially smaller than even that of prompt neutrinos,[38] and thus any observed tau neutrino would be astrophysical with high probability.

As discussed above, the detection signature of tau neutrino interactions is similar to electron neutrinos unless their energy exceeds about one PeV. Nonetheless, it is interesting to attempt the identification of a double-bang signature in the sample of observed starting events. This has been performed with a modification[39] of the reconstruction algorithm used for starting events.[26] No evidence of a double-bang signature has been found. However, this analysis, close to the detection threshold, is challenging because of systematic uncertainties of photon propagation through ice, and a substantial contribution of tau neutrinos to the observed events cannot be excluded. An alternative approach is to directly search for large double-pulse signatures in the recorded waveforms of optical sensors. An independent dedicated search[40] for this signature has not detected a clear double-bang event. However, the sensitivity of this search has not yet reached the observed astrophysical flux level, and more data is needed.

## 4. Comparison of observational results

### 4.1. *Energy spectrum*

The energy spectrum measured with the most recent data[24] for high-energy starting events is shown in Fig. 3 together with the spectrum obtained from the extension of the analysis towards lower energies.[25] Here, the normalization in each energy bin has been a free parameter in a maximum likelihood fit of the data-set. The best fit spectral index for Fig. 3(a) is $\gamma = 2.58 \pm 0.25$, slightly softer than results based on earlier data. This is consistent with the result in Fig. 3(b) of $\gamma = 2.46 \pm 0.12$ and is also consistent with measurements in the cascade channel,[28] which find similar soft indices. The spectrum depends slightly on the assumptions of atmospheric neutrinos

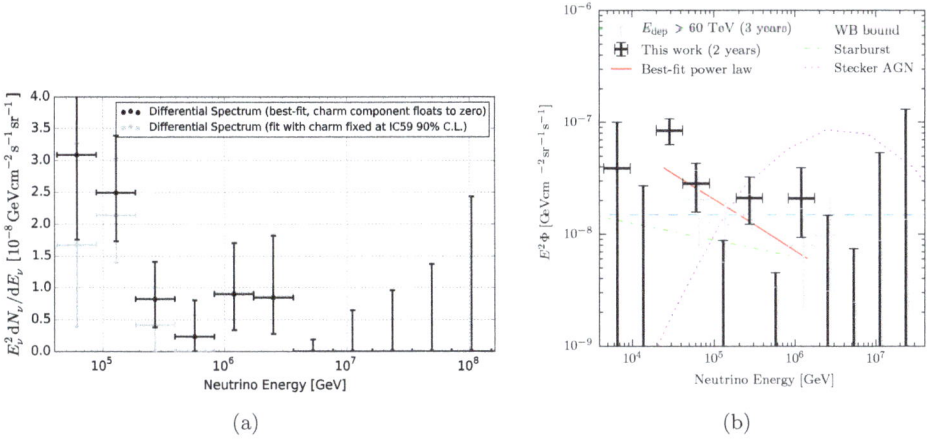

Fig. 3. The reconstructed energy spectrum for the analysis of starting events (left) and the analysis with lower energy threshold (right)

from charm decays. A larger charm component would lead to a harder astrophysical spectrum.

Two aspects are particularly interesting. One may wonder about the existence of a possible cut off at an energy of a few PeV or a possible spectral break in the spectrum. Note that the underfluctuation of events just below one PeV of deposited energy has previously caused some speculations. However, this fluctuation is not statistically significant, and it has decreased with the recently added data.

Clearly visible is the deviation from the hard $E^{-2}$ hypothesis. It seems that the steepness of the slope is dominated by data at lower energy of $\lesssim 50$ TeV. As shown in Ref. 25, above an energy of 100 TeV, the data would also be consistent with a hard spectrum $\gamma \simeq 2.26 \pm 0.35$.

For the question of a cut off, about three events would be expected for a hard $E^{-2}$ spectrum above 2 PeV, while none were observed. However, with a softer spectrum, as the best fit seems to indicate, this tension is strongly relaxed, and a cut off is not required to describe the observation.

All in all, the current statistics are not sufficient yet to answer questions concerning a possible spectral break or cut off, and future data will improve the picture.

While the starting event analysis is dominated by cascade-like events from the southern sky it is interesting to compare it to the energy distribution of up going muon neutrinos. As the measurement of uncontained muons does not directly allow for a good estimate of the neutrino energy, the spectral unfolding is difficult but benefits from higher statistics of events and well-controlled systematics and backgrounds. The measurement with two years of data[29] resulted in a spectral index slightly harder, $\gamma = 2.2 \pm 0.2$, than measured in the cascade-dominated channels. This tension has increased with recent data covering six years of IceCube,[30] as shown in Fig. 4. The left figure shows the reconstructed energy

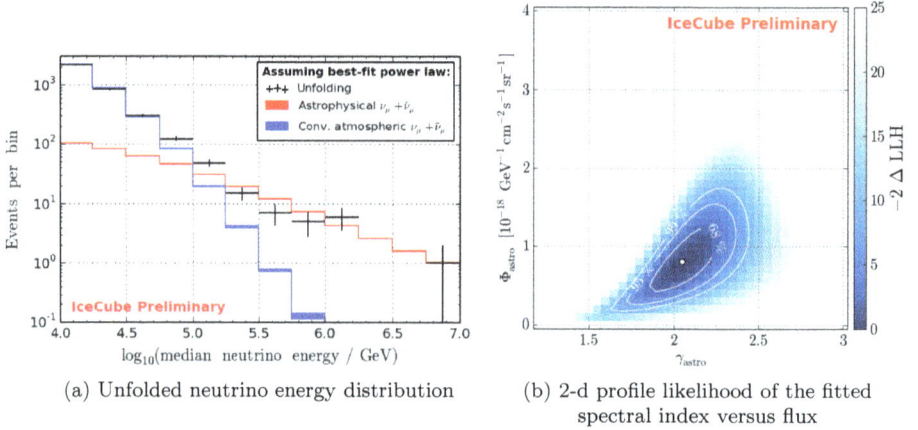

(a) Unfolded neutrino energy distribution

(b) 2-d profile likelihood of the fitted spectral index versus flux

Fig. 4.   Results of the analysis of up going muons based on six years data.

distribution, the right figure the profile likelihood for the extracted astrophysical flux parameters. Note that the reconstructed neutrino energy spectrum is model dependent. Shown here is the median expected neutrino energy calculated from the measured deposited energy assuming the best fit spectrum. The excess of data is clearly inconsistent with a pure atmospheric origin. With a best fit spectral index of $2.06 \pm 0.13$, the observed spectrum is harder than that of the cascade-dominated measurements.

For a quantitative discussion, the results for all analyses have to be compared by means of statistical confidence. It is found that the measured flux normalization is often correlated with the fitted spectral index. When analyzing the error contours for all analyses,[41] the apparent tension is reduced to about $2.5\sigma$, which is marginally statistically significant. For this discussion, it is furthermore important to highlight the systematic differences between these two measurements. The threshold for the up going muon signal is a few hundred TeV while astrophysical starting events are detected above a few tens of TeV. If only high-energy starting events were considered in the comparison, the spectra would be in agreement. Another important difference is the dominance of different hemispheres in both analyses. If the astrophysical flux was non-isotropic or, for example, composed of a galactic and an extra galactic component, a difference between the two analyses could be explained. Additional data will be needed to answer these questions. An extension to this discussion is found in Sec. 4.3.

### 4.2. *Global fit and flavor ratio*

Based on the different detection channels, a global fit of the combined data-sets can be attempted. In Ref. 42, such a global analysis has been performed using six different data-sets[13,14,21,22,25,29,43] consisting of up going muons, contained cascades, and

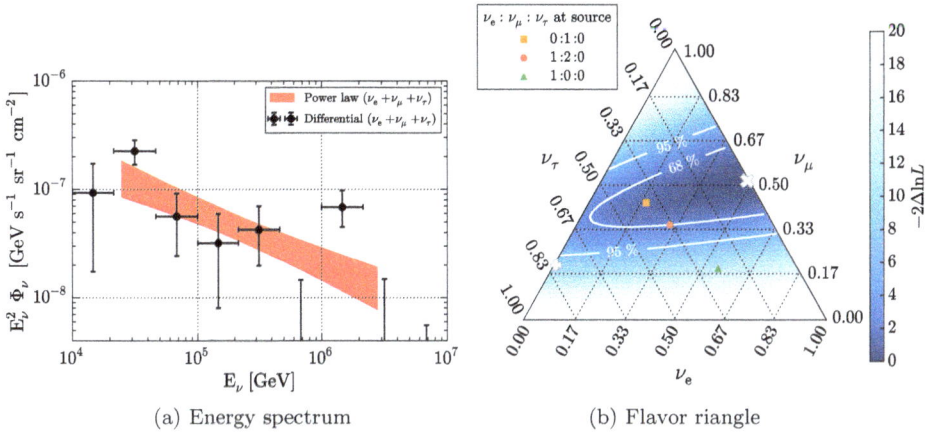

(a) Energy spectrum                              (b) Flavor riangle

Fig. 5.   Results of the global fit.

starting events. Special care was taken not to double count overlapping data-sets and
the combination of systematic uncertainties. For practical reasons, the fit does not
include the full systematics of the individual data-sets but combines them as generalized global parameters. These are the energy-scale uncertainty, the atmospheric
muon background normalization for each data-set, and the cosmic ray spectral
index, which affects the atmospheric neutrino background. Different hypotheses for
the astrophysical flux normalization and spectra are tested. The result for a single
power law is a spectral index of $\gamma = -2.50 \pm 0.09$, disfavoring a hard spectral index
of $\gamma = 2$ at the $3.8\sigma$ level. This significance is reduced to $2.1\sigma$ if an exponential cut
off is introduced. Similar to the discussions above, it can be seen in Fig. 5(a) that
this tension could be also relaxed without introducing a cut off if a transition from
a softer spectral index at lower energies to a harder spectrum at higher energies is
assumed.

The flavor ratio of the neutrino flux is particularly interesting because allows
constraining of the acceleration mechanism. The initial flavor composition at the
source is modified by neutrino oscillations and largely smeared out due to the long
baseline. Therefore, one expects to observe all neutrino flavors at Earth with a
similar flux. However, depending on the injection model of neutrino flavors at the
source and the assumed oscillation parameters, deviations from the exact $\nu_e : \nu_\mu :
\nu_\tau \approx 1 : 1 : 1$ mixing are expected at Earth. Despite not having directly identified
tau neutrino events, they contribute to the observed event rates of the considered
detection channels differently. Within the framework of the global fit, the flavor ratio
can be tested when fitting the flux normalization of each flavor separately in a joint
fit. The result is shown in Fig. 5. The measured data is consistent with a mixture
of all flavors and pure fluxes of $\nu_e$ and $\nu_\mu$ or $\nu_\tau$ are strongly disfavored. When
comparing to different injection scenarios the hypothesis of pure $\nu_e$ as expected
from sources with dominating muon decays can be excluded. Both scenarios of an

injected ratio $\nu_e : \nu_\mu : \nu_\tau \approx 0 : 1 : 0$ and $\nu_e : \nu_\mu : \nu_\tau \approx 1 : 2 : 0$ are compatible with the data. With more data, this measurement is expected to become increasingly important for the understanding of the source mechanism of the measured flux.

### 4.3. *Isotropy*

No analysis of the arrival directions of detected neutrinos has yet revealed a statistically significant clustering nor correlation with a known source.[24,44,45] However, tests for large-scale anisotropies allow investigating the important question of whether the assumption of isotropy of the cosmic signal is valid or if regions related to a few close sources dominate. In particular, whether the observed flux is of galactic or extra-galactic origins can be tested.

An important step is the recent observation of the astrophysical signal also in the up going muon analyses,[29,30] see Sec. 3.3. This confirms that the flux that was observed in the high-energy starting event analysis mostly by cascades from the Southern Hemisphere is accompanied by a roughly equally strong flux of muon neutrinos in the Northern Hemisphere. This indicates that at least a substantial fraction of the flux is isotropic and thus presumably extra-galactic.

A straightforward approach is to split the data samples into two separate regions of the sky and to compare the observed fluxes. This is indicated in Fig. 6.

The simplest test is to compare the Northern and Southern Hemispheres. As the Southern Hemisphere contains the central part of the galaxy, differences in the hemispheres may indicate differences between galactic and extra galactic components. This has been done within the analysis of starting events at a lower energy threshold,[25] see Fig. 7, the global fit,[42] and the analysis of

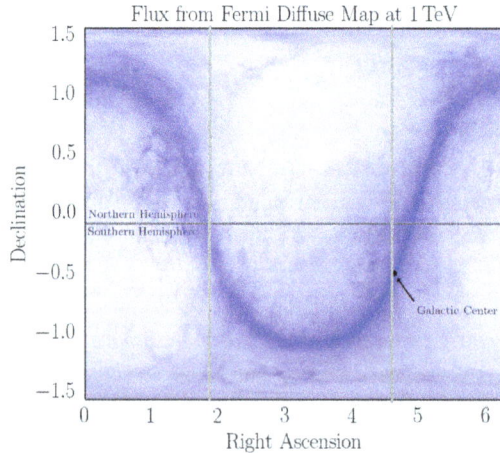

Fig. 6.    Splitting the data samples into regions of the sky: north-south and east-west. The figure shows the orientation of the galactic plane, indicated by the superimposed diffuse gamma emission measured by Fermi-LAT (data from Ref. 46). Vertical lines indicate a split by right ascension that results in quadrants in both hemispheres with and without the galactic plane.

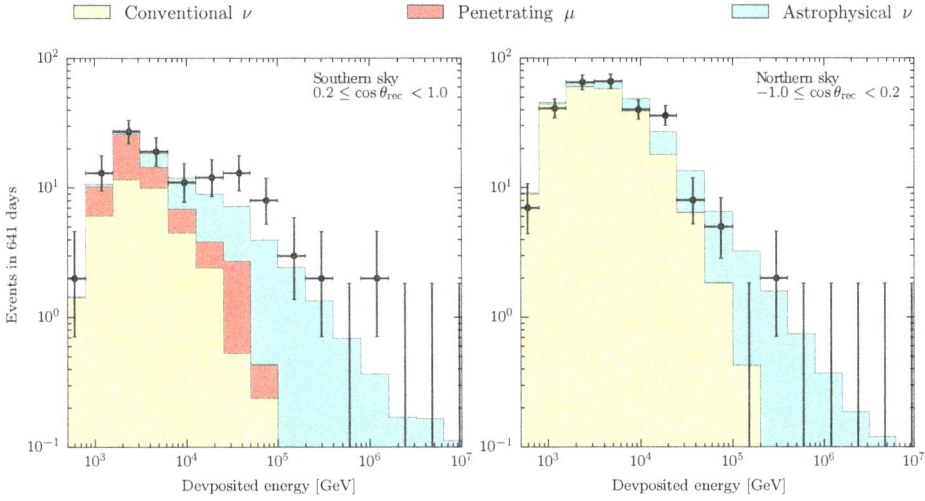

Fig. 7.  Energy spectra of starting events for northern and southern sky separately.

contained cascades.[28] Both hemispheres show a clear excess of an astrophysical signal over the atmospheric background, however, the southern excess seems stronger.

This question is more quantitatively addressed in the aforementioned global fit. It results in a spectral index of $\gamma = 2.0^{+0.3}_{-0.4}$ for the northern sky and $\gamma = 2.56 \pm 0.12$ for the southern sky, respectively. Note that the significance of a discrepancy is only $1.1\sigma$. The interpretation must be done carefully, as the observational conditions and systematic uncertainties between north and south strongly differ. The fit of the northern sky is dominated by the up going muon sample, which has a higher energy threshold compared to the cascade sample, which dominates the southern sky. Furthermore, detector systematics with all sensors directed downward, the absorption of high-energy events by Earth, and the rejection of atmospheric neutrino background differs strongly between both hemispheres.

This test has been repeated with contained cascades.[28] This measurement results in very similar spectral indices for north ($\gamma = 2.69^{+0.34}_{-0.34}$) and south ($\gamma = 2.68^{+0.20}_{-0.22}$). Also for this analysis, systematics differ between north and south and the energy range is slightly smaller. Obviously, at this point in time, the results are not conclusive yet.

A test that is not affected by these systematics has been performed in the northern sky with the six-year up going muon sample. The sample was split in right ascension instead of declination. The regions, indicated as vertical lines in Fig. 6, are chosen such that the two split samples are of similar statistics but complementary with respect to the galactic plane. The fit result, shown in Fig. 8, is a small but not statistically significant larger flux and softer spectrum than the region including the galactic plane.

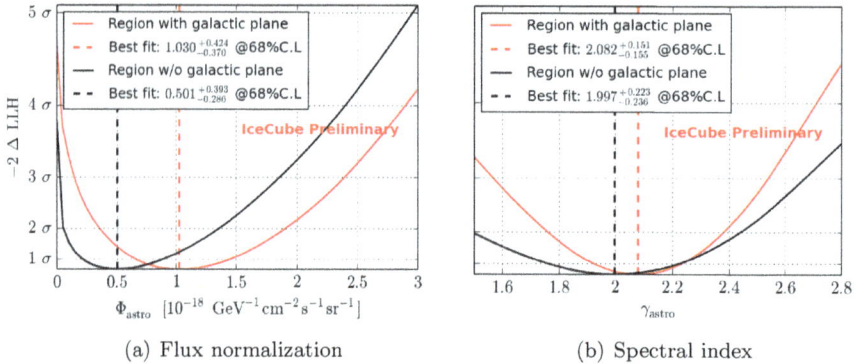

(a) Flux normalization    (b) Spectral index

Fig. 8. Profile likelihoods for the extracted flux parameters from the fit of the six-year up going muon sample, which is split by right ascension.

Again, a definitive answer on whether the flux from the galactic plane differs from the all-sky result cannot be deduced. Dedicated tests are underway, and additional data will certainly allow addressing this question further.

## 5. Summary and conclusions

In this contribution, we have reviewed the observational evidence and approaches to observing diffuse fluxes of astrophysical neutrinos. It is remarkable that only a few years after the initial discovery, the signal could be clearly identified in several detection channels. After the consolidation of the observational evidence, substantial progress has been made in characterizing the signal properties, in particular by improving analysis methods and by the comparison of the different results. Several important questions have been addressed, especially the characterization of the energy spectrum, the flavor composition, and a possible large-scale isotropy. While we are still awaiting the detection of point sources hopefully soon (see contribution by Ch. Finley), the above questions can be addressed with further improved analyses and new data.

Ultimately, IceCube will be limited in sensitivity. Therefore, the need for a next-generation instrument arises. The KM3Net[47] neutrino telescope in the Mediterranean Sea (see contribution by M. de Jong) and the BAIKAL-GVD neutrino telescope[48] in Lake Baikal (see contribution by Z. Djilkibaev) plan to install deep water instrumentation that exceed the size of the IceCube detector. Finally, the design of a next-generation instrument at the South Pole, IceCube-Gen2,[49] aims at a scale of $10 \, km^3$ volume (see contribution by E. Blaufuss and A. Karle). Also investigations are underway of possible improvements of the current IceCube performance for the search of cosmic neutrinos by extending the surface detector IceTop acting as a veto[18] to atmospheric signals.

## Acknowledgements

The author would like to thank Hans Niederhausen, Leif Rädel, Sebastian Schoenen Todor Stanev, Albrecht Karle and Tom Gaisser for careful reading of the manuscript as well as the whole IceCube Collaboration for their contributions.

## References

1. M. Markov, On high energy neutrino physics. In *Proceedings, 10th International Conference on High-Energy Physics (ICHEP 60)*, pp. 578–581, (1960).
2. C.V. Achar, M.G.K. Menon, V.S. Narasimham, P.V.R. Murthy, B.V. Sreekantan, *et al.*, Detection of muons produced by cosmic ray neutrinos deep underground, *Phys. Lett.* **18** (1965) 196–199. doi: 10.1016/0031-9163(65)90712-2.
3. F. Reines, M. Crouch, T. Jenkins, W. Kropp, H. Gurr, *et al.*, Evidence for high-energy cosmic ray neutrino interactions, *Phys. Rev. Lett.* **15** (1965) 429–433. doi: 10.1103/PhysRevLett.15.429.
4. T.K. Gaisser, F. Halzen, and T. Stanev, Particle astrophysics with high-energy neutrinos, *Phys. Rept.* **258** (1995) 173–236. doi: 10.1016/0370-1573(95)00003-Y.
5. I. Belolaptikov, *et al.*, The Baikal underwater neutrino telescope: Design, performance and first results, *Astropart. Phys.* **7** (1997) 263–282. doi: 10.1016/S0927-6505(97)00022-4.
6. V. Balkanov, *et al.*, Reconstruction of atmospheric neutrinos with the Baikal Neutrino Telescope NT-96. (1997).
7. E. Andres, P. Askebjer, S. Barwick, R. Bay, L. Bergstrom, *et al.*, The AMANDA neutrino telescope: Principle of operation and first results, *Astropart. Phys.* **13** (2000) 1–20. doi: 10.1016/S0927-6505(99)00092-4.
8. E. Andres, P. Askebjer, X. Bai, G. Barouch, S. Barwick, *et al.*, Observation of high-energy neutrinos using Cherenkov detectors embedded deep in Antarctic ice, *Nature.* **410** (2001) 441–443. doi: 10.1038/35068509.
9. A. Achterberg, *et al.*, Multi-year search for a diffuse flux of muon neutrinos with AMANDA-II, *Phys. Rev.* **D76** (2007) 042008. doi: 10.1103/PhysRevD.76.042008,10.1103/PhysRevD.77.089904.
10. M. Ageron, *et al.*, ANTARES: The first undersea neutrino telescope, *Nucl. Instrum. Meth.* **A656** (2011) 11–38. doi: 10.1016/j.nima.2011.06.103.
11. J. Aguilar, *et al.*, Search for a diffuse flux of high-energy $\nu_\mu$ with the ANTARES neutrino telescope, *Phys. Lett.* **B696** (2011) 16–22. doi: 10.1016/j.physletb.2010.11.070.
12. J. Ahrens, *et al.*, Sensitivity of the IceCube detector to astrophysical sources of high energy muon neutrinos, *Astropart. Phys.* **20** (2004) 507–532. doi: 10.1016/j.astropartphys.2003.09.003.
13. M. Aartsen, *et al.*, Search for neutrino-induced particle showers with IceCube-40, *Phys. Rev.* **D89**(10) (2014) 102001. doi: 10.1103/PhysRevD.89.102001.
14. M. Aartsen, *et al.*, Search for a diffuse flux of astrophysical muon neutrinos with the IceCube 59-string configuration, *Phys. Rev.* **D89**(6) (2014) 062007. doi: 10.1103/PhysRevD.89.062007.
15. P. Lipari, Proton and neutrino extragalactic astronomy, *Phys. Rev.* **D78** (2008) 083011. doi: 10.1103/PhysRevD.78.083011.
16. M. Kowalski, Status of high-energy neutrino astronomy (2014).

17. M. Ahlers and F. Halzen, Pinpointing extragalactic neutrino sources in light of recent IceCube observations, *Phys. Rev.* **D90** (4) (2014) 043005. doi: 10.1103/PhysRevD.90.043005.

18. J. Auffenberg, IceVeto: Extended PeV neutrino astronomy in the Southern Hemisphere with IceCube, *AIP Conf. Proc.* **1630** (2014) 50–53. doi: 10.1063/1.4902769.

19. S. Schonert, T.K. Gaisser, E. Resconi, and O. Schulz, Vetoing atmospheric neutrinos in a high energy neutrino telescope, *Phys. Rev.* **D79** (2009) 043009. doi: 10.1103/PhysRevD.79.043009.

20. J.G. Learned and S. Pakvasa, Detecting tau-neutrino oscillations at PeV energies, *Astropart. Phys.* **3** (1995) 267–274. doi: 10.1016/0927-6505(94)00043-3.

21. M. Aartsen, *et al.*, Evidence for high-energy extraterrestrial neutrinos at the IceCube detector, *Science.* **342** (2013) 1242856. doi: 10.1126/science.1242856.

22. M. Aartsen, *et al.*, Observation of high-energy astrophysical neutrinos in three years of IceCube data, *Phys. Rev. Lett.* **113** (2014) 101101. doi: 10.1103/PhysRevLett.113.101101.

23. R. Enberg, M.H. Reno, and I. Sarcevic, Prompt neutrino fluxes from atmospheric charm, *Phys. Rev.* **D78** (2008) 043005. doi: 10.1103/PhysRevD.78.043005.

24. C. Kopper, N. Kurahashi, *et al.*, Observation of astrophysical neutrinos in four years of IceCube data. In *Proceeding of the 34th International Cosmic Ray Conference*, The Hague, Netherlands, 30 July–6 August 2015, (2015).

25. M. Aartsen, *et al.*, Atmospheric and astrophysical neutrinos above 1 TeV interacting in IceCube, *Phys. Rev.* **D91** (2) (2015) 022001. doi: 10.1103/PhysRevD.91.022001.

26. M. Aartsen, *et al.*, Energy reconstruction methods in the IceCube neutrino telescope, *JINST.* **9** (2014) P03009. doi: 10.1088/1748-0221/9/03/P03009.

27. M.G. Aartsen, *et al.*, Measurement of the atmospheric $\nu_e$ spectrum with IceCube, *Phys. Rev.* **D91** (2015) 122004. doi: 10.1103/PhysRevD.91.122004.

28. M. Lesiak-Bzdak, H. Niederhausen, A. Stössl, *et al.*, High energy astrophysical neutrino flux characteristics for neutrino-induced cascades using IC79 and IC86-string IceCube configurations. In *Proceeding of the 34th International Cosmic Ray Conference*, The Hague, Netherlands, 30 July–6 August 2015, (2015).

29. M.G. Aartsen, *et al.*, Evidence for astrophysical muon neutrinos from the northern sky with IceCube, *Phys. Rev. Lett.* **115** (8) (2015) 081102. doi: 10.1103/PhysRevLett.115.081102.

30. S. Schoenen, L. Rädel, *et al.*, A measurement of the diffuse astrophysical muon neutrino flux using multiple years of IceCube data. In *Proceeding of the 34th International Cosmic Ray Conference*, The Hague, Netherlands, 30 July–6 August 2015, (2015).

31. S. Schoenen and L. Raedel, Detection of a multi-pev neutrino-induced muon event from the northern sky with icecube, *The Astronomer's Telegram.* **7856** (2015) 1.

32. F. Acero, Fermi large area telescope third source catalog, *Astrophys. J. Suppl.* **218**(2) (2015) 23. doi: 10.1088/0067-0049/218/2/23.

33. S.P. Wakely and D. Horan. TeVCat: An online catalog for very high energy gamma-ray astronomy. In *Proceedings, 30th International Cosmic Ray Conference (ICRC 2007)*, Vol. 3, pp. 1341–1344, (2007).

34. M. Aartsen, *et al.*, First observation of PeV-energy neutrinos with IceCube, *Phys. Rev. Lett.* **111** (2013) 021103. doi: 10.1103/PhysRevLett.111.021103.

35. A. Ishihara, *et al.*, A search for extremely high energy neutrinos in 6 years of IceCube data. In *Proceeding of the 34th International Cosmic Ray Conference*, The Hague, Netherlands, 30 July–6 August 2015, (2015).

36. V.S. Berezinsky and G.T. Zatsepin, Cosmic rays at ultrahigh-energies (neutrino?), *Phys. Lett.* **B28** (1969) 423–424. doi: 10.1016/0370-2693(69)90341-4.

37. J.F. Beacom, P. Crotty, and E.W. Kolb, Enhanced signal of astrophysical tau neutrinos propagating through earth, *Phys. Rev.* **D66** (2002) 021302. doi: 10.1103/PhysRevD. 66.021302.

38. P. Gondolo, G. Ingelman, and M. Thunman, Charm production and high-energy atmospheric muon and neutrino fluxes, *Astropart. Phys.* **5** (1996) 309–332. doi: 10.1016/0927-6505(96)00033-3.

39. P. Hallen. *On the measurement of high-energy tau neutrinos with IceCube.* PhD thesis, Master's thesis, RWTH Aachen, November 2013.

40. D.R. Williams, C.M. Vraeghe, D.L. Xu, *et al.*, A search for astrophysical tau neutrinos in three years of IceCube data. In *Proceeding of the 34th International Cosmic Ray Conference*, The Hague, Netherlands, 30 July–6 August 2015, (2015).

41. H. Niederhausen and S. Schoenen. Comparing independent multivariate measurements: The general case. IceCube Internal Report icecube/201510001-v2, Department Physics and Astronomy, Stony Brook and III.Physikalisches Institut RWTH Aachen University, (2015).

42. M.G. Aartsen, *et al.*, A combined maximum-likelihood analysis of the high-energy astrophysical neutrino flux measured with IceCube, *Astrophys. J.* **809** (1) (2015) 98. doi: 10.1088/0004-637X/809/1/98.

43. M.G. Aartsen, *et al.*, The IceCube neutrino observatory part II: Atmospheric and diffuse UHE neutrino searches of all flavors. In *Proceedings, 33rd International Cosmic Ray Conference (ICRC2013)*, (2013).

44. M.G. Aartsen, *et al.*, Searches for extended and point-like neutrino sources with four years of IceCube data, *Astrophys. J.* **796** (2) (2014) 109. doi: 10.1088/0004-637X/ 796/2/109.

45. M.G. Aartsen, *et al.*, Searches for small-scale anisotropies from neutrino point sources with three years of IceCube data, *Astropart. Phys.* **66** (2015) 39–52. doi: 10.1016/j. astropartphys.2015.01.001.

46. LAT Background Models, gll_iem_v05_rev1.fit. Data repository, Fermi Science Support Center, (2014).

47. A. Margiotta, The KM3NeT deep-sea neutrino telescope, *Nucl. Instrum. Meth.* **A766** (2014) 83–87. doi: 10.1016/j.nima.2014.05.090.

48. A.V. Avrorin, *et al.*, Current status of the BAIKAL-GVD project, *Nucl. Instrum. Meth.* **A725** (2013) 23–26. doi: 10.1016/j.nima.2012.11.151.

49. M. Aartsen, *et al.*, IceCube-Gen2: A vision for the future of neutrino astronomy in antarctica (2014).

Chapter 6

# IceCube and Astrophysical Neutrinos:
# The Search for Sources

Chad Finley

*Oskar Klein Centre*
*Physics Department, Stockholm University*
*10691 Stockholm, Sweden*
*cfinley@fysik.su.se*

The astrophysical neutrino flux discovered by IceCube is best known for the handful of spectacular events near and above one PeV. Less familiar are the hundreds of recorded astrophysical muon-neutrino events that are implied by the latest flux measurements but are hidden in the foreground of atmospheric neutrinos. Reconstructed with an angular resolution from 0.3° to 1.0°, they may make it possible to resolve sources of high energy neutrinos. Here we will review the most sensitive point source searches in IceCube and the connection with the measured astrophysical flux. We will review some of the tests that already constrain the sources of astrophysical neutrinos, and discuss the prospects for identifying their origins.

## 1. Introduction

The search for localized sources of neutrino emission is at the heart of neutrino astronomy, and it has taken on a new urgency with the discovery of the high-energy astrophysical neutrino flux. Here we will review the search results for point-like and extended sources by IceCube. We will consider how the non-detection to date of a neutrino source already provides some information about the origin of the astrophysical flux, and we will consider the implications for the prospects of source detection in the near future.

For clarity, **astrophysical neutrino flux** will refer to the totality of neutrinos arriving from space. In principle, one expects the high energy astrophysical flux to consist of many components, including: galactic sources, diffuse galactic plane emission, nearby or bright extragalactic sources, and diffuse emission from the ensemble of fainter, unresolvable extragalactic sources distributed throughout the universe. At present, IceCube measurements of the astrophysical flux seem to be adequately described by a single isotropic component, which therefore could be the extragalactic diffuse emission. This leads to the terms "astrophysical flux" and "diffuse flux" being used interchangeably, but it should not be taken to imply that the diffuse emission model has been "proven". Rather, in the context of searches for the sources of these neutrinos, the isotropic flux (or equivalently the diffuse extragalactic flux) serves as the null hypothesis from which we seek to identify deviations that require additional or alternative components.

In this review we will pay particular attention to the distinction between **track** and **shower** events, and between **starting-event** selections and **through-going event** selections. These selections represent complementary strategies for reducing the overwhelming background of events induced by cosmic ray air showers.

## 1.1. *Classes of neutrino events: Tracks and showers*

**Track events:** charged-current muon-neutrino interactions result in a muon that, at high energies, can travel for many kilometers through ice and bedrock. As the muon track crosses the detector, the long lever arm allows directional reconstruction with median angular resolution from 1.0° at 1 TeV down to 0.3° at 10 PeV neutrino energies.[1] The fact that the muon can come from a neutrino interaction occurring many kilometers away enlarges the effective area of the detector by roughly an order of magnitude at high energies. The combination of this effective area gain and good angular resolution is what makes track event samples generally optimal for point-source searches. The measured muon energy, however, provides only a lower bound on the primary neutrino energy, as the energy losses outside the detector volume are not known.

**Shower events:** for charged-current electron- and tau-neutrino interactions, as well as for all neutral current interactions, the shower of resulting particles has a short length ($\sim$10 m) compared to the detector array spacing (125 m between strings) and is, to first order, reconstructed as a sphere of light. At high energies, the location of the shower within the detector volume can be measured to approximately a few meters, and the direction of the primary neutrino can be reconstructed to 10° to 15° angular uncertainty, based on the detailed timing information of the recorded photoelectrons.[2] The energy deposited in the detector can be measured to 10–15% resolution, and for at least some types of events like $\nu_e$ charged-current electron neutrino interactions this is a measure of the neutrino energy itself.

## 1.2. *Event selections: Starting and through-going samples*

Cosmic rays interacting high in the atmosphere produce extensive air showers of particles, including muons and neutrinos that can reach the detector. These are usually known as atmospheric backgrounds, although they might better be called foregrounds. Atmospheric muons dominate, entering from above and triggering the detector at a rate of 2.5 kHz. At a rate approximately $10^6$ lower are atmospheric neutrinos, which are also caused by cosmic ray air showers but enter the detector from all directions since they pass through the Earth.

**Starting-event samples** use outer layers of the detector to veto incoming events. This veto technique has proven to be a powerful filter against the downward-going background. Starting event samples consist predominantly of shower event types, but have sensitivity to all neutrino flavors and interaction types. Here we will look at the High-Energy Starting Event (HESE) sample, which was the first event selection to isolate an astrophysical neutrino signal.[2,3]

**Through-going track samples** accept incoming as well as starting track events. As noted above, track samples provide larger effective area and better angular resolution, at the price of being sensitive only to muon neutrinos and of allowing more background from atmospheric neutrinos or muons. In principle the up-going direction is a pure, Earth-filtered sample of neutrino-induced events (both atmospheric and astrophysical). In practice, the much higher rate of down-going atmospheric muons means that a small fraction of these are mis-reconstructed as up-going and will enter the up-going event sample. Depending on the analysis goal, different levels of this mis-reconstructed muon background can be tolerated. Samples for measuring the diffuse astrophysical flux use tighter cuts to achieve 99.9% neutrino purity,[4] while samples for transient source searches such as gamma-ray bursts can relax the cuts to ~85% purity to achieve optimal sensitivity.[5]

## 2. Analysis results

### 2.1. *Point source searches with starting events*

We begin with the High-Energy Starting-Event sample shown in Fig. 1, which has been the subject of much attention. This three-year sample[3] contains 37 events; most of the events are showers. The estimated fraction of astrophysical neutrinos in

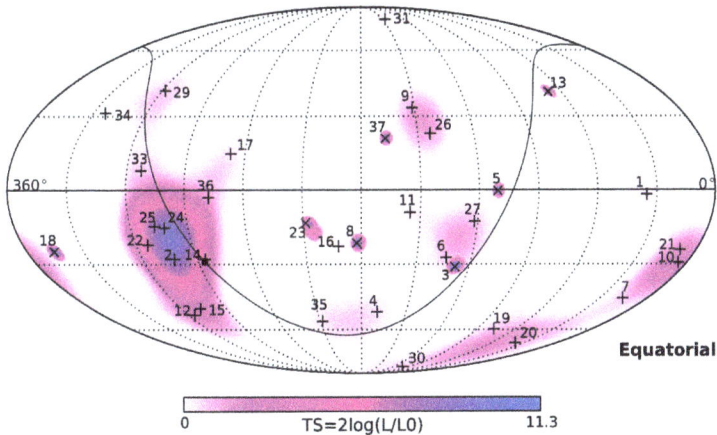

Fig. 1.   Skymap in equatorial coordinates showing the results (test statistic) of a maximum likelihood analysis of the 37 HESE events, looking for significant clustering at each position in the sky. The galactic plane is indicated by the solid curved line. From Ref. 3.

this sample is high, approximately half of the events. It can be seen that more events are recorded from the Southern sky than from the Northern sky; this is the result of the Southern sky muon background ($8 \pm 4$ events estimated to be down-going muons that evade the veto) and also of Earth-absorption of high energy neutrinos, which depletes the Northern sky events.

The figure shows a maximum-likelihood point-source analysis of the directions of the events. The color indicates the significance of the point-source plus background hypothesis relative to the background-only hypothesis at each location in the sky. (Note that for shower events, with large angular uncertainty, replacing the point-source hypothesis with an extended-source hypothesis of $\sim 10°$ width would have little impact on the result.) The most significant direction anywhere in the sky is the cluster of events near the galactic center, visible in the lower-left. The significance of this cluster, after correcting for trials related to searching the entire sky, is $1.5\sigma$.

Additional analyses have been performed with the starting-event sample.[3] A galactic plane analysis was performed that included a scan over different possible widths (in galactic latitude) of the emission region. The strongest correlation was found using a width of $\pm 7.5°$, with post-trial significance of $1.9\sigma$. A pre-defined candidate source list search, and a time-correlation analysis were also performed, with no indication of sources or clustering found.

## 2.2. *Point source searches with through-going track events*

The most recent published point source analysis based on through-going event selections was applied to four years of data.[1] This includes one year with the completed 86-string detector, and three previous years of the partially built detector. In the full detector, the sample consists of about 70 000 up-going events per year. This up-going sample is dominated by background atmospheric neutrinos, but it can be estimated to contain a fraction from 0.2% to 1% astrophysical neutrinos, depending on the whether the astrophysical flux measured at high energies continues down to lower energies as a hard ($E^{-2}$) or soft ($E^{-2.5}$) power-law, respectively.

The all-sky search is performed in both the northern and southern skies. In the southern sky, a very high energy cut is applied to reduce the muon background from $\sim 10^{11}$ to $\sim 10^5$ events per year. While this sample still consists predominantly of muons from cosmic ray air showers, rather than muons from neutrino interactions, at energies approximately above PeV the sample does provide some sensitivity to neutrino point sources in the southern sky.

The results of the all-sky search for point sources are shown in Fig. 2. Each pixel in the map shows the significance (p-value) of the best-fit to a point source at the given coordinates. The flux normalization and spectral index of the hypothesized neutrino point source are free parameters of the fit at each point.

Figure 2 indicates with boxes the locations of the most significant point source results in the northern and the southern skies. Given the large number of trials

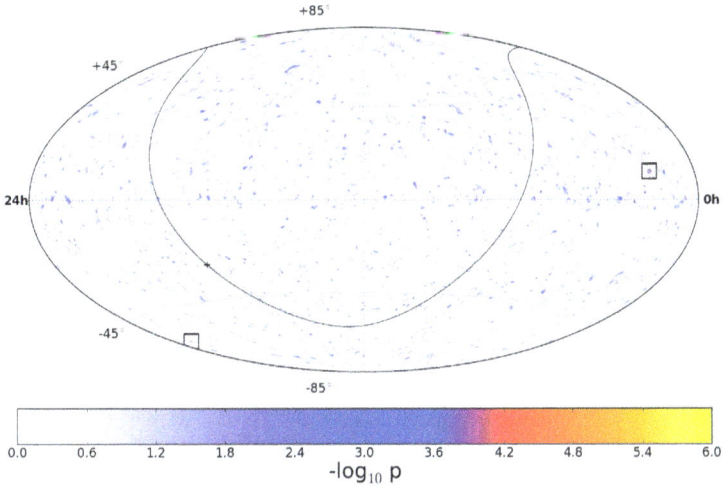

Fig. 2. Skymap in equatorial coordinates showing the results (logarithm of the p-value of the test-statistic) of a maximum likelihood analysis of the four-year point source sample, looking for significant clustering at each position in the sky. The point-source sample consists of approximately 180 000 atmospheric neutrino events (northern sky) and 215 000 atmospheric muon events (southern sky). The galactic plane is indicated by the solid curved line, and the galactic center by a cross. The two boxes enclose the locations of the most significant point source results for the northern and the southern sky searches. From Ref. 1.

associated with scanning the sky, neither these nor any other locations provide significant evidence for a neutrino point source after trial correction. A list of 44 preselected candidate sources was also tested. While sources on the list are implicitly tested as part of the all-sky search, the list provides a lower trial factor and lower threshold for detection. None of the 44 candidate sources tested was significant.

## 2.3. *Point source sensitivities and limits*

Figure 3 shows the sensitivity of neutrino point source searches. The lowest dashed line shows the four-year point source sample's sensitivity (median 90% CL upper limit expected if no signal is present) for a single point source, as a function of the source's declination. The sensitivity is given for a source with an $E^{-2}$ power-law energy spectrum. The Northern sky (right) allows for the most sensitive searches for neutrino sources to date. Here the Earth filter allows IceCube to be sensitive to point sources from TeV to PeV energies. In the Southern sky (left), the point source sample relies on a high energy threshold to reject most of the cosmic ray muon background. This means that the sensitivity to an $E^{-2}$ point source comes mainly from neutrinos with energies of PeV and above. Below 100 TeV, the point source sensitivity is better for most Southern declinations using the ANTARES detector, which is located in the Mediterranean Sea and can use the Earth as a filter for these directions.[6]

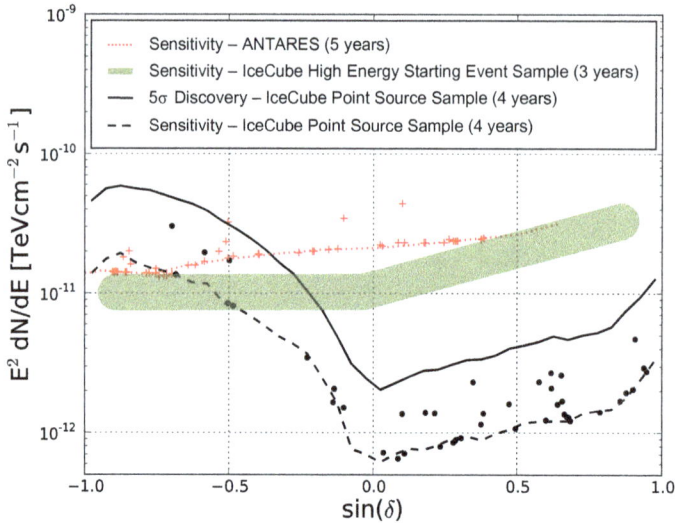

Fig. 3. Comparison of $E^{-2}$ point sources sensitivities as a function of declination for ANTARES (red dotted line), the IceCube High Energy Starting Event sample (green band), and the IceCube point source sample (black dashed line). Upper limits for selected source objects are also shown for ANTARES (crosses) and IceCube (circles). The solid black line is the IceCube point source sample discovery potential: the neutrino flux needed for the maximum likelihood analysis to detect a source at a specific location at $5\sigma$ significance 50% of the time. (Plot and IceCube data from Ref. 1, with sensitivity added for the High Energy sample estimated from the information in Ref. 3. ANTARES data from Ref. 6. All times in calendar years, not live time.)

It can be seen that the through-going track-event selection of muon neutrinos generally provides better sensitivity to point sources than the starting-event selection. The larger effective area and better angular resolution outweigh the disadvantage of higher atmospheric backgrounds. In the southern sky, however, the starting event selection allows sensitivity to sources at lower energies than IceCube could otherwise achieve, by vetoing the cosmic ray background from above the detector. The starting event selection also provides competitive sensitivity to extended sources, where good angular resolution is not needed.

### 2.4. Other searches

The point source sample has been used for extended-source searches, in which the point-source hypothesis is replaced with that of an extended source of radius from $1°$ up to $5°$. No evidence of an extended source was found in an all-sky search.[1] More than a dozen "stacking" searches have also been performed. In each of these, an *a priori* list of objects (point-like or extended) is tested simultaneously, with a relative-weighting based on theory and/or gamma-ray observations. No evidence was found for neutrino emission in these searches. The most significant result ($2\sigma$) was found for the set of six Milagro TeV sources on the galactic plane visible in the northern sky, which are hypothesized cosmic ray accelerators.[7,8] In some stacking analyses, notably the searches for neutrinos from gamma-ray bursts and from blazars, the lack of significant emission constrains their contribution to the astrophysical flux, see Sec. 3.2. Small-scale anisotropy studies, based on autocorrelation and spherical harmonic analyses, have placed generic constraints on ensembles of weakly-emitting, individually unresolved sources that could contribute to the diffuse flux.[9]

Time-dependent searches for individual flaring emission, including both searches correlated with multi-wavelength observations and untriggered searches, have been performed on four years of point source neutrino data.[10] The most significant result ($2.2\sigma$ post-trial) was found in an all-sky search for any time-dependent excess of events, and occurred at ($21°$ r.a., $0°$ dec.) over ~11 days in March 2010. No counterpart in photons has been identified.

It should be noted that a high-significance ($>6\sigma$) measurement of the Moon's shadow in cosmic ray muons was obtained already with two years of data from the partially-completed IceCube array. Without a standard candle neutrino point source, the moon shadow observation provides a verification of the astronomical pointing and allows a confirmation of the point-spread function that has been estimated in simulation of track events.[11]

## 3. Discussion of results

In this section we discuss what the absence of a significant source identification thus far can reveal about the origin of the astrophysical flux.

A good place to start is with the PeV muon neutrino observed by IceCube in 2014.[12] Being a track event, it has the best-reconstructed direction of any of the PeV events so far, with an angular uncertainty of $0.3°$. Its direction is about $150°$ away from the galactic center and $11°$ off the galactic plane, so the event does not appear to have an obvious association with the galaxy. Within $3°$ of the event, there are no known gamma-ray sources, either galactic or extragalactic. What possible source does this event come from?

### 3.1. *Galactic versus extragalactic origins*

Before considering extragalactic origins, we first note that there may be several hints of a galactic component in the observed astrophysical flux. These include the above-mentioned galactic plane correlation and the galactic center excess in the starting-sample analysis, and the Milagro candidate sources along the galactic plane in the through-going analysis. At present none of these is statistically significant evidence of galactic emission. Nevertheless, neutrinos from within the galaxy are expected, both from the (still unidentified) sources of galactic cosmic rays, and from interactions of these cosmic rays with the interstellar medium. The latter appears as a diffuse emission from along the galactic plane and is a guaranteed source of astrophysical neutrinos. Calculations of the expected neutrino flux from this process, based on the observed cosmic ray spectrum, predict a flux well below present sensitivities.[13] However, if the locally-measured cosmic ray spectrum is softer than in other parts of the galaxy, the neutrino flux may be higher.[14,15]

While the presence of a galactic component is uncertain, the presence of an extragalactic component is an inference more easily drawn from the existing data. The detection and measurement of the astrophysical neutrino flux in both the starting sample (dominated by southern sky sensitivity) and the northern sky muon neutrino sample,[4] as well as the distribution over the sky of the starting event directions shown in Fig. 1, together indicate that the galactic plane and sources within it cannot be the origin of *all* of the astrophysical neutrinos. A straightforward explanation is that these neutrinos come from an extragalactic source population distributed throughout the universe. This would yield a nearly isotropic flux, except for the nearer and brighter sources which would, with sufficient statistics, start to stand out.

Until a significant deviation from isotropy is detected, the simplest description consistent with the data is that *all* the astrophysical neutrinos are part of a diffuse flux from the ensemble of unresolved neutrino sources throughout the universe. With more events, better angular resolution, or correlations in time or direction with known astronomical phenomena, components of the astrophysical flux may start to be isolated, and resolved into extragalactic or galactic sources.

### 3.2. *Extragalactic source populations*

The challenge of identifying extragalactic sources is suggested by Fig 4. Here we consider a generic extragalactic population of sources that have similar neutrino

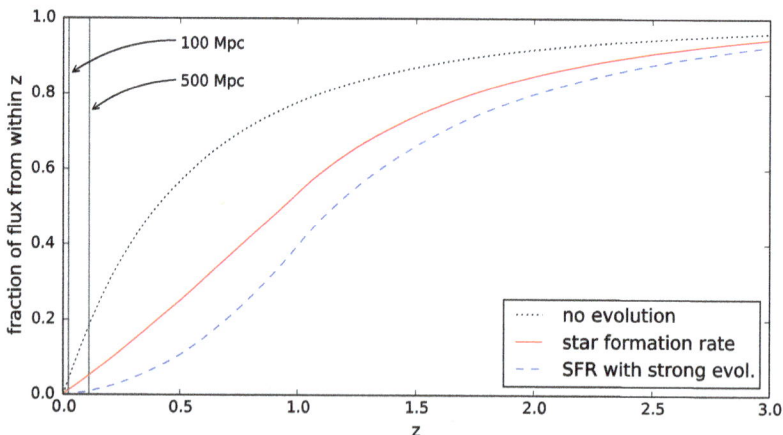

Fig. 4. Cumulative fraction of the astrophysical neutrino flux expected from sources closer than redshift $z$. Three models for how the population source density might have evolved over time are shown. The top curve (no evolution) assumes the comoving source density is constant; the middle curve assumes the source density traces the star formation rate (fitted from Ref. 16); in the bottom curve, the source density traces the star formation rate but with a suppression factor for $z < 1$.[17,18] The vertical lines indicate distances of 100 Mpc and 500 Mpc for reference.

luminosities and $E^{-2}$ power-law energy spectra. Three different models of the source density variation over the history of the universe are shown, corresponding to: a constant comoving density (no evolution), a density that follows the evolution of the star formation rate, and a third that follows the star formation rate for $z > 1$ but is an order of magnitude lower today at $z = 0$.[17,18] For each density model, the figure shows the fraction of the total neutrino flux from the source class that arrives from all the sources nearer than redshift $z$. A useful reference point is that for a source density following the star formation rate, approximately 1% of the extragalactic neutrinos detected at Earth originate from the sources within 100 Mpc ($z \approx 0.022$). (The cases of no evolution or strong evolution would yield about 4% or 0.1%, respectively.)

The challenge is that many candidate source classes, such as active galactic nuclei (AGNs) or starburst galaxies, have sufficiently high densities $\rho \sim 10^{-4}$ Mpc$^{-3}$ such that there are hundreds of objects within 100 Mpc and millions of objects within $z < 1$. Within one square degree (approximately the solid angle uncertainty of a well-reconstructed muon track event), there are still tens to hundreds of such objects out to a redshift of $z = 2$, a typical range from which such a neutrino might come. If the sources of the astrophysical flux are common, it can be difficult to associate a high-energy neutrino with its source, while at the same time the number of neutrinos expected from well-known nearby objects is small.

Classes of rare source objects such as blazars and gamma-ray bursts can be more readily constrained. For rare sources, more flux is concentrated in fewer individual objects making them brighter and easier to detect. Blazars, the sub-class of AGNs with jet emission pointed along our line of sight, are correspondingly rarer than AGNs as a whole. With densities $\rho \sim 10^{-7}$ Mpc$^{-3}$, blazars provide good prospects for detection or constraints using large stacking searches. The most powerful test of a blazar origin of the astrophysical flux has been performed using three years of the IceCube through-going point source sample.[19] All 862 blazars in the second Fermi LAT AGN Catalog (2LAC),[20] as well as four sub-samples based on different classifications of blazars, were tested in a maximum-likelihood point source stacking analysis. Only a non-significant excess ($\sim 1\sigma$) of neutrinos correlated with blazars was found.

Because the blazars are distributed more or less randomly across the whole sky, the flux upper limits from the stacked search can be converted to an isotropic flux per solid angle and compared with the diffuse flux measurement. Example results for two of the blazar samples are shown in Fig. 5. In all cases, the upper limit on the total flux from the tested blazars amounts to only a fraction ($\lesssim 20\%$) of the diffuse neutrino flux. (In Fig. 5, the comparison is with the diffuse flux fit in Ref. 21 which measures the astrophysical neutrino flux down to 10 TeV and fits a spectral index of $\gamma = 2.5$. For an $E^{-2}$ fit to the astrophysical neutrino flux, the constraints on the blazar catalog contribution are similarly at the fractional level.) It is moreover possible to transform the tested catalog limits to population limits, based on estimates of the fraction of the total gamma-ray emission from a blazar

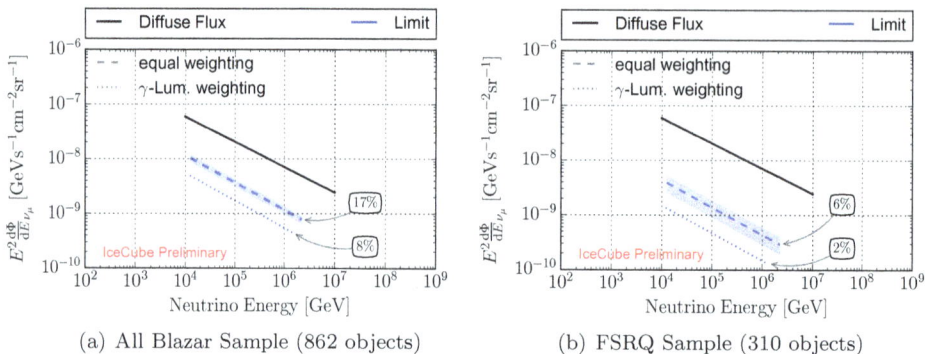

Fig. 5.    Comparison of the IceCube fit for the diffuse astrophysical flux in Ref. 21 (black, solid lines), and the upper limits on the flux from different blazar samples, converted to a diffuse flux. Left: all blazars in Fermi-2LAC. Right: flat-spectrum radio quasars (FSRQs) only. The dashed and dotted lines show limits obtained under different assumptions about whether the neutrino luminosity tracks the gamma-luminosity (and is therefore known on a source-by-source basis during the search), or not ("equal-weighting"). The percentage labels indicate the fraction of the diffuse neutrino flux that the blazar limits corresponds to. Figures from Ref. 19.

class that is resolved by Fermi and represented in the catalog. For example in the case of the rare blazars classified as flat spectrum radio quasars (FSRQs), it is estimated that the resolved sources comprising the catalog sample account for more than 50% of the total gamma-ray flux emitted by all FSRQs.[22]

Gamma-ray bursts (GRBs), with densities $\sim 10^{-9}$ Mpc$^{-3}$ per year, are sufficiently rare that their contribution to the astrophysical neutrino flux can be strongly constrained. These constraints are further improved because the bursts' brief duration means that searches have very low background. Significant limits on their neutrino emission were already placed before the discovery of the astrophysical flux.[23] A more recent analysis, using a four-year sample of through-going muon neutrino track events and 506 GRBs, constrains the contribution to the astrophysical neutrino flux by the total population of potentially observable GRBs to 1% or less.[5]

## 4.  Conclusions and prospects

The discovery of the high-energy astrophysical neutrino flux has initiated the field of neutrino astronomy, challenging us now to make the first identification of the sources of this flux. The sky distribution of the neutrinos observed so far suggests the origin likely includes a population of extragalactic sources and possibly includes galactic sources. The non-detection of sources to date already places strong constraints on several extragalactic source classes. Gamma-ray bursts and flat-spectrum radio quasars, for example, are sufficiently rare that neutrino correlation searches with these objects should have already produced significant detections if they were the dominant sources of the astrophysical flux. Next generation neutrino telescopes like IceCube-Gen2[24] and KM3NeT[25] with an order-of-magnitude improvement in point source sensitivity will be able to probe the more difficult case where extragalactic sources are common but weak, as well as identify and study the galactic sources responsible for the galactic cosmic rays. More uncertain, but especially exciting, is the possibility that the next generation detectors will be sensitive enough to observe time-dependent neutrino emission from one or more of the source classes hypothesized to be time-variable.

## Acknowledgements

The author thanks Imre Bartos and Maryon Ahrens for thoughtful discussions and is grateful for the financial support of the Swedish Research Council (VR), as well as from the Stockholm University Physics Department and the Oskar Klein Centre.

## References

1. M.G. Aartsen, *et al.*, Searches for extended and point-like neutrino sources with four years of IceCube data, *Astrophys. J.* **796** (2014) 109. doi: 10.1088/0004-637X/796/2/109.
2. M.G. Aartsen, *et al.*, Evidence for high-energy extraterrestrial neutrinos at the IceCube detector, *Science* **342** (2013) 1242856. doi: 10.1126/science.1242856.
3. M.G. Aartsen, *et al.*, Observation of high-energy astrophysical neutrinos in three years of IceCube data, *Phys. Rev. Lett.* **113** (2014) 101101. doi: 10.1103/PhysRevLett.113.101101.
4. M.G. Aartsen, *et al.*, Evidence for astrophysical muon neutrinos from the northern sky with IceCube, *Phys. Rev. Lett.* **115**(8) (2015) 081102. doi: 10.1103/PhysRevLett.115.081102.
5. M.G. Aartsen, *et al.*, Search for prompt neutrino emission from gamma-ray bursts with IceCube, *Astrophys. J.* **805**(1) (2015) L5. doi: 10.1088/2041-8205/805/1/L5.
6. S. Adrian-Martinez, *et al.*, Searches for point-like and extended neutrino sources close to the galactic centre using the ANTARES neutrino telescope, *Astrophys. J.* **786** (2014) L5. doi: 10.1088/2041-8205/786/1/L5.
7. F. Halzen, A. Kappes, and A. O'Murchadha, Prospects for identifying the sources of the galactic cosmic rays with IceCube, *Phys. Rev.* **D78** (2008) 063004. doi: 10.1103/PhysRevD.78.063004.
8. M.C. Gonzalez-Garcia, F. Halzen, and V. Niro, Reevaluation of the prospect of observing neutrinos from galactic sources in the light of recent results in gamma ray and neutrino astronomy, *Astropart. Phys.* **57–58** (2014) 39–48. doi: 10.1016/j.astropartphys.2014.04.001.
9. M.G. Aartsen, *et al.*, Searches for small-scale anisotropies from neutrino point sources with three years of IceCube data, *Astropart. Phys.* **66** (2015) 39–52. doi: 10.1016/j.astropartphys.2015.01.001.
10. M.G. Aartsen, *et al.*, Searches for time dependent neutrino sources with IceCube data from 2008 to 2012, *Astrophys. J.* **807**(1) (2015) 46. doi: 10.1088/0004-637X/807/1/46.
11. M.G. Aartsen, *et al.*, Observation of the cosmic-ray shadow of the moon with IceCube, *Phys. Rev.* **D89**(10) (2014) 102004. doi: 10.1103/PhysRevD.89.102004.
12. S. Schoenen and L. Raedel, Detection of a multi-PeV neutrino-induced muon event from the northern sky with IceCube, *The Astronomer's Telegram* **7856** (2015) 1.

13. V.S. Berezinsky, T.K. Gaisser, F. Halzen, and T. Stanev, Diffuse radiation from cosmic ray interactions in the galaxy, *Astropart. Phys.* **1** (1993) 281–288. doi: 10.1016/0927-6505(93)90014-5.

14. A. Neronov and D. Semikoz, Neutrinos from extra-large hadron collider in the Milky way, *Astropart. Phys.* **72** (2016) 32–37. doi: 10.1016/j.astropartphys.2015.06.004.

15. D. Gaggero, D. Grasso, A. Marinelli, A. Urbano, and M. Valli, The gamma-ray and neutrino sky: A consistent picture of Fermi-LAT, H.E.S.S., Milagro, and IceCube results. (2015). arXiv: 1504.00227.

16. L.-X. Li, Probing the cosmic metallicity evolution with gamma-ray bursts, *Mon. Not. Roy. Astron. Soc.* **388** (2008) 1487. doi: 10.1111/j.1365-2966.2008.13488.x.

17. M. Ahlers and F. Halzen, Pinpointing extragalactic neutrino sources in light of recent IceCube observations, *Phys. Rev.* **D90**(4) (2014) 043005. doi: 10.1103/PhysRevD.90.043005.

18. T.A. Thompson, E. Quataert, E. Waxman, and A. Loeb, Assessing the starburst contribution to the gamma-ray and neutrino backgrounds (2006). arXiv: 0608699.

19. T. Glüsenkamp. Analysis of the cumulative neutrino flux from Fermi-LAT blazar populations using 3 years of IceCube data. In *5th Roma International Conference on Astro-Particle physics (RICAP 14)*, Noto, Sicily, Italy, September 30-October 3, 2014, (2015). arXiv: 1502.03104.

20. M. Ackermann, *et al.*, The second catalog of active galactic nuclei detected by the Fermi Large Area Telescope, *Astrophys. J.* **743** (2011) 171. doi: 10.1088/0004-637X/743/ 2/171.

21. M.G. Aartsen, *et al.*, Atmospheric and astrophysical neutrinos above 1 TeV interacting in IceCube, *Phys. Rev.* **D91**(2) (2015) 022001. doi: 10.1103/PhysRevD.91.022001.

22. M. Ajello, M.S. Shaw, R.W. Romani, C.D. Dermer, L. Costamante, O.G. King, W. Max-Moerbeck, A. Readhead, A. Reimer, J.L. Richards, and M. Stevenson, The luminosity function of Fermi-detected flat-spectrum radio quasars, *Astrophys. J.* 751: (2012) 108. doi: 10.1088/0004-637X/751/2/108.

23. R. Abbasi, *et al.*, An absence of neutrinos associated with cosmic-ray acceleration in $\gamma$-ray bursts, *Nature* **484** (2012) 351–354. doi: 10.1038/nature11068.

24. M.G. Aartsen, *et al.*, IceCube-Gen2: A vision for the future of neutrino Astronomy in Antarctica (2014).

25. A. Margiotta, Status of the KM3NeT project, *JINST.* **9** (2014) C04020. doi: 10.1088/ 1748-0221/9/04/C04020.

# Chapter 7

# Recent Results from the ANTARES Neutrino Telescope

Paschal Coyle

*Centre de Physique des Particules de Marseille,*
*163 Avenue de Luminy, Case 902, Marseille 13288 cedex 09, France*
*coyle@cppm.in2p3.fr*

Clancy W. James

*Friedrich-Alexander-Universität Erlangen-Nürnberg,*
*Erlangen Centre for Astroparticle Physics,*
*Erwin-Rommel-Str. 1, 91058 Erlangen, Germany*
*Clancy.James@physik.uni-erlangen.de*

The ANTARES deep sea neutrino telescope has been taking data continuously since its completion in 2008. With its excellent view of the galactic plane and good angular resolution the telescope can constrain the origin of the diffuse astrophysical neutrino flux reported by IceCube. Assuming various spectral indices for the energy spectrum of neutrino emitters, the Southern sky and in particular central regions of our galaxy have been studied searching for point-like objects, for extended regions of emission, and for signal from transient objects selected through multi-messenger observations. For the first time, cascade events are used for these searches.

ANTARES has also provided results on searches for hypothetical particles (such as magnetic monopoles and nuclearites in the cosmic radiation), and multi-messenger studies of the sky in combination with various detectors. Of particular note are the searches for dark matter: the limits obtained for the spin-dependent WIMP-nucleon cross section surpassing those of current direct-detection experiments.

## 1. Introduction

Being weakly interacting, the neutrino is unique and complementary to other astrophysical probes such as multi-wavelength light and charged cosmic rays. The neutrino can escape from regions of dense matter or radiation fields and can travel cosmological distances without being absorbed. Neutrinos make it possible to distinguish unambiguously between hadronic and electronic acceleration and to localise these acceleration sites more precisely than cosmic rays detectors. High-energy neutrinos may also be produced by the annihilation of dark matter particles which may have accumulated in the cores of massive astrophysical objects such as the Sun. Since the recent observation of a diffuse flux of cosmic neutrinos by the IceCube Collaboration, an understanding of its origin has become a top priority for the astroparticle physics community.

The ANTARES neutrino telescope being located in the Northern Hemisphere has a good visibility for the galactic plane and, due to the exceptional optical properties of the deep seawater, provides an excellent angular resolution on the neutrino direction. The ANTARES detector (see Ref. 1 for details), is located 40 km offshore from Toulon at a depth of 2475 m. It was completed on 29 May 2008, making it the largest neutrino telescope in the northern hemisphere and the first to operate in the deep sea. The telescope is optimised to detect upgoing high energy neutrinos (>10 GeV) by observing the Cherenkov light produced in seawater from secondary charged leptons that originate in neutrino interactions near the vicinity of the instrumented volume. Due to the long range of the muon, neutrino interaction vertices tens of kilometres away from the detector can be observed thereby increasing the effective volume. Other neutrino flavours are also detected, although with lower efficiency and degraded angular precision because of the shorter range of the corresponding leptons.

The detector infrastructure comprises 12 mooring lines hosting light sensors. Due to its location in the deep sea, the infrastructure also provides opportunities for innovative measurements in Earth and sea sciences (see for example Ref. 3). Another project benefiting from the deep sea infrastructure is an R&D system of hydrophones which investigates the detection of ultra-high energy neutrinos using the sound produced by their interaction in water. This system called AMADEUS (Antares Modules for the Acoustic Detection Under the Sea)[4] is a feasibility study for a prospective future large-scale acoustic detector.

The decommissioning of the ANTARES telescope is planned for 2017, at which point the KM3NeT neutrino telescope[2] will have surpassed ANTARES in sensitivity.

## 2. General description of the detector

A schematic view of the ANTARES neutrino telescope is given in Fig. 1. The basic detection element is the Optical Module which houses a 10-inch photo-multiplier tube inside a pressure-resistant glass sphere. Each node of the three-dimensional telescope array is the assembly of a mechanical structure, which supports three optical modules, looking downwards at 45°, and a titanium container housing the offshore electronics and embedded processors. In its nominal configuration, a detector line comprises 25 storeys linked together with an electro-mechanical cable. The distance between storeys is 14.5 m with the first storey starting 100 m from the seabed. The line is anchored on the seabed and is held vertical by a submerged buoy.

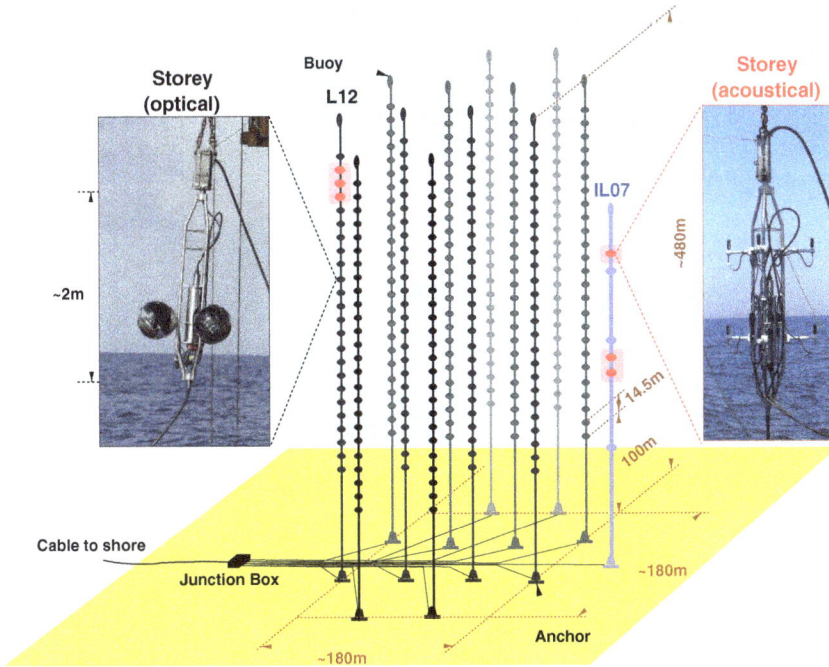

Fig. 1.   Schematic view of the ANTARES detector.

The full neutrino telescope comprises 12 lines, 11 with the nominal configuration, the twelfth line being equipped with 20 storeys and completed by devices dedicated to acoustic detection. Thus, the total number of the OMs installed in the detector is 885. The lines are arranged in an octagonal configuration, with a typical interline spacing of 60–70 m. The infrastructure is completed by the Instrumentation Line, which supports the instruments used to perform environmental measurements.

The detection lines are flexible and variations in the intensity and direction of the sea current can induce a coherent displacement, typically of the order of a few meters for the uppermost storeys. In addition, the storeys may rotate about the vertical axis. For the ultimate precision on the neutrino directions it is important to follow such movements. An acoustic positioning system[5] and compasses within the electronics container provide a precision of certimetres and degrees on the Optical Module position and orientation.

The data communication and the power distribution to the lines are provided via a network of interlink cables and a junction box on the sea floor. Power is transmitted in alternating current from shore, via a single conductor in a 42 km long electro-optic cable, to a transformer in the junction box; the power return is via the sea.

The data acquisition system is based on the "all-data-to-shore" concept. In this mode, all signals from the PMTs that pass a preset threshold (typically 0.3 single photo electrons) are digitised in a custom built ASIC chip and all digital data are sent to shore, via an optical Dense Wavelength Division Multiplexing system. On shore the data are processed in real-time by a farm of commodity PCs. The data flow is typically a couple of Gb/s and is dominated by the light generated by K40 decays of the salt in the seawater (typically a 50 kHz baseline singles rate per photo-multiplier).

## 3. Event reconstruction

The main channel for the search for astrophysical point-like sources of neutrinos is the muon neutrino. The high rate of downgoing $\mu$ from the interactions of cosmic rays in the atmosphere restricts such searches to events coming from below, or only a few degrees above, the horizon. The remaining background is then the flux of atmospheric $\nu_\mu$ and those few remaining atmospheric $\mu$ events misreconstructed as upgoing. The long scattering length of blue light in seawater provides an excellent directional resolution on the $\nu_\mu$ primary of 0.38° for an $E^{-2}$ source.[6] Thus, angular clustering requirements yield a strong suppression of both backgrounds.

The effective area of neutrino telescopes such as ANTARES and IceCube to cascade events (neutral-current (NC) interactions, and $\nu_e$ and $\nu_\tau$ charged-current (CC) interactions) is generally lower than that of $\nu_\mu$ CC interactions, due to the shorter range of the outgoing lepton. Additionally, the angular resolution of the cascade channel is inferior. Nevertheless, it has several advantages: neutrino events are more easily distinguished from the background of atmospheric muons, allowing both up- and down-going events to be studied. Furthermore, the energy deposited in the detector is better correlated with the energy of the primary neutrino.

The performance of the current ANTARES cascade reconstruction algorithm[7] yields a median angular resolution of typically 3° for the energy range 10–100 TeV (Fig. 2). The corresponding energy resolution is about 5% (Fig. 3) and thus is limited by the current systematic uncertainty of 10%. Below 10 TeV, the resolution degrades due the reduced number of detected photons, while above 300 TeV the events start to saturate the detector.

## 4. Point source searches for astrophysical neutrinos

The inclusion of the cascade channel has allowed for the first time a combined point-source search using both muon-track and cascade events using 1690 days of effective livetime from 2007 to 2013.[7] After cuts, the sample consisted of 6261 muon-track events, and 156 cascade events, with an estimated contamination of 10% mis-reconstructed atmospheric muons in each. The resulting skymap is shown in Fig. 4.

While the atmospheric background produces predominantly muon-track events, an $E^{-2}$ point source with a flavour-uniform flux would be expected to produce a cascade-to-track ratio of 3:10, significantly increasing the sensitivity of the

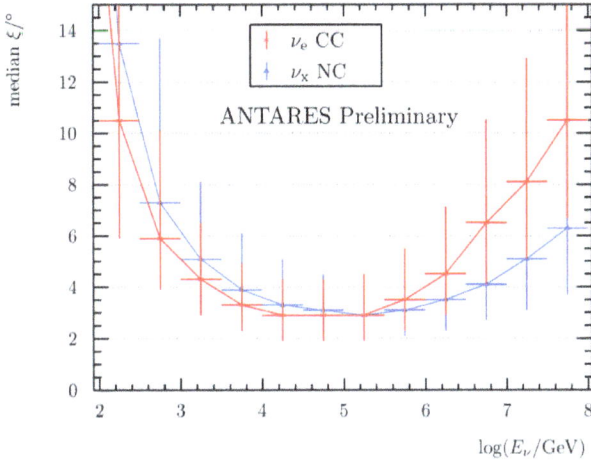

Fig. 2.  Angular resolution for cascade topology events.

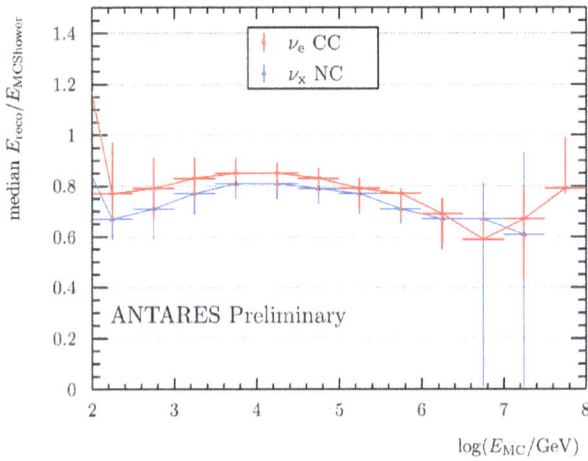

Fig. 3.  Energy resolution for cascade topology events.

search. The achieved search sensitivity was approximately $10^{-8}\,\mathrm{GeV}^{-1}\mathrm{cm}^{-2}\mathrm{s}^{-1}$ for $\delta < -40°$ as shown in Fig. 5. An untargeted point-source search, a search over a list of pre-specified candidates, and a search using the origins of the IceCube events reported in Ref. 8 were applied to this data. No significant excess was observed.

## 4.1.  *Joint ANTARES/IceCube point source search*

A joint analysis using ANTARES and IceCube data is detailed in Ref. 9. The fractional number of source events expected to be present in each data set is

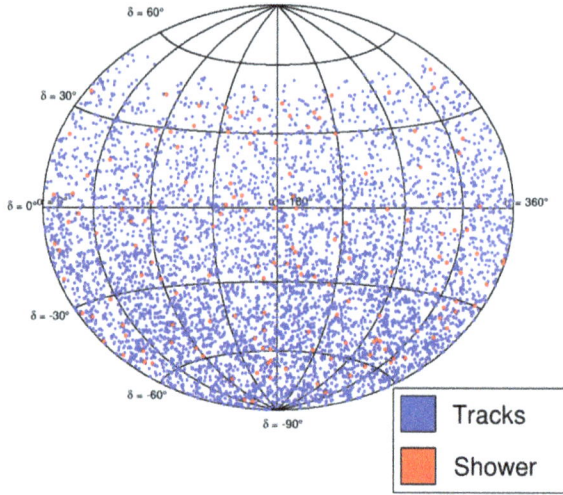

Fig. 4.   Arrival directions of events in the all-sky point source analysis.

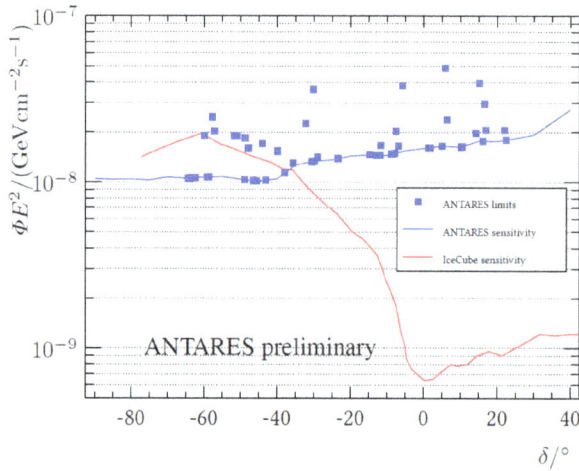

Fig. 5.   Limits and sensitivity of the ANTARES targeted source for flavour uniform neutrino points sources with an $E^{-2}$ spectra in terms of flux per flavour.

shown in Fig. 6 for the current best-fit to the IceCube flux. The results of the combined search are shown in Fig. 7, for an $E^{-2.5}$ source spectrum. The ANTARES contribution is dominant for declination $< -15°$. No significant cluster is found, with the most significant source on the candidate list being the blazar 3C 279, with a pre-trial $p$-value of 5%. The combined analysis improves on the results from both experiments, indicating the complementarity of the two instruments.

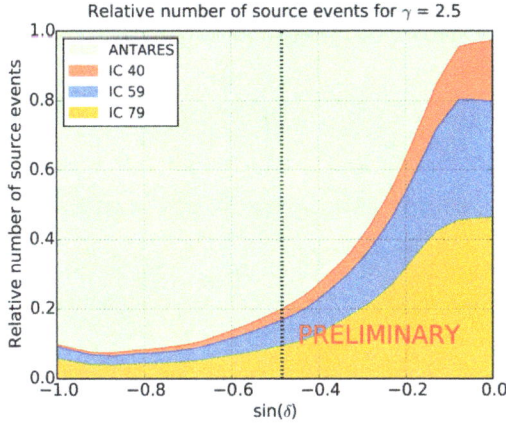

Fig. 6.    Fractional contribution of each data set to the total number of signal events passing cuts in the joint ANTARES–IceCube analysis for an $E^{-2.5}$ spectra with no cutoff as a function of $\delta$.

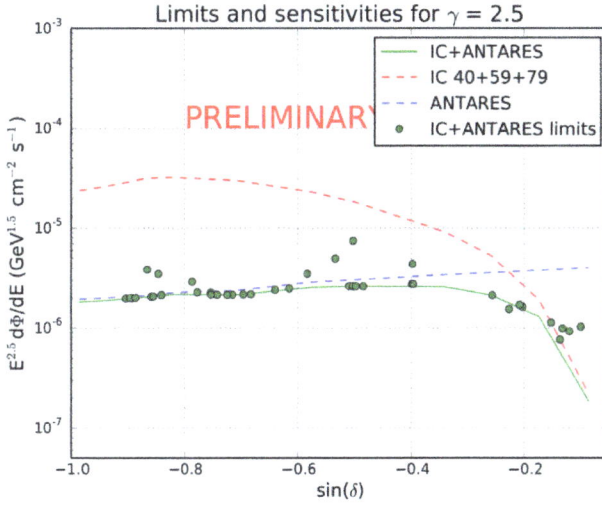

Fig. 7.    Sensitivity (lines) and limits (dots) of the joint ANTARES–IceCube analysis for an $E^{-2.5}$ spectra with no cutoff as a function of $\delta$. ANTARES (blue), IceCube (red), combined (green).

## 4.2.   *Limits on a point source origin of the IceCube signal*

It has been proposed[10] that the cluster of IceCube events seen in Ref. 8 could be due to a single point-like source, which is not detectable due to the poor angular resolution. The non-detection of an ANTARES point-like source in this region, as reported by Ref. 11, limits the flux of such a source as a function of spectral index, shown by the solid lines and y-axis of Fig. 8. The flux required to produce a given number of events in the HESE analysis (x-axis) is also shown. The range where the latter is greater than the former rules out a corresponding contribution from any

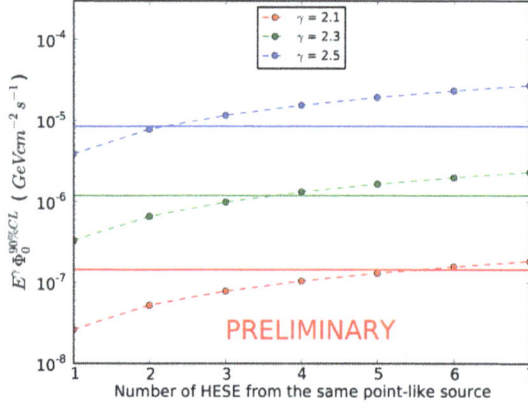

Fig. 8.  ANTARES limits (solid lines) at 90% C.L. on the contribution of point-like sources to the IceCube HESE sample for various spectral indices, shown for $\delta = -29°$. These are compared with the flux required to produce a given number of HESE events. Similar results are obtained for other declinations around the galactic centre.

single point-like source with that spectral index at 90% C.L. The result above is particularly relevant because the current best-fit spectrum (between 25 TeV and 2.8 PeV) of the IceCube flux has a spectral index of $-2.5 \pm 0.09$.[12] ANTARES can thus rule out any single point-source of neutrinos in the region of the galactic centre with spectral index of –2.5 as having a flux corresponding to more than 2 HESE events.

## 5. Transient sources

### 5.1. *Flares from AGN and X-ray binaries*

Active galactic nuclei (AGN) have long been proposed as a source of high-energy cosmic rays and, hence, neutrinos.[13] Blazars, AGNs with jets pointing towards the line-of-sight, exhibit bright flares which dominate the extragalactic gamma-ray sky observed by Fermi-LAT.[14]

Using multi-wavelength observations, several bright blazars have been reported by the TANAMI Collaboration[15] to lie within the 50% error bounds of the reconstructed arrival directions of the PeV-scale events IC 14 and IC 20 observed by Ice-Cube. ANTARES observes signal-like events from the two brightest blazars, both in the field of IC 20,[16] although this is also consistent with background fluctuations. A lack of such events from the field of IC 14 excludes a neutrino spectrum softer than $E^{-2.4}$ as being responsible for this event. The highest-energy event IC 35 ("Big Bird") was detected during an extremely bright flare from the blazar PKS B1424-418, which lies within the 50% error region of the IC 35 arrival direction. ANTARES finds only one event within 5° of this source during the flaring

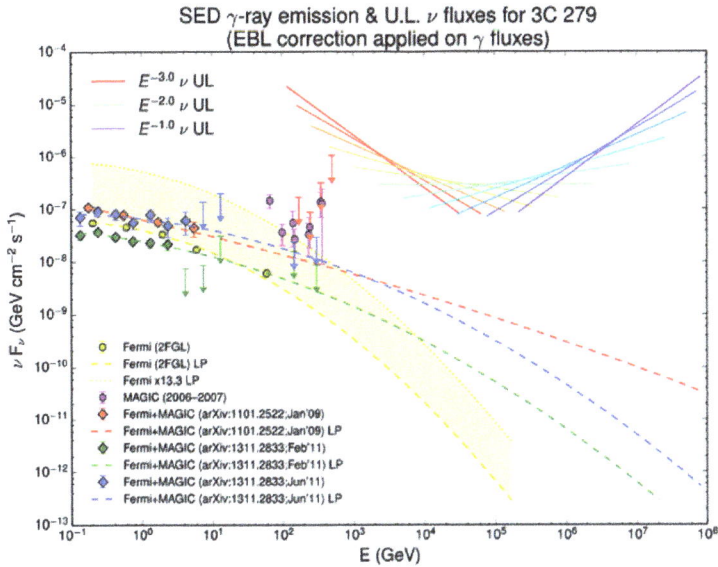

Fig. 9. Limits on the neutrino flux from the blazar 3C279 as a function of spectral index (solid lines), compared to the observed (points) and extrapolated (dashed lines) gamma-ray spectra observed by FERMI and IACTs.

period, whereas approximately three would be expected from random background fluctuations alone.

In another analysis,[17] ANTARES targets a sample of 41 blazar flares observed by Fermi LAT and 7 by the IACTs H.E.S.S., MAGIC, and VERITAS. The lowest pre-trial p-value of 3.3% was found for the blazar 3C 279, which comes from the coincidence of one event with a 2008 flare previously reported in Ref. 18. However, the post- trial p-value is not significant. The resulting limits are given in Fig. 9.

Similar methods were also used to search for neutrino emission during the flares from galactic X-ray binaries.[19] A total of 34 X-ray- and gamma-ray-selected binaries were studied, with no significant detections, allowing some of the more optimistic models for hadronic acceleration in these sources to be rejected at 90% C.L.

## 5.2. *Gamma-ray bursts*

Long-duration gamma-ray bursts (GRBs) have been proposed as a source of the highest-energy cosmic rays.[23] ANTARES has searched for a neutrino flux from GRBs considering two models of the emission processes: the NeuCosmA description of Ref. 20 and the "photospheric" model of Ref. 24. In each case, the expected signal is simulated on a burst-by-burst basis, and the detector response and background are modelled using the exact detector conditions at the time of the burst. The ANTARES analysis using the NeuCosmA model was developed and applied to a sample of 296 bursts in Ref. 25, with no coincident neutrino events detected.

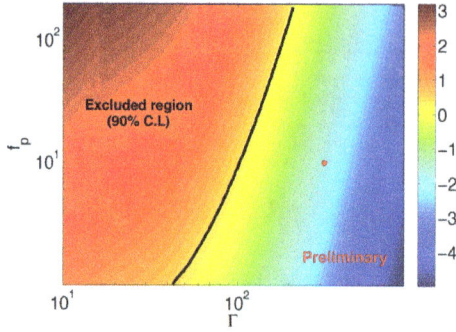

Fig. 10.   Range of jet Γ-factor and baryonic loading fp excluded by ANTARES in the case of
GRB110918A using the NeuCosmA model of Ref. 20, as described in Ref. 21. The assumed values
of $\Gamma = 316$ and $fp = 10$ are shown by the red point, while the colour-coding gives the expected
number of observable neutrinos.

Since then, one especially powerful burst GRB110918A, and the nearby burst
GRB130427A, have been identified as promising candidates for neutrino detection,
and studied in detail in Ref. 21. No coincident events are observed from either
burst. The predicted $\nu$ emission scales with $\Gamma^{-5}$ and linearly with $f_p$, allowing limits to be set on the bulk gamma-factor and baryonic loading of the jet, as shown
in Fig. 10.

A search using the photospheric models is developed in Ref. 26, and will shortly
be unblinded. The GRB search methods are also being extended to test Lorentz
invariance violation,[21] which would delay the arrival times of TeV neutrinos compared to GeV photons.

### 5.3. Optical and X-ray follow-up

The TAToO (telescopes-ANTARES target-of-opportunity) program[27] performs
near-real-time reconstruction of muon-track events. If a sufficiently high energy
event is reconstructed as coming from below the horizon (i.e. those events most
likely to be of astrophysical origin), an alert message is generated to trigger robotic
optical telescopes, and, with a higher threshold, the Swift-XRT. The very short alert-
generation time (a few seconds) and half-sky simultaneous coverage of ANTARES
makes it ideal for detecting transient signals, and optical and X-ray follow up observations have been initiated within 20 s and one hour respectively.

Results from 42 optical and 7 X-ray alerts have been analysed. While no associated transient event was detected, this non-observation can be used to place limits on the astrophysical origin of the detected neutrinos,[22] as shown in Fig. 11.
The steep fall-off of the light-curves emphasises the need for a rapid alert generation and follow-up: observations within one minute can rule out a GRB origin
with high confidence, while those after one day would be unlikely to detect even a
bright GRB.

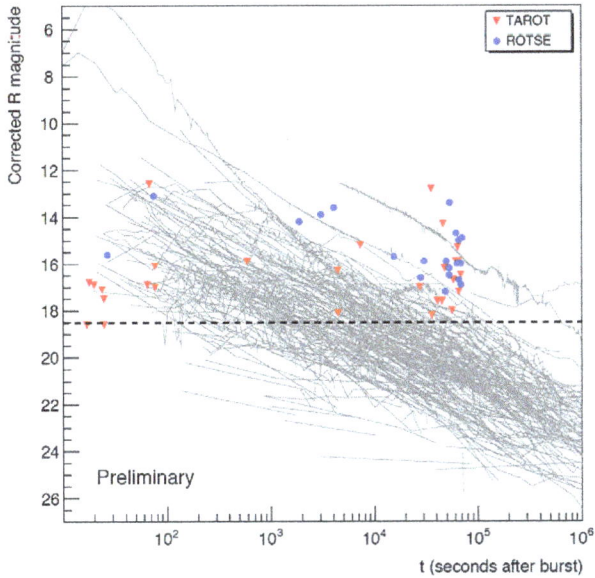

Fig. 11. Limiting magnitudes and delay times of optical follow-up observations to ANTARES alerts with ROTSE and TAROT[22] compared to (grey lines) the light-curves from measured GRBs.

## 6. Diffuse flux searches

A diffuse flux search in ANTARES has been developed that makes optimal uses of both muon track and cascade events.[28] Since any explicit selection of muon-like and cascade-like events inevitably discards events with topologies falling between the two classes, no such selection was made. The procedure was applied to 913 days of effective livetime between 2007 and 2013. The expected number of events from the standard and prompt atmospheric background[30,31] was 9.5±2.5, composed of 5.5 $\nu_\mu$ CC, 1 atmospheric $\mu$, and 2.9 $\nu$ NC and $\nu_e$ events. The expectation from the IceCube neutrino flux reported by Ref. 8 was 5.0 ± 1.1 events. After unblinding, 12 events passed the selection cuts–consistent with both background only, and background and IceCube diffuse flux expectations. The resulting limits on an $E^{-2}$ flux are given in Fig. 12.

## 7. Extended source searches

In addition to the numerous point-like candidate neutrino sources, several extended regions have been proposed as hadronic acceleration sites. ANTARES searches for an excess neutrino flux from these regions using "on-zones" defined by specific templates, which are compared to "off-zones" of exactly the same size and shape, but offset in right ascension. Thus the off-source regions give an unbiased estimate of the background in the source region in a way that is independent of simulations.

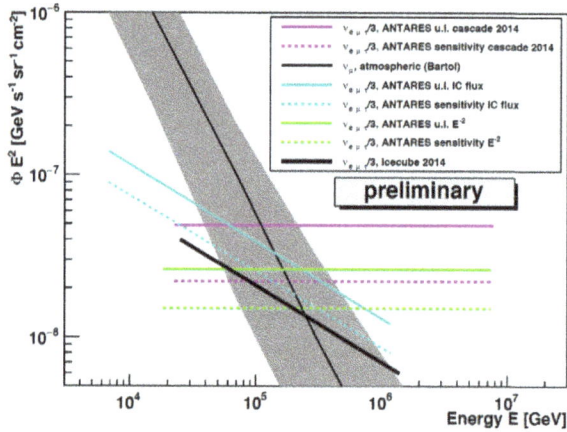

Fig. 12.   ANTARES sensitivity to (dotted), and limits on (solid), a diffuse astrophysical neutrino measured by IceCube.[28] shown are (pink) the previous ANTARES limit on an $E^{-2}$ spectrum,[29] and current results on (blue) the flux (thick black line) observed by IceCube.[12] and (green) an $E^{-2}$ spectrum. This is compared to the conventional atmospheric background flux (thin black line),[30] with associated error (grey shading).

Results for the Fermi Bubbles, galactic plane, and the IceCube cluster are described below.

## 7.1. *Fermi bubbles*

The Fermi Bubbles[35] are large regions of $\gamma$-ray emission extending perpendicular to the Galactic Centre and have been proposed as sites of hadronic acceleration,[34] with neutrinos expected from p-p collisions. A first search in ANTARES data from 2008–2011 for emission from these regions was presented in Ref. 33 — here, an update is presented adding 2012–2013 data.

The on- and off-zone regions used in the Fermi Bubble analysis are shown in Fig. 13. Flavour-uniform $E^{-2}$ and $E^{-2.18}$ neutrino fluxes are assumed, where the latter is motivated by the best-fit proton spectrum of $E^{-2.25}$ reported by Ref. 34. Exponential cut-offs at energies of 500, 100, and 50 TeV are also tested. A slight excess is found in the source region, corresponding to a $1.9\sigma$ significance. The corresponding upper limits on an $E^{-2.18}$ neutrino flux are compared in Fig. 14 to the expectations from Ref. 34.

## 7.2. *Galactic plane*

Cosmic rays in our galaxy will collide with the interstellar medium to produce pions and, hence, neutrinos. Direct evidence for these processes comes from observations by Fermi-LAT[39] of the diffuse galactic gamma-ray background. It is also interesting that the number of IceCube high energy starting events (HESE) in the $E > 100$ TeV

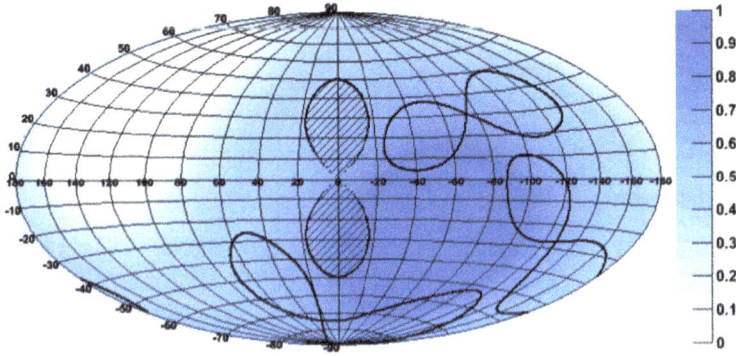

Fig. 13.   On- and off-zone search regions for the Fermi Bubble search of Ref. 32, compared to the ANTARES visibility (blue shading).

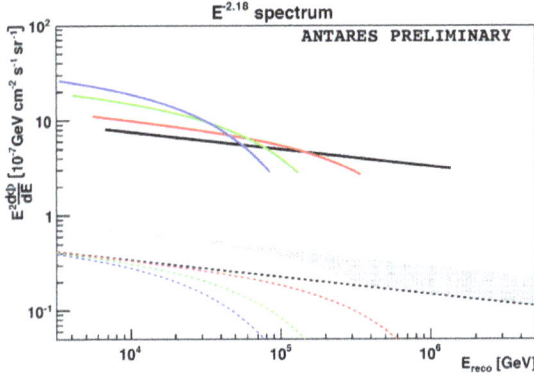

Fig. 14.   90% C.L. upper limits (lines) on the neutrino flux from the Fermi Bubbles, compared to (shaded regions) expectations[34] for different spectral shapes.

range with angular directions consistent with this region corresponds to a flux consistent with that observed in $\gamma$-rays,[36] as shown in Fig. 15. The large uncertainty in the arrival directions of cascade-like HESE, and their low number, makes this comparison difficult however.

The ANTARES northern latitude is ideally suited for studying the expected neutrino flux from the inner galactic plane. A search has been performed in the regions of galactic longitude $|l| < 40°$ and latitude $|b| < 3°$, as reported in Ref. 37. The search used nine off-zones and one on-zone, and found no excess in the on-zone region (one event compared to an average of 2.5 for the off-zones). The resulting limits are shown in Fig. 15. In particular, the hypothesis of a 1 to 1 relation between the $\gamma$-ray and neutrino flux from the Galactic Ridge is ruled out at 90% confidence, indicating that ANTARES is already testing the

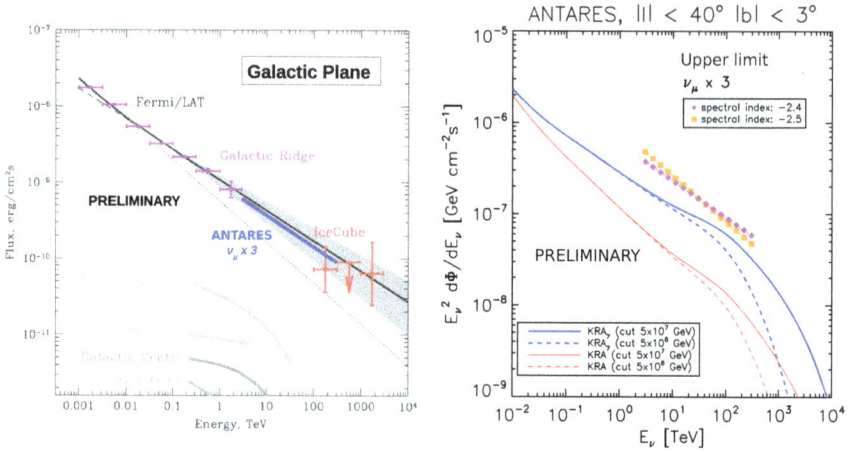

Fig. 15.   ANTARES 90% C.L. upper limits for the search for an excess of events from the central galactic region,[37] Left: for an $E^{-2.4}$ neutrino spectra, compared to the expected neutrino flux (solid black line) extrapolated from the FERMI-LAT diffuse gamma-ray up to high energies. Right: for $E^{-2.4}$ and $E^{-2.5}$ neutrino spectra, compared to the predicted neutrino flux from Ref. 38.

well-established multimessenger $\gamma$-$\nu$-CR paradigm in our galaxy. The limits, how-ever, cannot rule out models from more-detailed simulations of galactic cosmic-ray propagation.

### 7.3. IceCube cluster

The same techniques employed in the galactic plane search were used to probe the origin of the cluster of IceCube events reported in Ref. 8. The analysis of Ref. 37 used twelve off-zones and one on-zone to search for an excess of events. One event passing the selection cuts is observed in both the on-zone and the average off-zone, i.e. no excess is observed. Resulting limits on the maximum number of HESE events produced by a source with different spectral indices are presented in Fig. 16. For the best-fit IceCube diffuse spectral index $\Gamma = 2.5$,[12] ANTARES rejects at 90% confidence a flux from this region expected to produce three or more of the IceCube events in the cluster. This extends the results of Refs. 6 and 11 for this region, which limit the existence of point-like and mildly extended sources in this region.

### 8.  Dark matter and exotics

ANTARES has placed limits on different WIMP dark-matter scenarios by search-ing for high-energy neutrino emission from WIMP annihilation in the Sun, Earth, galactic centre, and dwarf galaxies. Since the dark-matter density is expected to be strongly peaked near the centres of these objects the excellent

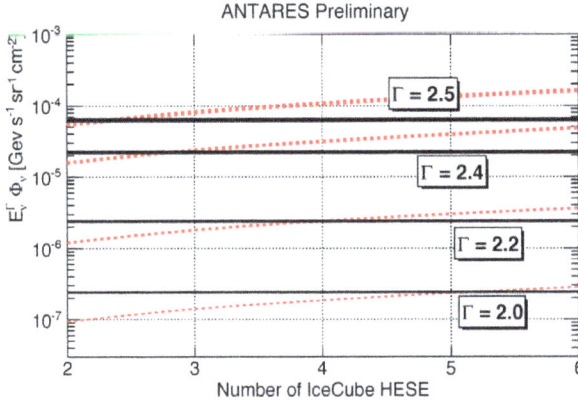

Fig. 16.   ANTARES upper limits at 90% C.L. (black) on a flavour-uniform neutrino flux from the IceCube cluster region as a function of the spectral index $\Gamma$, compared to (red) the flux required to produce an expected number of events in the IceCube HESE analysis.[40] The maximum number of IceCube events allowed at 90% C.L. is indicated by the crossing points of the red and black lines for a given spectral index. See Ref. 37 for details.

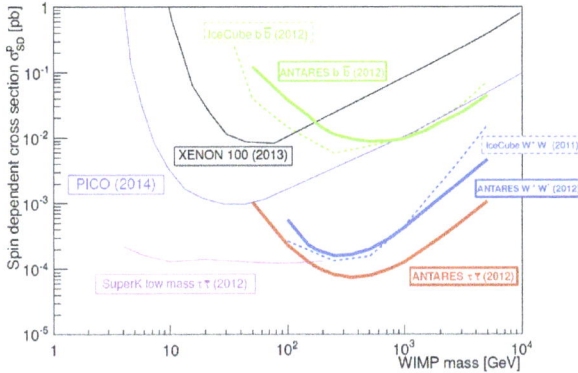

Fig. 17.   ANTARES limits on $\sigma_{SD}^p$ from the Sun as a function of the WIMP mass.

angular resolution of ANTARES, yields competitive limits for WIMP masses above 50 GeV.

Limits on the spin-dependent (WIMP-proton) interaction cross section $\sigma_{SD}^p$ from ANTARES observations of the Sun (Fig. 17)[42] and on the WIMP-WIMP velocity-averaged self-annihilation cross section $\sigma_A \nu$ from the galactic centre using the $\tau\tau$ channel[4] are given in Fig. 18, and are described in further detail in ref. 43.

Dark-matter analyses by ANTARES also includes a search for a WIMP signature from the centre of the Earth[44] and a test of secluded dark-matter models in the Sun.[45,46]

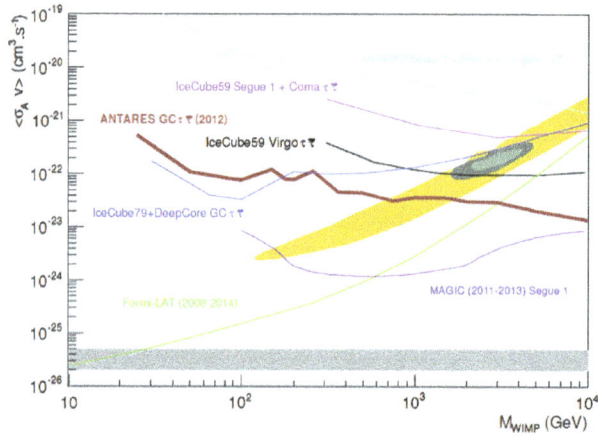

Fig. 18.   ANTARES limits on $\sigma_A \nu$ from the galactic centre as a function of the WIMP mass.

ANTARES also places limits on beyond-the-standard-model physics, with searches for magnetic monopoles and nuclearites. Updates to existing limits are presented in Ref. 47.

## 9.  Conclusions

The ANTARES neutrino telescope has proved to be a highly successful instrument, performing a wide range of physics analyses. In particular, its excellent angular resolution on both muon-track and cascade events, facilitated by the excellent optical properties of deep-sea water, is well suited to studying point-like sources of neutrinos. This capability has allowed relevant constraints to be placed upon the the origin of the astrophysical neutrino flux reported by IceCube and in particular any possible galactic contribution. ANTARES will continue data-taking until end of 2016, thus most of the analyses reported here will be extended with an additional three years of data.

## References

1. M. Ageron, *et al.*, *Nucl. Instrum. Meth. A* **656** (2011) 11.
2. S. Adrián-Martínez, *et al.*, *J. Phys. G* **43** (2016) 084001.
3. C. Tamburini, *et al.*, PLoS ONE 8(7) e67523 (2013) doi:10.1371/journal.pone.0067523.
4. J.A. Aguilar, *et al.*, *Nucl. Instrum. Meth. A* **626** (2011) 128.
5. S. Adrián-Martínez, *et al.*, *JINST* **7** (2012) T08002.
6. S. Adrián-Martínez, *et al.*, *Astrophys. J.* **786** (2014) L5.
7. T. Michael, PoS (ICRC2015) 1078.[a]; Light at the end of the shower; An all-flavour neutrino point-source search with the ANTARES neutrino telescope, PhD Dissertation, NIKHEF, The Netherlands (2016).

[a] PoS (ICRC2015): online at http://pos.sissa.it/cgi-bin/reader/conf.cgi?confid=236

8. M.G. Aartsen, *et al.*, *Phys. Rev. Lett.* **113** (2014) 101101.
9. S. Adrián-Martínez *et al.*, *Astrophys. J.* **823** (2016) 65.
10. M.C. Gonzalez-Garcia, F. Halzen and V. Niro, *Astropart. Phys.* **57** (2014) 39.
11. J. Barrios-Martí, *PoS* (ICRC2015) 1077.
12. M.G. Aartsen, *et al.*, *Astrophys. J.* **809** (2015) 98.
13. T.K. Gaisser, F. Halzen and T. Stanev, *Phys. Rep.* **258** (1995) 173.
14. W. B. Atwood *et al.*, *Astrophys. J.* **697** (2009) 1071.
15. F. Krauß, *et al.*, *A&A* **566** (2014) L7.
16. S. Adrián-Marínez, *et al.*, *A&A* **576** (2015) L8.
17. S. Adrián-Marínez, *et al.*, *J. Cosmol. Astropart. Phys.* **12** (2015) 014.
18. S. Adrián-Marínez, *et al.*, *Astropart. Phys.* **36** (2012) 204.
19. D. Dornic & A. Sánchez-Losa, *PoS* (ICRC2015) 1046.
20. S. Hümmer, M. Rüger, F. Spanier and W. Winter, *Astrophys. J.* **721** (2010) 630.
21. J. Schmid & D. Turpin, *PoS* (ICRC2015) 1057.
22. A. Mathieu, *PoS* (ICRC2015) 1093.
23. E. Waxman and J. Bahcall, *Phys. Rev. Lett.* **78** (1997) 2292.
24. S. Gao, K. Asano, and P. Mészáros, *J. Cosmol Astropart. Phys.* **11** (2012) 58.
25. S. Adrián-Marínez, *et al.*, *A&A* **559** (2013) A9.
26. M. Sanguineti, *PoS* (ICRC2015) 1068.
27. M. Ageron *et al.*, *Astropart. Phys.* **35** (2012) 530.
28. J. Schnabel & S. Hallmann, *PoS* (ICRC2015) 1065.
29. J. A. Aguilar, *et al.*, *Phys. Lett. B* **696** (2011) 16.
30. M. Honda, T. Kajita, K. Kasahara, S. Midorikawa, and T. Sanuki, *Phys. Rev. D* **75** (2007) 043006.
31. R. Enberg, M.H. Reno, and I. Sarcevic, *Phys. Rev. D* **78** (2008) 043005.
32. S. Hallmann, *PoS* (ICRC2015) 1059.
33. S. Adrián-Marínez, *et al.*, *Eur. Phys. J. C* **74** (2014) 2701.
34. C. Lunardini, S. Razzaque, and L. Yang, *Phys. Rev. D* **92** (2015) 021301.
35. M. Su, T.R. Slatyer, and D.P. Finkbeiner, *Astrophys. J.* **724** (2010) 1044.
36. A. Neronov, D. Semikoz, and C. Tchernin, *Phys. Rev. D* **89** (2014) 103002.
37. L. Fusco, *PoS* (ICRC2015) 1055.
38. D. Gaggero, D. Grasso, A. Marinelli, A. Urbano, and M. Valli, *Astrophys. J. Lett.* **815** (2015) L25 *PoS* (ICRC2015).
39. M. Ackermann, *et al.*, *Astrophys. J.* **750** (2012) 3.
40. M. Spurio, *Phys. Rev. D* **90** (2014) 103004.
41. S. Adrián-Marínez, *et al.*, *JCAP* **10** (2015) 68.
42. S. Adrián-Marínez, *et al.*, *Phys. Lett. B.* **759** (2016) 69.
43. C. Tönnis, *PoS* (ICRC2015) 1207.
44. A. Gleixner & C Tönnis, *PoS* (ICRC2015) 1110.
45. S. Adrián-Marínez, *et al.*, *JCAP* **5** (2016) 16.
46. M. Ardid & C. Tönnis, *PoS* (ICRC2015) 1212.
47. I. El Bojaddaini & G.E. Păvălas, *PoS* (ICRC2015) 1060.

# Chapter 8

# The Baikal Neutrino Project

V. M. Aynutdinov and Zh.-A. M. Dzhilkibaev

*Institute for Nuclear Research,*
*$60^{th}$ October Anniversary prospect 7a, Moscow 117312*
*djilkib@yandex.ru*

The Baikal neutrino project addresses a wide spectrum of scientific problems, most notably the detection of neutrinos of astrophysical origin. The project was started in 1980 with the first site investigations in Lake Baikal. Between 1993–1998, the world's first deep underwater neutrino telescope, NT200, was constructed, and it has taken data for more than 2 decade. The second-generation neutrino telescope, Baikal-GVD, will be a research facility of cubic-kilometer scale consisting of independent subarrays (clusters of light sensors). The prototyping/early construction phase of the Baikal-GVD project concluded in April 2015 with deployment and commissioning of the first cluster of Baikal-GVD.

## 1. Introduction

The Baikal neutrino project's roots trace to 1959–1960, when the Soviet scientist M. Markov[1] proposed the idea of large-scale deep-sea experiments for studying the properties of neutrinos and the astrophysical sources of their origin. This idea proved to be extremely fruitful. Investigations performed under conditions of a drastically reduced background of penetrating cosmic radiation, when huge volumes of surrounding rock or water themselves serve as targets for high-energy neutrinos, made it possible to achieve experiments within extremely high level of sensitivity, allowing the study of rare processes that cannot be detected in laboratories at the surface.

Following a suggestion by Soviet physicist A. Chudakov, Lake Baikal was considered as a site to test and deploy future large-scale neutrino telescopes. The choice of Lake Baikal was determined by the high transparency of deep fresh water, the depth of the lake, and the ice cover that allowed installing deep underwater equipment during two winter months. On October 1, 1980, regarded as the start of the Baikal neutrino experiment, the Laboratory of High-Energy Neutrino Astrophysics was established at the Institute for Nuclear Research AS USSR (now INR RAS). Later it became the core of the Baikal collaboration which included, at different periods, the Joint Institute for Nuclear Research (Dubna), the Irkutsk State University, the Moscow Lomonosov State University, the research center DESY-Zeuthen (Germany), the Nizhny Novgorod State Technical University, the St. Petersburg State Marine Technical University, and other scientific institutions.

The design, deployment, and startup of the full-scale neutrino telescope were preceded by experimental investigations of hydro-optical, hydro-physical, and hydrological conditions in Baikal, which were performed for about a decade. In these experiments, for the first time, an intrinsic emission of deep waters in the lake was discovered, which is caused by the oxidation of particles a few micrometers in size. Deep-underwater pilot Cherenkov detectors were installed for long-term operation to test methods of data acquisition and to perform the first physical experiments. One of the notable results was an upper limit on the flux of slowly moving GUT monopoles catalyzing proton decay, obtained from data collected by stationary strings of light sensors, operating during 1984–1986.[2] An important step towards the project's realization were the development and production of the hybrid photodetector QUASAR-370 exclusively for the Baikal neutrino telescope (see below).

## 2.  The Baikal neutrino telescope NT200

The site chosen for the Baikal neutrino telescope is in the southern basin of Lake Baikal. Here, the combination of hydrological, hydrophysical, and landscape factors is optimal for the deployment and operation of such a detector. The water depth is about 1360 m at distances larger than about three km from the shore. The water transparency is characterized by an absorption length of about 20–25 m and a scattering length of 30–50 m.[4] Water luminescence is moderate at the detector site.

The first-generation neutrino telescope NT200 comprises 192 optical modules (OMs) placed at a depth of 1100–1200 m. The optical modules are carried by eight strings that are attached to an umbrella-like frame. The strings are anchored by weights at the lake floor and held in a vertical position by buoys at various depths. The configuration spans 69 m in height and 43 m in diameter. The detector was deployed (and is hauled up for maintenance) within a period of about 6 weeks between February and April, when the lake is covered with a thick ice layer providing a stable working platform. It is connected to shore by several copper cables on the lake floor, which allows operation over the full year.

The optical modules[3] are glass spheres equipped with QUASAR-370 phototubes; they are grouped pairwise along a string. In order to suppress accidental hits from dark noise (about 30 kHz) and bioluminescence (typically 50 kHz but seasonally up to hundreds of kHz), the two photomultipliers of a pair are switched in sync, defining a channel, with typically only 0.1 kHz noise rate. The basic cell of NT200 consists of a *svjaska* (Russian for "bundle"), comprising two optical module pairs and an electronics module for time and amplitude conversion and slow control functions. A majority trigger is formed if more than 3 channels are fired within a time window of 500 ns (this is about twice the time a relativistic particle needs to cross the NT200 array). Trigger and interstring synchronization electronics are housed in an array electronics module at the top of the umbrella frame. This module is less than 100 m away from the optical modules, allowing for nanosecond synchronization over copper cable.

The QUASAR-370 is a hybrid device. Photoelectrons from a large hemispherical cathode (K2CsSb) with $>2\pi$ viewing angle are accelerated by 25 kV to a fast, high-gain scintillator that is placed near the center of the glass bulb. The light from the scintillator is read out by a small conventional photomultiplier tube. One photoelectron from the hemispherical photocathode yields typically 20 photoelectrons in the conventional photomultiplier. This high multiplication factor results in an excellent single-electron resolution of 70%. Furthermore, the QUASAR-370 is characterized by a small time jitter (2 ns) and a small sensitivity to the Earth's magnetic field.

The first deep underwater neutrino telescope NT200 was constructed in the years 1993–1998. The experimental data of 1994 already showed two candidate neutrino events — the first in the worldwide race of deep-underwater and deep-ice experiments.[5] The small spacing of modules leads to a comparably low energy threshold of about 10–15 GeV for muon detection. About 400 upward muon events were collected over 1034 days of detector livetime. A wide program of scientific research was implemented, and important results were obtained in searches for a diffuse cosmic neutrino flux.[6] Limits were derived on the flux of magnetic monopoles[7] and of muons from dark matter annihilation in the center of the Earth[8] and the Sun,[9] as well as the neutrino flux associated with GRBs.[10] In order to improve the

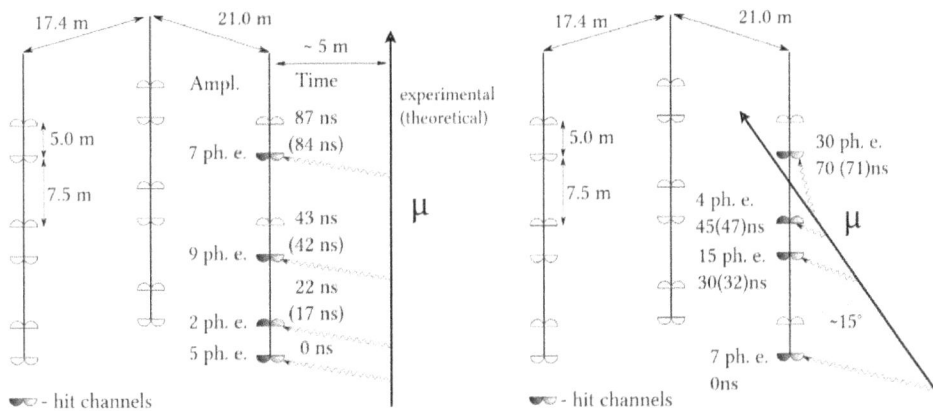

Fig. 1. The two neutrino candidates in NT36. The hit PMT pairs (channels) are marked in black. Numbers give the measured amplitudes (in photoelectrons) and arrival times with respect to the first hit channel. Times in parentheses are those expected for a vertically upward going muon (left) and upward muon passing the string under 15° (right).

sensitivity to high-energy cascade events, NT200 was surrounded in 2005–2007 with three sparsely instrumented outer strings (6 optical module pairs per string). This configuration is named NT200+.

## 3. The gigaton volume detector in Lake Baikal

The long-term operation of NT200 and the obtained physics results have demonstrated the effectiveness of deep underwater neutrino detection in the fresh water of Lake Baikal. The next-generation neutrino telescope Baikal-GVD will be a research facility aimed mainly at studying astrophysical neutrino fluxes and mapping the high-energy neutrino sky in the Southern Hemisphere, including the region of the galactic center.

The concept of Baikal-GVD is based on a number of evident requirements for the design and architecture of the recording system of the new array: taking the utmost advantage of the ice cover of Lake Baikal for deployment, the extendability of the facility, the effective operation even in the first deployment stages, and the possibility of realizing different spatial configurations of light sensors with the same measuring system.

### 3.1. *Design of Baikal-GVD*

Baikal-GVD consists of an array of optical modules — photomultiplier tubes enclosed in pressure-resistant glass spheres.[11] The OMs are arranged on vertical load-carrying cables, faced downward and spaced by 15 m. Every 12 OMs on such strings are combined in a section. Strings are grouped in cluster: functionally independent subarrays that are synchronized with each other and connected to shore by individual electro-optical cables. Each cluster comprises eight strings — seven peripheral strings uniformly arranged at 60 m distance around a central one. The distances between the central strings of neighboring clusters are 300 m.

Two telescope configurations with 2304 OMs (GVD) and 10368 OMs (GVD*4) have been optimized. The first configuration consists of 12 clusters of 345 m long strings. Each string consists of two sections and comprises 24 OMs at depths of 950 m to 1300 m. The second configuration consists of 27 clusters of 705 m long strings. Each string of this configuration consists of four sections and comprises 48 OMs at depths of 600 m to 1300 m. The clusters for GVD*4 are spaced over an area of approximately 2 km$^2$; the instrumented water volume is about 1.4 km$^3$.

### 3.1.1. *The optical module*

The optical module is the basic element of the Baikal-GVD neutrino telescope. Each OM consists of a pressure-resistant glass sphere with 43.2 cm diameter that houses the electronics and the PMT; the latter is surrounded by a high-permittivity alloy cage for shielding it against the Earth's magnetic field. The chosen PMT is Hamamatsu R7081-100, with a 10-inch hemispherical photocathode and a quantum efficiency up to 35%. The electronics in the OM includes a high-voltage power supply unit, a fast two-channel preamplifier, a controller, and two LEDs for time and amplitude calibration of the measuring channel. The OM controller is intended for HV control, for monitoring the PMT noise, and for time and amplitude calibration.

### 3.1.2. *The string section*

The optical modules are grouped into sections.[12] Each section includes 12 OMs and the central module (CeM). PMT signals from all OMs are transmitted through 90 m long coaxial cables to the CeM of the section, where they are digitized by custom-made ADC boards with 200 MHz sampling rate. The waveform information is read out by the master board, also located in the CeM. The master board is connected via local Ethernet to the cluster DAQ-center and control of the section operation and the section trigger logic. A request analyzer forms the section trigger request (local trigger) on the basis of channel requests L (low channel threshold, ∼0.3 p.e.) and H (high threshold, ∼3 p.e.) from 12 ADC channels. This unit contains a programmable coincidence matrix (12H×12L), which provides a simple way to generate the section trigger request. The basic trigger modes are: coincidences of >N L-requests within a selectable time window (L>N-trigger), and coincidences of L and H requests from any neighbouring OMs within a section (L&H-trigger). A request of the section trigger is transferred from the master board through a string communication module (CoM) to the cluster DAQ-center, where a global trigger for all sections is generated. Data from the strings are transferred through a DSL-modem Ethernet channel to the cluster center. The data between the cluster DAQ-center and the shore station are transmitted through optical fibers of about 6 km length.

## 3.2. *Performance of Baikal-GVD*

The objective of the optimization of the GVD design was to reach a large cascade detection volume while also effectively recording high-energy muons. The muon effective areas for the two optimized GVD configurations are shown in Fig. 2 (left). The area for the GVD*4 configuration (10368 OMs) rises from 0.3 km$^2$ at 1 TeV to about of 1.8 km$^2$ asymptotically. The resolution of the muon direction (median mismatch angle) is about 0.25 degrees. Fig. 2 (right) shows the shower effective volumes for the two configurations, reaching 0.4–2.4 km$^3$ for GVD*4 at energies above 10 TeV. The accuracy of energy reconstruction is about of 20–35% depending

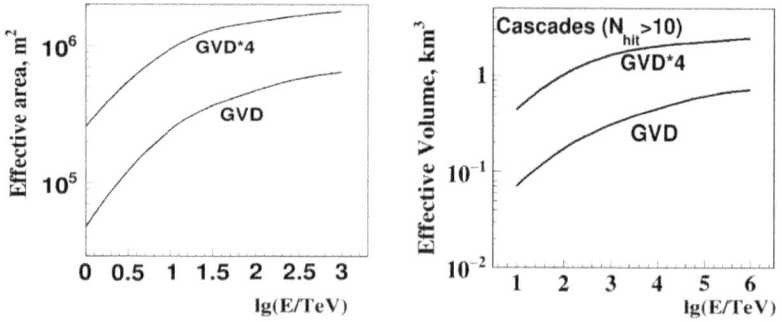

Fig. 2.   Left: Muon effective area. Right: Effective volume for cascade detection. The curves labeled GVD*4 and GVD relate to configurations with 10368 and 2304 OMs, respectively.

on the shower energy. The accuracy of the shower direction reconstruction is 3.5–6.5 degrees (median value).

### 3.3. *Project implementation*

The first prototype with the GVD electronics was installed in Lake Baikal in April 2008. It was a reduced-size section with six OMs. This unit allowed studying basic elements of the future detector: new optical modules and a measuring system based on Flash Analog-to-Digital Converters (FADC). During the following two years, different versions of prototype strings were tested as a part of the NT200+ detector. The 2009 prototype string consisted of 12 OMs, with six Hamamatsu R8055 PMTs and six Photonis XP1807. In April 2010, a string with eight R7081HQE PMTs and four R8055 PMTs (both Hamamatsu) was deployed in Lake Baikal. The operation of these prototype strings in 2009 and 2010 allowed a first assessment of the DAQ performance.[11]

The prototyping/early construction phase of the Baikal-GVD project was aimed at construction and operation of the first GVD-cluster. It was started in 2011 with the deployment and operation of autonomous engineering arrays. The next important step in the realization of the GVD project was made in 2013 with the deployment of an enlarged engineering array — the first stage of the first GVD-cluster, which comprised 72 OMs arranged on three 345 m long strings and was operated from April 2013 to February 2014.[13] In 2014, the second stage of the demonstration cluster was deployed, with 112 OMs arranged on five strings and an instrumentation string with array calibration and environment monitoring equipment. During 2014, it was operated in various testing and data-taking modes. One of the main goals of the operation in testing mode was to study the quality of in-situ calibration procedures. The performance and quality of these procedures have been verified by reconstructing position and intensity of a distant calibration laser. This laser fires five series of 480 nm light pulses at five fixed intensity levels ranging from approximately $10^{12}$ to $6 \cdot 10^{13}$ photons/pulse. This corresponds to shower energies from about 10 PeV to 600 PeV and allows testing the quality of high-energy cascade reconstruction. The position of the laser source was reconstructed using the arrival times of photons detected by the PMTs, taking into account the timing calibration of the PMTs by LEDs. Results of the reconstruction were compared with the laser coordinates obtained by the acoustic positioning system. The reconstruction accuracy (median value) is about 3 m, confirming the expected performance.

## 4. DUBNA — the first cluster of Baikal-GVD

In April 2015, the first cluster of Baikal-GVD was deployed and put into operation. A schematic view of the cluster is shown in Fig. 3. It was named DUBNA and encloses 1.7 megatons of water. The configuration comprises a total of 192 OMs, arranged on eight 345 m long strings, as well as an acoustic positioning system[14] and an instrumentation string with equipment for array calibration and monitoring of environmental parameters. Each string comprises 24 OMs spaced by 15 m at depths of 900 m to 1250 m below the surface. In 2015, seven side strings were located at a

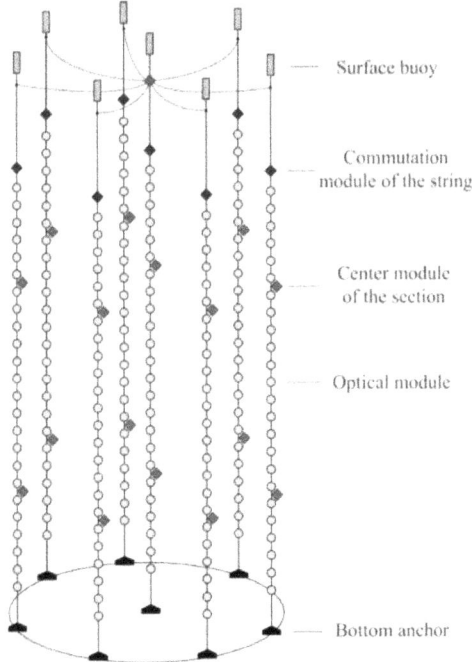

Surface buoy

Commutation module of the string

Center module of the section

Optical module

Bottom anchor

Fig. 3.   Schematic view of cluster.

reduced radius of 40 m around a central string (compared to 60 m for the baseline configuration). The reason was to increase the sensitivity to low-energy atmospheric muons and neutrinos, which are used for array calibration. Next year, strings will be moved to the baseline distances.

The cluster of Baikal-GVD in its baseline configuration will have the potential to detect astrophysical neutrinos with a flux value measured recently by IceCube.[15,16] The search for such high-energy neutrinos is based on the selection of cascade events generated by neutrino interactions in the sensitive volume of the array. After applying an iterative procedure of vertex reconstruction followed by the rejection of hits contradicting the cascade hypothesis on each iteration stage,[6] events with a final multiplicity of hit OMs, $N_{\text{hit}} > 20$, are selected as high-energy neutrino events. With this event selection, the neutrino effective areas for each flavor, assuming an equal flux of neutrinos and antineutrinos and averaged over all arrival angles, have been derived (see Fig. 4 left). These areas are about ten times smaller than the corresponding areas of IceCube.[15] The accuracy of shower direction reconstruction is about 4 degrees (median value), which is substantially better than the 10–15 degrees accuracy for IceCube.[15] The fraction of shower events ($E_{\text{sh}} = 100\,\text{TeV}$) with mismatch angles between generated and reconstructed muon directions less than a given value is shown in Fig. 4 (right). Energy distributions of the expected number of shower events per year for the IceCube astrophysical fluxes, for different flavors and all-flavor, as well as the distribution of expected background events from atmospheric neutrinos are shown in Fig. 5 (left). The expected background events from atmospheric neutrinos are strongly suppressed for energies higher than 100 TeV. From an all-flavor astrophysical flux with the normalization $E^2 F = 3.6 \cdot 10^{-8}\,\text{GeV}\,\text{cm}^{-2}\text{s}^{-1}\text{sr}^{-1}$, we expect about one event per year with $E_{\text{sh}} > 100\,\text{TeV}$ in the cluster, compared to about 10 events in IceCube. The sensitivity per flavor for one cluster for a neutrino flux with an $E^{-2}$ spectrum and flavor ratio 1:1:1 as a function of the observation years is shown in Fig. 5 (right). Three years of observation allows sensitivity at a level of flux value measured by IceCube.

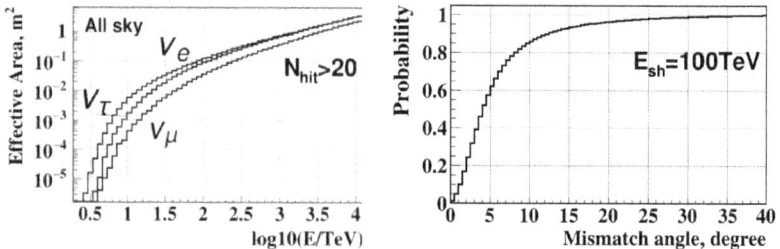

Fig. 4.   Left: Neutrino effective areas for each flavor averaged over all arrival angles. The effective area includes effects from attenuation of neutrinos in Earth. Right: The fraction of reconstructed showers with mismatch angles less than a given value.

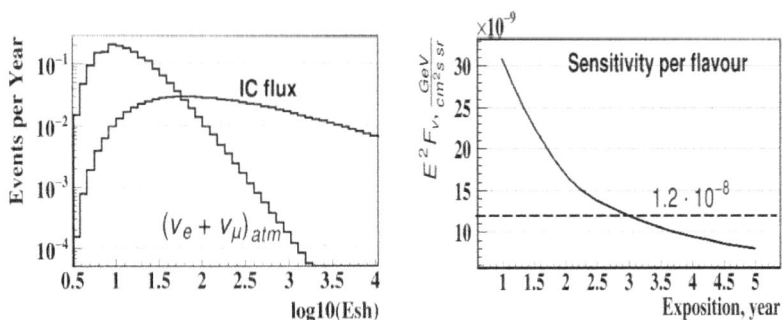

Fig. 5. Left: Expected distribution of events per year from the astrophysical flux obtained by IceCube. Also shown is the expected distribution of background events from atmospheric neutrinos. Right: Cluster sensitivity per flavor for a neutrino flux with an $E^{-2}$ spectrum as a function of the observation years. The long-dashed line indicates the one flavor neutrino flux value obtained by IceCube.

## 5. Conclusion

The ambition of the Baikal collaboration is the construction of a km$^3$-scale neutrino telescope — the Gigaton Volume Detector in Lake Baikal. During the R&D phase of the GVD project in 2008–2010, its basic elements — new optical modules, FADC readout units, underwater communications, and trigger systems — were developed, produced and tested *in situ* by long-term operating prototype strings. The prototyping phase of the GVD project began in April 2011 with the aim of comprehensive *in situ* testing of all elements and systems of the future detector. This phase concluded with deployment and commissioning of the first cluster of GVD in April 2015. For the years 2016–2020, the deployment of 10–12 clusters with about 2000 OMs is envisaged.

## Acknowledgements

This work was supported by the Russian Fund for Basic Research (grants 16-29-13032, 14-02-00972).

## References

1. M.A. Markov, On high energy neutrino physics, In *Proc. 10th Ann. Int. Conf. on High Energy Physics at Rochester (ICHEP)*, Rochester, USA, August, 1960, pp. 578–581.
2. L.B. Bezrukov, *et al.*, Search for superheavy magnetic monopoles in deep underwater experiments at Lake Baikal, *Sov. J. Nucl. Phys.* **52**(1) (1990) 54–59.
3. R. Bagduev, *et al.*, The optical module of the Baikal deep underwater neutrino telescope, *Nucl. Instr. Meth.* **A420** (1999) 877–881.
4. A. Avrorin, *et al.*, Asp-15 — A stationary device for the measurement of the optical water properties at the NT200 neutrino telescope site, *Nucl. Instr. Meth.* **A693** (2012) 186–194.

5. L.B. Bezrukov, *et al.*, Preliminary results on a search for neutrinos from the center of the Earth with the Baikal underwater telescope, In *Proc. of $2^{nd}$ Workshop on The dark side of the Universe,* 221–225, Rome, Italy (1995).

6. A. Avrorin, *et al.*, Search for high-energy neutrinos in the Baikal neutrino experiment, *Astr. Lett.* **35**(10) (2009) 651–662.

7. V.M. Aynutdinov, *et al.*, Search for relativistic magnetic monopoles with the Baikal neutrino telescope, *Astropart. Phys.* **29** (2008) 366–372.

8. K. Antipin, *et al.*, Nearly vertical muons from the lower hemisphere in the Baikal neutrino experiment, In *Proc. of $1^{st}$ Workshop on Exotic Phys. With Neutrino Telescopes,* Uppsala, Sweden, September 2006, pp. 34–38.

9. A.D. Avrorin, *et al.*, Search for neutrino emission from relic dark matter in the sun with Baikal neutrino telescope, *Astropart. Phys.* **62** (2015) 12–20.

10. A.V. Avrorin, *et al.*, Search for neutrinos from gamma-ray bursts with the Baikal neutrino telescope NT200, *Astr. Lett.* **37**(10) (2011) 692–698.

11. A.V. Avrorin, *et al.*, Status of the Baikal-GVD project, *Nucl. Instr. Meth. A* **692** (2012) 46–52.

12. A. Avrorin, *et al.*, Data acquisition system of the NT1000 Baikal neutrino telescope, *Instr. Exp. Tech.* **57** (2014) 262–273.

13. A. Avrorin, *et al.*, The prototyping/early construction phase of the Baikal-GVD project, *Nucl. Instr. Meth.* **742** (2014) 82–88.

14. A. Avrorin, *et al.*, A hydroacoustic positioning system for the experimental cluster of the cubic-kilometer-scale neutrino telescope at Lake Baikal, *Instr. Exp. Tech.* **56** (2013) 449–458.

15. M.G. Aartsen, *et al.*, Evidence for high-energy extraterrestrial neutrinos at the IceCube detector, *Science* **342** (2013) 1242856.

16. M.G. Aartsen, *et al.*, Observation of high-energy astrophysical neutrinos in three years if IceCube data, *Phys. Rev. Lett.* **113**(10) (2014) 101101.

# Chapter 9

# The Dawn of Multi-Messenger Astronomy

Marcos Santander

*Barnard College, Columbia University*
*3009 Broadway, New York, NY, USA*
*santander@nevis.columbia.edu*

The recent discoveries of high-energy astrophysical neutrinos and gravitational waves have opened new windows of exploration to the Universe. Combining neutrino observations with measurements of electromagnetic radiation and cosmic rays promises to unveil the sources responsible for the neutrino emission and to help solve long-standing problems in astrophysics such as the origin of cosmic rays. Neutrino observations may also help localize gravitational-wave sources, and enable the study of their astrophysical progenitors. In this work we review the current status and future plans for multi-messenger searches of neutrino sources.

## 1. Introduction

The continuing study of the high-energy sky has revealed a large number of powerful astrophysical objects capable of emitting radiation across the entire electromagnetic spectrum. The high magnetic fields and strong astrophysical shocks observed in some of these objects are expected to be responsible for the acceleration of cosmic rays up to the highest observable energies, in the $10^{20}$ eV range. As they are being accelerated, or during their propagation, cosmic rays can interact with ambient material or radiation fields, leading to the production of high-energy neutrinos and gamma rays through the decay of charged and neutral mesons.[53,54] Fast transient sources which are potential cosmic-ray accelerators, such as gamma-ray bursts (GRBs), may also emit gravitational waves due to bulk motion of matter in the source progenitor.

A complete understanding of the most energetic phenomena in the Universe therefore, calls for a joint study of the different "cosmic messengers" they emit: cosmic rays, neutrinos, photons, and gravitational waves. Multi-messenger astronomy is an emerging subfield of high-energy astrophysics that aims at combining observations from instruments sensitive to these different messenger particles. This approach may solve the long-standing mystery of the sources of cosmic rays, further our understanding of particle acceleration in astrophysical shocks, increase the sensitivity of searches for gravitational-wave emitters, and potentially unveil new types of sources through the detection of spatial and temporal correlations between two or more messenger channels.

A major result for the field has been the detection of an astrophysical flux of neutrinos by the IceCube observatory at the level of $E_\nu^2 J_\nu(E_\nu) \sim 10^{-8}$ GeV cm$^{-2}$ s$^{-1}$ sr$^{-1}$ per neutrino flavor in the energy range between a few tens of TeV and a few PeV.[7] While no point sources have been detected so far, the flux upper limits set by IceCube and ANTARES searches[9] are at the level of 1–10% of the all-sky astrophysical flux, hinting at a large population of neutrino sources. No significant correlation has been found with the galactic plane, which tends to favor an extragalactic origin of the astrophysical neutrinos, with a potential sub-dominant contribution from galactic sources.

The search for neutrino sources should benefit from the boost in sensitivity provided by the multi-messenger approach. Over the last few years, data from a large network of astrophysical observatories around the world (shown in Fig. 1) has been used to study correlations between neutrino events and other messenger signals. As more facilities go online in coming years, a drastic increase in the sensitivity of these searches is expected, which may finally reveal these elusive sources.

In the following sections, we review recent results from multi-messenger searches for high-energy neutrino sources. Searches for electromagnetic counterparts to the neutrino emission are described in Sec. 2. Section 3 covers results from correlation studies between neutrinos and high-energy cosmic rays. In Sec. 4 we describe joint

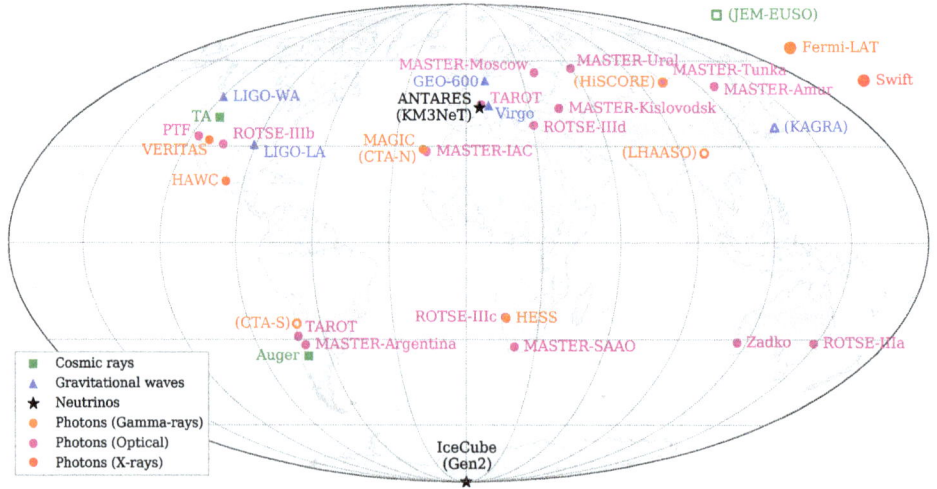

Fig. 1.   Global distribution of some of the observatories (past, present, and future) dedicated to multi-messenger studies. Current and past facilities are shown with filled markers. Some representative future detectors are shown with empty markers with their names in parentheses.

searches of neutrino and gravitational-wave emitters. We describe current efforts to build a multi-messenger observatory network to provide rapid communications of high-energy astrophysical events in Sec. 5. We present conclusions and an outlook for future studies in Sec. 6.

## 2. Photons

The high-energy hadronic interactions responsible for the creation of the astrophysical neutrinos observed by IceCube should also produce gamma rays through the decay of neutral pions. Lower energy photons may also be detectable for those cases where the source or the propagation medium is opaque to gamma rays. Studies of spatial and temporal correlations between neutrinos and EM radiation rely on searching for neutrino emission from known EM sources where cosmic-ray acceleration is expected, or on performing follow-up EM observations of high-energy neutrino positions that are likely astrophysical in origin. The first approach has been used to set upper limits on the neutrino flux from GRBs, AGNs, SNRs, and galaxy clusters. In the case of GRBs, null neutrino detections have set important constraints on models that postulate these objects as sources of cosmic rays with energies above $10^{18}$ eV (see Ref. 13 for a recent update on this work). In the following subsections, we summarize recent EM–$\nu$ searches that use the second approach.

While temporal correlations can be explored down to the microsecond level (given the precision in the determination of the neutrino arrival time), the main challenge for spatial correlation studies is presented by the limited angular resolution of neutrino directional reconstructions. In IceCube, the angular resolution of "cascade" events produced by charged-current $\nu_{e,\tau}$, or neutral-current interactions of any flavor, is about $15°$ at energies above $100 \, \text{TeV}$. Charged-current $\nu_\mu$ interactions produce km-long muon "tracks" that can be typically reconstructed to within $1°$. The muon reconstruction capability of IceCube has been validated by the observation of the cosmic-ray shadow of the Moon within $0.2°$ of its expected position.[8] Deep-sea neutrino telescopes benefit from the longer scattering length of Cherenkov light in water to reconstruct neutrino events to higher precision. The angular resolution of ANTARES[24] for muon tracks is believed to be better than $0.3°$ above $10 \, \text{TeV}$ and about $4°$ for cascade events, while for the future KM3NeT detector[60] the angular resolution for cascades is expected to improve to $2°$.[41] Recent correlation studies have concentrated on searching for EM emission coincident with muon track positions to benefit from the better angular resolution of these events and reduce the probability of accidental correlations.

### 2.1. *Gamma-rays*

At production, the flux of TeV–PeV astrophysical neutrinos should be associated with a flux of gamma-rays of similar spectral characteristics. Photons in this energy range are attenuated during propagation by pair-producing on background radiation fields, with the extragalactic background light (EBL) dominating at $E_\gamma < 10\,\mathrm{TeV}$ and the cosmic microwave background (CMB) for $E_\gamma > 10\,\mathrm{TeV}$. At PeV energies, the photon attenuation length is below $10\,\mathrm{kpc}$, which restricts correlation studies of PeV photons and neutrinos to our galaxy. Past gamma-ray observations have been used to test the association of the astrophysical neutrinos with the Galactic Plane,[30,55] the Galactic Halo,[66] and the *Fermi* "bubbles".[58] The sensitivity of these tests will be greatly improved by observations from current and future air-shower arrays, such as IceTop,[16] HAWC,[21] LHAASO[40] and HiScore.[68]

Neutrino correlations with sources of extragalactic gamma-rays can be investigated at GeV–TeV energies, where absorption is not as severe, if the hadronic gamma emission extends to this energy range. The main instruments in this band are the *Fermi* Large Area Telescope (LAT),[36] the H.E.S.S.,[29] MAGIC,[4] and VERITAS[71] ground-based telescopes, and the HAWC array. The sensitivities of current and future gamma-ray telescopes are shown in Fig. 2.

The connection between the neutrino flux and extragalactic radiation backgrounds has been explored in recent studies. Simple extrapolations of the astrophysical neutrino flux down to GeV energies lead to an associated photon flux that can account for a significant fraction or even overflow (depending on the assumed

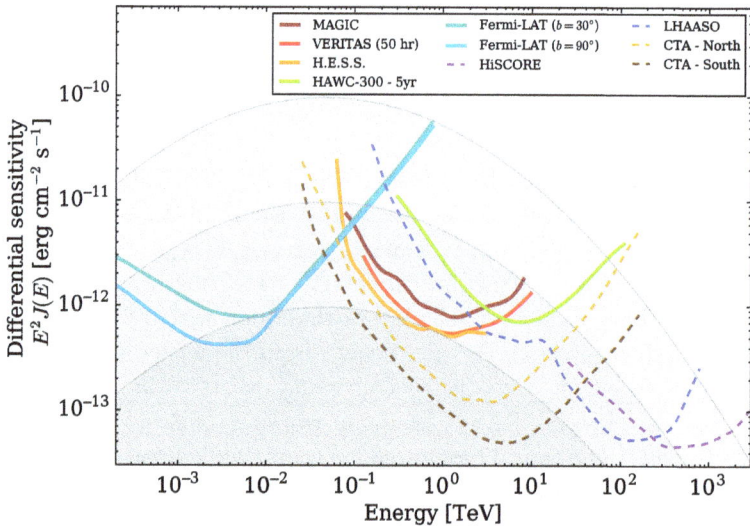

Fig. 2.   Differential $5\sigma$ sensitivity of current (solid lines) and future (dashed lines) gamma-ray observatories. The *Fermi*-LAT[35,46] sensitivity curve is given for a 10 year exposure at two galactic latitudes (30° and 90°). The *Fermi*-LAT and HAWC curves are given for quarter-decade energy bins. The VERITAS,[63] MAGIC,[31] H.E.S.S.,[50] and CTA[39] curves are given for 50 hours of observation and 5 energy bins per decade. The HAWC 300 sensitivity,[21] and that of the future HiScore[68] and LHAASO[49] arrays, is given for a five-year exposure. For reference, the shaded grey regions indicate, from the top, 100%, 10%, and 1% levels of the gamma-ray spectrum of the Crab nebula.

neutrino spectral index) the isotropic gamma-ray background (IGRB) measured by *Fermi*-LAT.[23] However, *Fermi* source population studies[42] indicate that the IGRB is dominated by unresolved AGNs (typically assumed to be leptonic sources) which results in a lower fraction of the IGRB that could be connected to neutrinos. While significant uncertainties remain on these extrapolations, current measurements are starting to probe the role that proton-proton sources (such as the archetypical star-forming galaxies) play in diffuse gamma-ray backgrounds.[33,37]

The large field of view and high duty cycle of the LAT provides temporal and spatial gamma-ray coverage for a large fraction of the neutrino events detected by IceCube. Data from the LAT has been analyzed to search for new sources, or flux enhancements in known ones, in coincidence with IceCube neutrino events. No spatially-coincident gamma emission has been found in correlation with muon track events[38] with the exception of a neutrino candidate event near the location of the gamma-ray blazar PKS 0723-008 (see Fig. 3), likely due to a chance alignment ($p = 37\%$). It has been recently reported[51] that a 2 PeV cascade event detected by IceCube occurred in relative temporal and spatial coincidence with an extended high-fluence flare of the blazar PKS B1424-418, although also in this case the association does not appear to be statistically significant ($p = 5\%$).

Searches for neutrino gamma-ray counterparts in the very-high-energy range (VHE, $E_\gamma > 100$ GeV) are underway using the H.E.S.S., MAGIC, and VERITAS Imaging Air Cherenkov Telescopes (IACTs) and the HAWC air shower array. For IACTs, these searches are limited to the observation of muon track positions, given the 3.5°–5° field-of-view of current generation telescopes. The VERITAS and H.E.S.S. telescopes have observed muon track positions published by ANTARES[64]

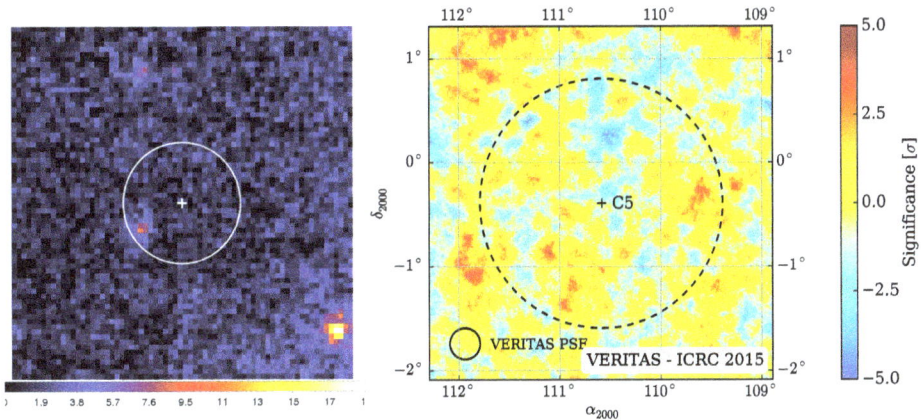

Fig. 3. Gamma-ray observations of an IceCube muon-track position (event #5 in Ref. 7). *Left:* *Fermi*-LAT test-statistic sky map of gamma rays in the 1–300 GeV range (see Ref. 38). *Right:* VERITAS significance sky map (see Ref. 59). The white circle in the *Fermi* sky map and the dashed circles in the VERITAS map show the <1.2° median angular resolution claimed for this neutrino event.

and IceCube,[45,59] and null results from these studies have set constraints on the steady VHE gamma-ray flux associated with each neutrino position. Increasing the sensitivity to transient gamma-ray sources requires a system that can issue prompt alerts to VHE instruments if a hint of an increase in neutrino activity is detected. Since 2012, IceCube operates a program that sends triggers to MAGIC and VERITAS whenever the number of neutrino events detected over a certain period around a VHE source crosses a predefined significance threshold,[4] so far with no significant gamma-ray detections.

A golden channel for follow-up observations is the sample of high-energy through-going muon events used by IceCube to measure the astrophysical $\nu_\mu$ flux[12] given its good angular resolution and high astrophysical purity. At energies above a few hundred TeV, where the astrophysical neutrino flux dominates the atmospheric background, the IceCube effective area to through-going muon neutrinos is more than ten times larger than for "starting" neutrino events, where the first interaction occurs in the detector volume. The highest energy neutrino detected so far (recently reported by IceCube[57]) comes from this sample and has an energy of $2.6 \pm 0.3$ PeV with an atmospheric $p$-value of $<0.01\%$. An archival analysis of HAWC data around the time of the event showed no gamma-ray emission at the neutrino location. Current efforts are underway to promptly circulate the positions of high-energy starting and through-going muon neutrino events to partner instruments using the AMON network (see Sec. 5), which would boost the sensitivity of EM follow-ups to transient neutrino sources. Future searches for VHE gamma-ray neutrino counterparts will receive a significant boost from the construction of the Cherenkov Telescope Array (CTA),[22] which will provide an order-of-magnitude improvement in sensitivity with respect to current IACTs (see Fig. 2). At MeV energies, proposed missions such as the ComPair satellite[62] can provide a large field-of-view coverage of the sky with an angular resolution in the $1°$ to $10°$ range.

## 2.2. *Multi-wavelength observations*

Besides gamma rays, other wavelength bands are being explored to search for transient EM emission associated with neutrino events. Realtime alerts from ANTARES and IceCube are currently sent to a network of optical and X-ray telescopes to search for transient sources such as GRBs and core-collapse supernovae (CCSNe) in correlation with interesting neutrino positions. Since 2008, IceCube operates optical (OFU,[15]) and X-ray (XFU) follow-up programs in parallel to the gamma-ray follow-up program described in the previous subsection. The rate of false trigger alerts is reduced by requiring that two or more spatially-coincident neutrinos are detected within a certain time window.[56] OFU alerts have been sent to the ROTSE telescope network (which has since stopped operations) and the Palomar Transient Factory (PTF), and have been supplemented by retrospectives searches through the Pan-STARRS1 $3\pi$ survey data. The XFU program triggers

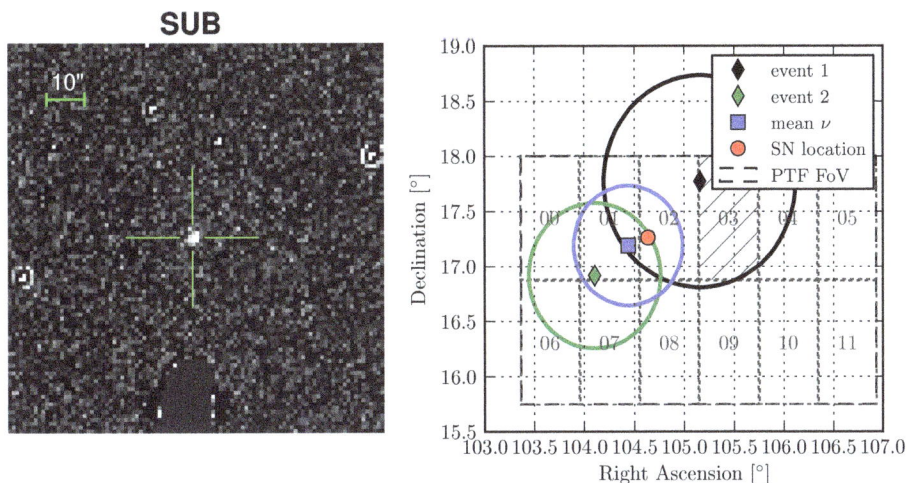

Fig. 4.   *Left*: PTF reference-subtracted discovery image of the CCSN *PTF12csy* taken on April 9, 2012 at the location of an IceCube neutrino doublet detected on March 30, 2012. *Right:* Map showing the two neutrino event positions and uncertainties compared to the field of view of the PTF camera CCDs. (From Ref. 11.)

observations in the 0.3–10 keV band using the XRT X-ray instrument onboard the *Swift* satellite.

The number of neutrino "doublets" observed so far by these programs agrees with the rate expected from the atmospheric neutrino background. Although not a statistically significant correlation ($2\sigma$), the capability of the OFU program to detect optical transients has been demonstrated by the discovery[11] of the CCSN *PTF12csy* (Fig. 4) in PTF follow-up observations of a neutrino doublet position.

Results from *Swift* observations performed as part of the IceCube XFU program are presented in Ref. 44. Seven 1–2ks exposures are required to cover the $\sim 0.5°$ muon error circle with the $0.4°$ XRT field-of-view. Given their increased sensitivity, XFU observations performed so far have unveiled more than 100 previously uncatalogued X-ray sources, although none of them appear to be clear counterparts for the neutrino events. These studies would greatly benefit from a deeper all-sky X-ray catalog, such as the one to be created by the *eRosita*[61] mission, which could be compared to new sources detected in triggered observations. The sensitivity to short transients would also be improved by the operation of an all-sky (or large field-of-view) X-ray telescope, which would reduce the hour-scale delay between the neutrino trigger and the start of X-ray observations.

Results from a similar X-ray and optical follow-up program in ANTARES are covered in Ref.[26] The Telescopes-ANTARES Target of Opportunity (TAToO) alert system[28] started operations in 2009, and it has triggered optical observations using the TAROT and ROTSE telescopes, and in X-rays with *Swift* XRT. Also in this case, no significant optical or X-ray[26] counterparts to the neutrino events have been detected so far. On September 1, 2015, *Swift* observations triggered by an

ANTARES alert revealed a variable, and previously unknown X-ray source in the 18-arcmin error circle of the neutrino event.[a] While this detection appears to have been caused by a chance alignment with an X-ray source, the strong multiwavelength follow-up campaign triggered by this detection highlights the interest of the astrophysical community in contributing to the discovery of the first neutrino source.

## 3. Cosmic rays

Cosmic rays are scattered by galactic (GMF) and intergalactic (IGMF) magnetic fields during propagation, limiting the applicability of spatial correlation studies with neutrino positions to the ultra-high energy cosmic ray range (UHECRs, $E_{CR} \gtrsim 10^{18}$ eV). Simulations indicate that $10^{20}$ eV protons are deflected only a few degrees by the galactic magnetic field (GMF) during propagation, although significant uncertainties remain on this figure given our incomplete knowledge of the chemical composition of the UHECR flux and the strength and structure of the GMF and IGMF. If neutrons are present in the cosmic-ray flux, their limited decay range of ~900 kpc at $10^{20}$ eV restricts the reach of neutrino-neutron correlation searches to our immediate galactic vicinity. The range of UHECRs with energies above $10^{20}$ eV is also limited to a few tens of Mpc by the Greisen-Zatsepin-Kuzmin (GZK) energy-loss mechanism.

The two main facilities dedicated to the study of UHECRs are the Pierre Auger Observatory (Auger)[2] in Mendoza, Argentina, and the Telescope Array[47] (TA) in Utah, USA, with instrumented areas of 3000 and 800 km$^2$, respectively. No strong evidence for UHECR point sources has been found so far, although a recent analysis[17] of Telescope Array data shows hints of a 20° "hotspot" in the northern sky at energies above $5.7 \times 10^{19}$ eV with a significance of 3.4$\sigma$. Possible indications of a dipole anisotropy have been reported in Auger data at the 4$\sigma$ level.[1]

A neutrino-UHECR connection is favored by the similarity between the astrophysical neutrino flux level measured by IceCube and the Waxman-Bahcall (WB) flux,[70] which represents an upper bound on the neutrino flux from UHECR sources, assuming they accelerate protons that convert most of their energy to pions. However, IceCube data currently favors a softer spectral index[10] than the $\propto E^{-2}$ spectrum assumed by the WB model. Extending the energy range of neutrino observations may elucidate the role that UHECRs play in the neutrino spectrum. As neutrinos carry about 5% of the parent cosmic-ray proton energy, the TeV–PeV neutrino sample used for these searches is too low in energy to be directly produced in UHECR interactions, which are expected to be observed at $E_\nu > 10^{16}$ eV.

---

[a]Coincidentally, the ANTARES trigger position was found in the vicinity of the star Antares ($\alpha$ Sco).

Rather, the TeV–PeV neutrinos, including those associated with the astrophysical flux detected by IceCube, are used as tracers of cosmic-ray acceleration that could be responsible for UHECR emission.

Correlation studies using UHECR data from TA and Auger, and neutrinos events from ANTARES and IceCube, have been conducted in recent years, so far with no statistically significant detection. An ANTARES search[27] used a sample of up-going candidate neutrino events and Auger UHECR showers with energies above $5.5 \times 10^{19}$ eV. No significant correlation was found at several angular scales and an upper limit was derived on the neutrino flux from each UHECR direction. More recently, the IceCube, Auger, and TA collaborations presented results[14] from an analysis that compared UHECR directions to both cascade and muon track neutrino events (Fig. 5). No significant deviation from the isotropic expectation was found using high-energy muon tracks. For cascade events, a post-trial $p$-value of $5 \times 10^{-4}$ was found for a typical angular separation of $22°$ from UHECR events. An *a posteriori* test where the UHECR positions were fixed gives a $p$-value of $8.5 \times 10^{-3}$. The statistical significance appears to be driven by event pairs in the region of the TA "hotspot" and it will be interesting to follow how the correlation evolves as the data set of both UHECRs and high-energy neutrinos continues to grow over the coming years.

Further enhancements to Auger,[3] the planned expansion of TA to almost $2600 \, \text{km}^2$, and the future launch of the JEM-EUSO mission[43] will provide a significant boost to UHECR statistics above $10^{20}$ eV, where GMF deflections are reduced. Joint searches using data from these upgraded instruments and next-generation neutrino telescopes will help explore current hints of an UHECR-neutrino correlation.

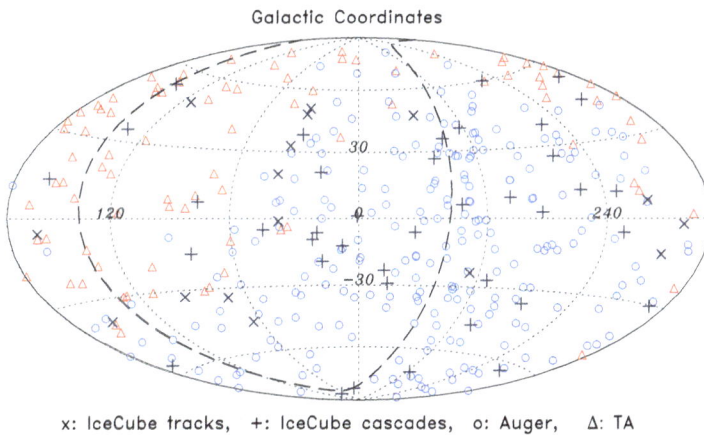

Fig. 5. Galactic-coordinates sky map of IceCube high-energy cascades (plus signs), high-energy tracks (crosses), and UHECRs detected by Auger (circles) and TA (triangles). The super-galactic plane is indicated by a dashed line. (From Ref. 14.)

## 4. Gravitational waves

The recent announcement of the first detection of gravitational waves by LIGO[20] represents a groundbreaking result that opens yet another channel to study the extreme universe in addition to photons, cosmic rays, and neutrinos. It has been proposed that energetic sources such as GRBs, CCSNe, and soft gamma repeaters (SGR) are emitters of neutrinos and gravitational waves (GW). While the neutrinos are produced by particle interactions in the relativistic outflows from these objects, GWs are related to the bulk matter dynamics of the source progenitor. Searches for spatial and temporal correlations between GW and neutrino signals have a higher sensitivity to these type of sources than those performed separately on each channel. A combined study also enables searches for more exotic phenomena such as "choked" GRBs, where the jet is not able to break out from the progenitor and therefore no gamma rays are emitted. Other events that are too faint or obscured to be detected by EM telescopes, or that are missed by the limited sky coverage of these detectors, may also be observed through this approach. Besides providing insights on the GW source progenitor, a coincident GW-neutrino detection can drastically shrink the source confidence region from several hundreds of square degrees (as in the case of the first GW detection) to the sub-square-degree level for neutrinos enabling targeted EM follow-up observations. The status of combined neutrino-GW searches leading to the first GW detection is presented in Ref. 32.

The most sensitive GW observatories currently in operation are km-scale Michelson laser interferometers. The LIGO observatory[18] operates detectors in two locations in the USA: Livingston, Louisiana, and Hanford, Washington. Both detectors were recently upgraded and in September 2015 the observatory started science operations for its "Advanced LIGO" (aLIGO) phase,[48] during which the first GW signals were detected. A LIGO site in India has also been proposed, which would significantly improve the ability of the observatory to locate sources in the sky. In Europe, the Virgo detector (near Cascina, Italy) is currently being upgraded[69] to its "Advanced Virgo" configuration (AdV), while the GEO 600 observatory[72] (near Sarstedt, Germany) is being used as a test-bench for advanced technology concepts. KAGRA,[65] a future observatory in Kamioka, Japan, is currently under construction and the beginning of scientific operations is expected towards the end of the decade.

Previous to the GW discovery, several searches for GW-neutrino spatial and temporal correlations were performed using GW data from LIGO and Virgo, and neutrino events from ANTARES[25] and IceCube,[6] with no significant correlations detected. The first GW event, named *GW150914*, was detected by LIGO on September 14th, 2015 and was produced by the merger of two $\sim 30 M_\odot$ black holes at a distance of 410 Mpc. Although no neutrino signal is expected in a black hole-black hole merger, a search for coincident neutrino events was performed using data from ANTARES and IceCube. Three neutrino events (all from IceCube) were found in an *a priori*-defined $\pm 500\,$s window around the GW detection (Fig. 6), in good

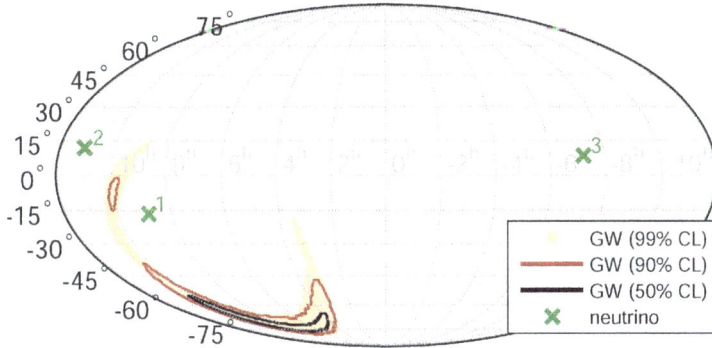

Fig. 6.  Error regions for the GW event *GW150914* for 50%, 90%, and 99% containment levels with three IceCube neutrinos detected in a ±500s time window around the GW event.[20]

agreement with atmospheric background expectations. Additionally, none of the neutrinos were spatially coincident with the GW uncertainty region and all-sky upper limits were derived on any potential associated neutrino source.

A second LIGO GW event was recently announced,[19] also detected during the first run of the aLIGO configuration, while a second run is expected to start towards the end of 2016. In addition to the continuing operation of aLIGO, the start of operations of AdV, the completion of KAGRA, the possible construction of the LIGO India site and the outstanding performance of the LISA Pathfinder mission[34] promise a bright future for GW astronomy and for correlation studies with neutrinos.

## 5. Transient searches

Most transient searches described so far have been performed on the basis of individual agreements between neutrino observatories and one or more multi-messenger detectors. These studies have been aimed at exploring a particular detection channel or augmenting the sensitivity to certain types of sources. However, as the sources of astrophysical neutrinos remain unknown, a better approach is to combine all available measurements in order to search for temporal and spatial correlations. A realtime detection of an interesting correlation between two or more channels can be used to trigger follow-up observations using pointed instruments. The superior angular and energy resolution of these instruments can increase the significance of a potential transient signal.

The Astrophysical Multimessenger Observatory Network (AMON)[b] is the current effort to realize this strategy. AMON provides the computational infrastructure to interconnect neutrino and gravitational-wave observatories with EM partners so

---

[b]http://amon.gravity.psu.edu/

that transient alerts can be transmitted without delay. A database of past alerts also enables archival coincidence searches of multi-messenger signals.

Observatories are classified as "triggering" or "follow-up" instruments. Triggering instruments, such as IceCube and HAWC, have large fields of view and high duty cycles. These observatories send event information to AMON including the event detection time, its position and uncertainty, an estimate of the false-trigger probability associated with this event, and additional detector-dependent quantities. AMON will forward interesting triggers (such as PeV neutrino event locations) directly to follow-up observatories, while lower-significance "sub-threshold" events will be used by the AMON online analysis to search for realtime coincidences and issue alerts if a correlation is found. The distribution of the alerts is performed using the Gamma-ray Coordinates Network (GCN).

AMON has recently started realtime operations,[67] and analyses of archival data sets made public by different partners have been performed.[52] For EM follow-ups of neutrinos associated with GRBs, a >1000 increase in efficiency is expected from conducting correlation studies between single neutrino events and several EM streams with respect to the current EM follow-up "status quo" that requires significant detections in each channel.

## 6. Conclusions and outlook

The high-energy astrophysical neutrino flux revealed by IceCube opens exciting possibilities to explore the extreme universe using multiple cosmic messengers. The impact of this breakthrough result has sparked a large number of observational programs and theoretical studies aimed at unveiling the sources of astrophysical neutrinos using a multi-messenger approach which involves photons, cosmic rays and gravitational waves.

An optimal scenario for multi-messenger searches consists of several neutrino telescopes with multi-km$^3$ effective volumes that are capable of detecting a large number of astrophysical events, reconstructing them with good angular resolution, and broadcasting their positions in near-realtime to partner multi-messenger observatories. Work is currently underway in different fronts to achieve many of these desired goals. Two next-generation neutrino telescopes are in the works: IceCube-Gen2,[5] a 10 km$^3$ extension to IceCube, and KM3NeT, the first km$^3$-scale detector to be built in the northern hemisphere. The joint operation of both detectors will greatly improve our sensitivity to neutrino point-sources, while providing a large high-purity astrophysical neutrino sample to use in follow-up observations. Even for current generation instruments, the sensitivity of counterpart searches can be significantly improved by exploiting the high-energy through-going muons detected by IceCube, which constitute the "golden channel" for multi-messenger searches given their sub-degree angular resolution and high astrophysical probability. Significant improvements in reconstruction techniques over the coming years will boost the angular resolution of cascade events in IceCube from its current value of ~15°, and it is feasible that KM3NeT will be able to deliver an angular resolution for these events at the level of $\sim 2°$.

While the directional uncertainty for cascades prevents most follow-up observations from using pointed instruments given their limited sky coverage, large field of view detectors with high-duty cycles (such as *Fermi*-LAT and HAWC in gamma rays, Auger and TA in cosmic rays, and the aLIGO and AdV in gravitational waves) can be used to search for temporal and spatial correlations. The sensitivity to counterparts will continue to increase thanks to the construction and operation of next-generation instruments, such as CTA, LHAASO, and HiScore (gamma-rays); JEM-EUSO (cosmic rays); and KAGRA (gravitational waves). The last remaining step is to interconnect this vast observational network. In this sense, the AMON network is currently starting operations and will provide an avenue for the rapid dissemination of neutrino alerts.

In summary, we foresee that over the next few years a greatly-improved global network of multi-messenger observatories will enter regular operations, and we look forward to the revolutionary discovery of the first point sources of astrophysical neutrinos as the crowning achievement for this joint international effort.

## Acknowledgements

The author would like to thank Imre Bartos, Jon Dumm, Geraldina Golup, Jamie Holder, Azadeh Keivani, Thomas Kintscher, Gernot Maier, Aurore Mathieu, Reshmi Mukherjee, Daniel Nieto, Michael Shaevitz, Fabian Schüssler, Ignacio Taboada, Gordana Tešić and Stefan Westerhoff for their helpful discussions and comments, and for providing some of the material used in this work.

The author acknowledges support from the US National Science Foundation through NSF grant PHYS-1505811.

## References

1. A. Aab, *et al.*, Large scale distribution of ultra high energy cosmic rays detected at the Pierre Auger Observatory with zenith angles up to 80. *Astrophys. J.* **802**(2) (2015) 111.
2. A. Aab, *et al.*, The Pierre Auger Cosmic Ray Observatory. *Nucl. Instrum. Meth.* **A798** (2015) 172–213.
3. A. Aab, *et al.*, The Pierre Auger Observatory: Contributions to the 34th International Cosmic Ray Conference (ICRC 2015). *PoS*, ICRC2015:593, 2015.
4. M.G. Aartsen, *et al.*, The IceCube neutrino observatory Part I: Point source searches. *ArXiv e-prints*, 2013.
5. M.G. Aartsen, *et al.*, IceCube-Gen2: A vision for the future of neutrino astronomy in Antarctica. *arXiv preprint*, 2014.
6. M.G. Aartsen, *et al.*, Multimessenger search for sources of gravitational waves and high-energy neutrinos: Initial results for LIGO-Virgo and IceCube. *Phys. Rev.* **D90**(10) (2014) 102002.

7. M.G. Aartsen, *et al.*, Observation of high-energy astrophysical neutrinos in three years of IceCube data. *Phys. Rev. Lett.* **113** (2014) 101101.

8. M.G. Aartsen, *et al.*, Observation of the cosmic-ray shadow of the moon with icecube. *Phys. Rev. D* **89** (2014) 102004.

9. M.G. Aartsen, *et al.*, Searches for extended and point-like neutrino sources with four years of IceCube data. *ArXiv e-prints*, 2014.

10. M.G. Aartsen, *et al.*, A combined maximum-likelihood analysis of the high-energy astrophysical neutrino flux measured with IceCube. *Astrophys. J.* **809**(1) (2015) 98.

11. M.G. Aartsen, *et al.*, Detection of a Type IIn Supernova in optical follow-up observations of IceCube neutrino events. *Astrophys. J.* **811**(1) (2015) 52.

12. M.G. Aartsen, *et al.*, Evidence for astrophysical muon neutrinos from the northern sky with IceCube. *Phys. Rev. Lett.* **115**(8) (2015) 081102.

13. M.G. Aartsen, *et al.*, Search for prompt neutrino emission from gamma-ray bursts with IceCube. *Astrophys. J. Lett.* **805** (2015) L5.

14. M.G. Aartsen, *et al.*, Search for correlations between the arrival directions of IceCube neutrino events and ultrahigh-energy cosmic rays detected by the Pierre Auger Observatory and the Telescope Array. *JCAP* **1601**(01) (2016) 037.

15. R. Abbasi, *et al.*, Searching for soft relativistic jets in core-collapse supernovae with the IceCube optical follow-up program. *Astron. Astrophys.* **539** (2012) A60.

16. R. Abbasi, *et al.*, IceTop: The surface component of IceCube. *Nucl. Instrum. Meth.* **A700** (2013) 188–220.

17. R.U. Abbasi, *et al.*, Indications of intermediate-scale anisotropy of cosmic rays with energy greater than 57 EeV in the northern sky measured with the surface detector of the Telescope Array Experiment. *Astrophys. J.* **790** (2014) L21.

18. B.P. Abbott, *et al.*, LIGO: The laser interferometer gravitational-wave observatory. *Rep. Prog. Phys.* **72**(7) (2009) 076901.

19. B.P. Abbott, *et al.*, GW151226: Observation of gravitational waves from a 22-solar-mass binary black hole coalescence. *Phys. Rev. Lett.* **116** (2016) 241103.

20. B.P. Abbott, *et al.*, Observation of gravitational waves from a binary black hole merger. *Phys. Rev. Lett.* **116** (2016) 061102.

21. A.U. Abeysekara, *et al.*, Sensitivity of the high altitude water Cherenkov detector to sources of multi-TeV gamma rays. *Astropart. Phys.* **50–52** (2013) 26–32.

22. B.S. Acharya, *et al.*, Introducing the CTA concept. *Astropart. Phys.* **43** (2013) 3–18.

23. M. Ackermann, *et al.*, The spectrum of isotropic diffuse gamma-ray emission between 100 MeV and 820 GeV. *Astrophys. J.* **799** (2015) 86.

24. S. Adrian-Martinez, *et al.*, Search for cosmic neutrino point sources with four year data of the ANTARES telescope. *Astrophys. J.* **760** (2012) 53.

25. S. Adrian-Martinez, *et al.*, A first search for coincident gravitational waves and high energy neutrinos using LIGO, Virgo and ANTARES data from 2007. *JCAP* **1306** (2013) 008.

26. S. Adrian-Martinez, *et al.*, Optical and X-ray early follow-up of ANTARES neutrino alerts. *arXiv preprint*, 2015.

27. S. Adrián-Martnez, *et al.*, Search for a correlation between ANTARES neutrinos and Pierre Auger Observatory UHECRs arrival directions. *Astrophys. J.* **774** (2013) 19.

28. M. Ageron, *et al.*, The ANTARES Telescope neutrino alert system. *Astropart. Phys.* **35** (2012) 530–536.

29. F. Aharonian, *et al.*, Observations of the Crab Nebula with H.E.S.S. *Astron. Astrophys.* **457** (2006) 899–915.

30. M. Ahlers and K. Murase. Probing the galactic origin of the IceCube excess with gamma-rays. *Phys. Rev.* **D90**(2) (2014) 023010.

31. J. Aleksić, *et al.*, The major upgrade of the MAGIC telescopes, Part II: A performance study using observations of the Crab Nebula. *Astropart. Phys.* **72** (2016) 76–94.

32. S. Ando, *et al.*, Colloquium: Multimessenger astronomy with gravitational waves and high-energy neutrinos. *Rev. Mod. Phys.* **85**(4) (2013) 1401–1420.

33. S. Ando, I. Tamborra, and F. Zandanel. Tomographic constraints on high-energy neutrinos of hadronuclear Origin. *Phys. Rev. Lett.* **115**(22) (2015) 221101.

34. M. Armano, *et al.*, Sub-femto-*g* free fall for space-based gravitational wave observatories: Lisa pathfinder results. *Phys. Rev. Lett.* **116** (2016) 231101.

35. W. Atwood, *et al.*, Pass 8: Toward the full realization of the Fermi-LAT scientific potential. *2012 Fermi Symposium proceedings — eConf C121028*, 2013.

36. W.B. Atwood, *et al.*, The large area telescope on the Fermi Gamma-Ray Space Telescope Mission. *Astrophys. J.* **697** (2009) 1071–1102.

37. K. Bechtol, M. Ahlers, M. Di Mauro, M. Ajello, and J. Vandenbroucke. Evidence against star-forming galaxies as the dominant source of IceCube neutrinos. 2015.

38. A.M. Brown, J. Adams, and P.M. Chadwick. $\gamma$-ray observations of extraterrestrial neutrino track events. *Mon. Not. Roy. Astron. Soc.* **451**(1) (2015) 323–331.

39. CTA Consortium. CTA Performance. https://portal.cta-observatory.org/Pages/CTA-Performance.aspx, 2015.

40. S. Cui, Y. Liu, Y. Liu, and X. Ma. Simulation on gamma ray astronomy research with LHAASO-KM2A. *Astropart. Phys.* **54** (2014) 86–92.

41. D. Stransky, *et al.*, (for the KM3NeT Collaboration). Reconstruction of cascade-type neutrino events in KM3NeT/ARCA. *PoS*, ICRC2015:1108, 2015.

42. M. Di Mauro and F. Donato. Composition of the Fermi-LAT isotropic gamma-ray background intensity: Emission from extragalactic point sources and dark matter annihilations. *Phys. Rev.* **D91**(12) (2015) 123001.

43. T. Ebisuzaki, *et al.*, The JEM-EUSO mission to explore the extreme universe. *AIP Conf. Proc.* **1238** (2010) 369–376.

44. P.A. Evans, *et al.*, Swift follow-up of IceCube triggers, and implications for the Advanced-LIGO era. *Mon. Not. Roy. Astron. Soc.* **448**(3) (2015) 2210–2223.

45. F. Schüssler, *et al.* for the H. E. S. S. Collaboration. The H.E.S.S. multi-messenger program. *ArXiv e-prints*, 2015.

46. Fermi-LAT Collaboration. Fermi-LAT performance. http://www.slac.stanford.edu/exp/glast/groups/canda/lat_Performance.htm, 2015.

47. M. Fukushima. Telescope Array Project for extremely high energy cosmic rays. *Prog. Theor. Phys. Suppl.* **151** (2003) 206–210.

48. G.M. Harry and the LIGO Scientific Collaboration. Advanced LIGO: The next generation of gravitational wave detectors. *Class. Quant. Grav.* **27**(8) (2010) 084006.

49. H. He, *et al.*, Design highlights and status of the LHAASO project. *PoS*, ICRC2015:1010, 2015.

50. M. Holler, A. Balzer, R. Chalmé-Calvet, M. de Naurois, and D. Zaborov. Photon Reconstruction for H.E.S.S. Using a semi-analytical shower model. In *Proceedings, 34th International Cosmic Ray Conference (ICRC 2015)*, 2015.

51. M. Kadler, *et al.*, Coincidence of a high-fluence blazar outburst with a PeV-energy neutrino event. *Nat. Phys.* 2016.

52. A. Keivani, D.B. Fox, G. Tešić, D.F. Cowen, and J. Fixelle. AMON searches for jointly-emitting neutrino + gamma-ray transients. *PoS*, ICRC2015:786, 2015.

53. S.R. Kelner and F.A. Aharonian. Energy spectra of gamma-rays, electrons and neutrinos produced at interactions of relativistic protons with low energy radiation. *Phys. Rev.* **D78** (2008) 034013 [Erratum: *Phys. Rev.* **D82** 099901 (2010)].

54. S.R. Kelner, F.A. Aharonian, and V.V. Bugayov. Energy spectra of gamma-rays, electrons and neutrinos produced at proton-proton interactions in the very high energy regime. *Phys. Rev.* **D74** (2006) 034018 [Erratum: *Phys. Rev.* **D79** (2009) 039901].
55. M.D. Kistler. On TeV gamma rays and the search for galactic neutrinos. 2015.
56. M. Kowalski and A. Mohr. Detecting neutrino-transients with optical follow-up observations. *Astropart. Phys.* **27** (2007) 533–538.
57. L. Rädel and S. Schoenen for the IceCube Collaboration. ATel #7856. http://www.astronomerstelegram.org/?read=7856, 2015.
58. C. Lunardini, S. Razzaque, and L. Yang. Multimessenger study of the Fermi bubbles: Very high energy gamma rays and neutrinos. *Phys. Rev.* **D92**(2) (2015) 021301.
59. M. Santander, for the VERITAS and IceCube Collaborations. Searching for TeV gamma-ray emission associated with IceCube high-energy neutrinos using VERITAS. *ArXiv e-prints*, 2015.
60. A. Margiotta. The KM3NeT deep-sea neutrino telescope. *Nucl. Instrum. Meth.* **A766** (2014) 83–87.
61. A. Merloni, *et al.*, eROSITA Science Book: Mapping the Structure of the Energetic Universe. *MPE document (online)*, 2012.
62. A.A. Moiseev, *et al.*, Compton-Pair Production Space Telescope (ComPair) for MeV gamma-ray astronomy. 2015.
63. N. Park. Performance of the VERITAS experiment. In *Proceedings, 34th International Cosmic Ray Conference (ICRC 2015)*, 2015.
64. F. Schüssler, *et al.*, Multiwavelength study of the region around the ANTARES neutrino excess. In *Proceedings, 33rd International Cosmic Ray Conference (ICRC2013)*, 2013.
65. K. Somiya. Detector configuration of KAGRA: The Japanese cryogenic gravitational-wave detector. *Class. Quant. Grav.* **29** (2012) 124007.
66. A.M. Taylor, S. Gabici, and F. Aharonian. Galactic halo origin of the neutrinos detected by IceCube. *Phys. Rev.* **D89**(10) (2014) 103003.
67. G. Tešić and A. Keivani. AMON: Transition to real-time operations. *PoS*, ICRC2015:329, 2015.
68. M. Tluczykont *et al.*, The HiSCORE concept for gamma-ray and cosmic-ray astrophysics beyond 10 TeV. *Astropart. Phys.* **56** (2014) 42–53.
69. Virgo Collaboration. Advanced Virgo Baseline Design. Report No. VIR-027A-09, 2009.
70. E. Waxman and J. Bahcall. High energy neutrinos from astrophysical sources: An upper bound. *Phys. Rev. D* **59** (1998) 023002.
71. T.C. Weekes, *et al.*, VERITAS: The Very Energetic Radiation Imaging Telescope Array System. *Astropart. Phys.* **17** (2002) 221–243.
72. B. Willke, *et al.*, The GEO 600 gravitational wave detector. *Class. Quant. Grav.* **19** (2002) 1377–1387.

# Chapter 10

## Neutrinos from Core-Collapse Supernovae

Marek Kowalski

*Humboldt-University zu Berlin & DESY,*
*marek.kowalski@desy.de*

## 1. Introduction

When a massive star with $M > 8M_\odot$ runs out of fuel to sustain the fusion in its center, it triggers arguably the most violent type of explosion the Universe has to offer, a core-collapse supernova. The gravitational energy released during the collapse is easily estimated:

$$\Delta E_{\mathrm{SN}} \approx G \frac{M^2}{R_{\mathrm{NS}}} \approx 2.7 \times 10^{53} \mathrm{erg} \left( \frac{M}{M_\odot} \right)^2,$$

assuming $R_{\mathrm{NS}} = 10$ km for the radius of the neutron star that forms at the center. The collapse occurs within a second, over which the system will heat-up from the outer shells bouncing of the stiff neutron star matter, initially cooling only through the emission of quasi-thermal neutrinos. This initial phase lasts for a few seconds determined by the diffusion time $t_d$ of the neutrinos out of the neutron star. The temperature of the neutrinos $T_\nu$ can be estimated using the Stefan–Bolzmann law for fermions, $T_\nu \sim (0.2\Delta E_{\mathrm{SN}}/t_d/4\pi\sigma R_{\mathrm{NS}}^2)^{1/4}$, where the prefactor 0.2 accounts for the fermionic degrees of freedom of the neutrinos. This corresponds to about $\langle E_\nu \rangle = 10\,\mathrm{MeV}$ average energy per neutrino for a SN near the mass threshold. Since 99% of the total energy is emitted through neutrinos, their total number is enormous, $\Delta E_{\mathrm{SN}}/\langle E_\nu \rangle \sim 10^{58}$. This basic picture of a core-collapse was confirmed in 1987 with the observation of Supernova 1987A (SN 1987A), which exploded in the Large Magellanic Cloud and from which 19 neutrinos were detected in total by Kamiokande[1] and IMB.[2] However, with more events the above picture could be examined in detail, opening up a window to the physics of neutrinos and to the explosion process itself. Today, Super-Kamiokande[3] would record $\sim 10^4$ individual neutrino events from a supernova in our galaxy, while IceCube would see the signal from as many as one million neutrinos, increasing the collective count rate of all IceCube sensors.[4] In particular IceCube, with its huge volume would measure the neutrino lightcurve with uncontended precision. This will be discussed further in Sec. 3.

How is the remaining energy released during a core-collapse supernova? Depending on the progenitor star, the observational consequences can be very different, ranging from optical light curves that, powered by shock interaction with the circumstellar medium (CSM), persist over several months to the very prompt emission of X-rays associated with gamma-ray bursts. Here we focus on the question of high energy neutrino production. The kinetic energy of the shells bouncing back off the neutron star is

$$E_{\text{kin}} \approx 2 \times 10^{51} \text{erg} \times \left( \frac{M_{\text{shell}}}{M_\odot} \right) \left( \frac{v_{\text{shell}}}{10^4 \text{km/s}} \right)^2. \tag{1}$$

The conditions provided in a supernova allow for shock acceleration, so a fraction of the kinetic energy is expected to be transferred to high energy protons and heavier nuclei. High energy neutrinos are produced in the collision of these protons with ambient matter or radiation, which in principle exists in abundant quantities in a SN environment. Assuming $10^{51}$ erg is transferred in this process to TeV neutrinos, it would yield $6 \times 10^{50}$ neutrinos. The IceCube effective area for muon neutrinos at 1 TeV is nearly $\sim 1 \, \text{m}^2$ (and $\sim 10 \, \text{m}^2$ for 10 TeV[5]), hence one would detect on average one TeV neutrino from SNe up to 100 Mpc distances. Within such a distance, more than 1000 SNe explode every year, including some very energetic ones with massive progenitors. Observing individual supernovae in connection with high energy neutrinos appears within reach and correspondingly, searches have been performed to detect high energy neutrinos from supernovae. The available models, search strategies and current constraints will be reviewed in the following section.

## 2. High energy neutrinos

The significant release in energy (see Eq. (1)) during the collapse of a massive star offers many conceivable mechanisms for high energy neutrino production, a few of which will be discussed in the following. Current constraints and strategies for neutrino detection from Supernovae are discussed thereafter.

### 2.1. *High energy neutrino predictions*

We start with the latest phase of a supernova first: In supernova remnants (SNR) particles are accelerated across shocks produced by the expanding matter shells propagating into interstellar space. Galactic SNRs have been studied in much detail using multi-wavelength observations ranging up to 100 TeV energies. They are the prime suspect for galactic cosmic-ray production up to the knee of the CR energy spectrum. For efficient neutrino production, the source needs not only to accelerate hadrons but also be surrounded by sufficiently abundant target material, usually in the form of molecular clouds. Taking a model prediction for SNR RXJ1713 as a benchmark,[6] the KM3NeT detector would allow testing the predictions for hadronic acceleration within 2–3 years at $3\sigma$ significance.[7] The IceCube experiment, located at the South Pole, is not optimally positioned too observe the most promising SNR candidates while the ANTARES detector appears to be to small.

Moving to earlier phases, i.e. minutes to hours after the explosion, the picture becomes less well constrained for the simple reason that one has to rely on the observation of extragalactic objects combined with the fact that for the denser environments modeling becomes more challenging. An important transition in a supernova turning into a remnant is the shock breakout. In the optically-thick interior of the progenitor, the shock is dominated by radiation but when the optical depth ahead of the shock drops below unity, the radiation decouples from the gas. In this environment, collective plasma instabilities can lead to magnetic field amplification, transforming the shock to become collisionless.[8] These are ideal conditions for TeV neutrino production. We discuss in the following a specific scenario that can lead to a sizable number of high-energy neutrinos. Supernovae of Type IIn — where the "n" denotes the presence of narrow width hydrogen emission lines — are of special interest, as their optical lightcurves appear not to be powered by radioactive decay but by fast ejecta interacting with the circumstellar medium (CSM). These objects provide the right conditions for efficient neutrino production due to the very large mass losses (either stellar winds or eruptions) associated with their progenitors (e.g. mass loss rates of[9] up to $10^{-1} M_\odot \mathrm{yr}^{-1}$ decades before the explosion). Such an environment allows for efficient collisionless shock acceleration.[10–12] Moreover, as a consequence of magnetic fields amplifications in the dense medium, the resulting energy spectrum of CRs is predicted to extend to $10^{17}$ eV. The CSM also provides a natural target for neutrino production through $pp$ interaction. The time profile of the neutrino emission depends on the radial profile of the wind and will generally start declining after a few months.[13] The prediction for the neutrino flux from all SNe IIn is consistent with the observations of the diffuse flux by IceCube.[13] To test the model, one can search for a spacial and temporal correlation of known SNe IIn with the detected neutrinos. Note that as a proxy for the neutrino emission one can use the optical flux, which is also assumed to be powered by the interaction of the shells with the CSM. While Type IIn objects are a unique type of supernovae, the same mechanisms might be at play for other types, provided that sufficient amounts of target material are available.

Can one also expect significant emission on minute to seconds time scales? Burst-like emission in a range of stellar explosion scenarios is motivated by the observation of long gamma ray bursts (GRBs),[a] that have been associated with

---

[a]See the contribution[14] from Peter Mészáros for a detailed discussion.

Type Ic core-collapse SNe.[b] GRBs require narrow jets containing highly relativistic colliding shocks to explain both the energy output, short duration and the observed afterglows. Due to the strong beaming, the objects only appear bright when the observer is located in the emission cone. In this case, however, GRBs become detectable throughout the visible Universe. That GRBs might be formidable emitters of high-energy neutrinos was first suggested by Waxman and Bahcall in 1997.[15] They have shown that if the emission of X-rays is explained by colliding shocks, it would also allow for shock acceleration of protons to the highest energies. Neutrinos are then produced in the collision of the high-energy protons with the ambient photon fields. By now the IceCube experiment has ruled out the initial prediction of Waxman and Bahcall.[16] However, more refined modeling of the same processes has resulted in a downward-correction of the flux by more than an order of magnitude,[17] below the sensitivity of IceCube.

Between GRBs and regular core-collapse SNe one finds various other representatives of stellar explosions, including low-luminosity GRBs (LLGRBs) and choked-jet SNe, which are further described in the contribution of Mészáros.[14] In these objects, a less collimated and mildly relativistic jet is launched that can be absorbed by the stellar envelope or emerge with much reduced energies. Correspondingly, these objects might be dark in GeV–TeV gamma rays.[18] Their occurrence, though not as easily detected, should be more frequent compared to GRBs. Hence it is conceivable that these objects significantly contribute to the diffuse flux of astrophysical TeV–PeV neutrinos observed by IceCube.[18]

So far, the search for neutrino emission from individual objects, such as SN2008D,[19] has not led to a detection of neutrinos, providing weak constraints on the parameters of the model.[20] Further constraints from an early IceCube analysis[21] (discussed in more detail below) on SNe with relativistic jets are shown in Fig. 1. Generally, however, the models remain uncertain and not yet strongly constrained.

Finally, another interesting scenario producing short bursts of neutrinos is offered by very young, fast spinning magnetized pulsars born in a supernova. These objects have been used to explain the highest energy cosmic rays[24,25] and would also predict a short burst of high energy neutrinos peaking at $\sim 10^{17}$ eV energies. However, this scenario is starting to be constrained by the non-observation of the very high energy neutrinos expected from the collective population of such pulsars.[26]

### 2.2. *Strategies towards detection of neutrinos from SNe*

Several strategies to detect high energy neutrinos from SNe exist, each being sensitive to different emission scenarios. Because the objects of interest are transient by nature, both time and directional information can be used. By cross-correlating

---

[b]Short GRBs with a duration of <2 s are assumed to be produced in compact object binary mergers.

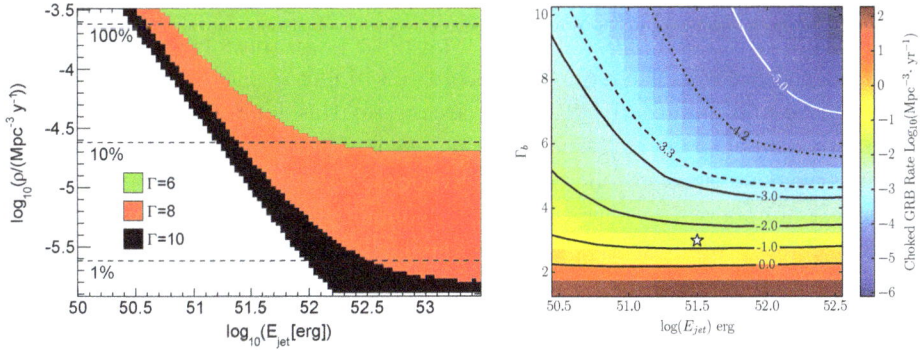

Fig. 1. Left: IceCube constraints[21] on the choked jet SN model[22] for different Lorentz boost factors $\Gamma$ as a function of the rate of SNe with jets $\rho$ and the jet energy $E_{jet}$. The colored regions are excluded at 90% confidence level. Horizontal dashed lines indicate a fraction of SNe with jets of 100%, 10% or 1% (relative to a CCSN rate of 1 per year within a distance of radius 10 Mpc). The limits only apply for a burst duration of 100 s or shorter. Right: Constraint from an independent analysis using IceCube/DeepCore data,[23] targeting lower energies and making weaker assumption on the burst duration.

the information with the information about SNe observed in the optical or other wavelength bands, the search gains further sensitivity.

The most prominent example of a multi-messenger search is that for neutrinos from GRBs. GRBs are very rare and short transients, hence the background of atmospheric events (e.g. $10^{-5}$ atmospheric neutrino events per second and square degree for IceCube, and less than a hundredth of that for ANTARES) accumulated in the time window of the prompt emission of a GRB is very low. This allows efficient stacking of the objects, and the sensitivity rises almost linearly with the number of GRBs. So far, no significant correlation was observed, leading to stringent limits on neutrino production: the maximum fraction that GRBs contribute to the diffuse astrophysical neutrino flux is currently $\sim 1$.[27]

In case of regular core-collapse SNe, there is no bright $\gamma$-ray burst to mark the stellar explosion. Extrapolating the optical light curve, including a possible detection of the shock break out, allows one to narrow down the explosion to a few hours (see e.g. Ref. 28). However, this requires ideally nightly observations of the full sky with sufficient depth to catch the supernovae soon after the explosion and current surveys all have limitations. Examples of current efforts are the dedicated Supernova survey ASASSN[c] that surveys the full sky every 2–3 days to about 16–17th magnitudes V-band depth,[29] the iPTF[d] survey at Palomar Observatory surveys several thousand square degrees on a variable cadence (1–3 days) to a depth of about 20–21st magnitude[30] and the PanStarrs survey at Mount

---

[c]http://www.astronomy.ohio-state.edu/~assassin/
[d]http://www.ptf.caltech.edu/iptf

Haleakala, Hawaii,[e] performing a $3\pi$ steradian survey to about 22nd magnitude depth, returning to each part of the sky typically 8 to 10 times per year.[31] No single survey today covers the requirements of being simultaneously complete, having high cadence and sufficiently deep. To circumvent this limitation, neutrino telescopes are searching for transient neutrino event signatures in near real time in order to trigger subsequent follow-up observations.[32] Both ANTARES and IceCube send target-of-opportunity alerts when neutrino multiplets — two or more neutrinos coincident in time and direction — are observed.[21,33] (ANTARES also sends alerts for the most energetic neutrino events, a program that IceCube operates as well since in 2016, see below). Follow-up observations are performed with the ROTSE (switched off in Nov 2014) and the TAROT network of small robotic telescopes,[21,33] with the Samuel Oschin telescope of iPTF,[34] as well as with the SWIFT X-ray satellite.[33,35]

While the first implementation for follow-up in IceCube and ANTARES has only produced upper limits,[21,34] the extension of the IceCube program to PTF has delivered a first interesting observation: PTF12csy, a bright SN in coincidence with the most significant neutrino alert so far (two neutrinos detected only 1.6 s apart).[34] Spectroscopic observations showed that it is a SN of Type IIn detected at a redshift of 0.068, i.e. for a prospective source rather distant. The lightcurve of PTF12csy is shown in Fig. 2. The probability for this to be a chance coincidence, two atmospheric neutrinos accidentally aligning with a SN, is estimated at 1.6%. However, based on archival data from Pan-STARRS as well as spectral information, the SN was established to be already more than 170 days old at the time of the neutrino detection, ruling out most models for neutrino production.[f]

But even the early search with 2008/2009 data from the IceCube experiment has already produced interesting constraints.[21] From the non-observation of a clear significant neutrino-multiplet (with or without SN counterpart), one can derive constraints on the emission parameters. For each $\Gamma$-value the 90% confidence region in the $E_{\mathrm{jet}}$-$\rho$-plane is displayed in Fig. 1. The colored regions are excluded with 90% confidence. The limits include the optical information, i.e. that no optical counterpart was found. This improved the limit and allows tests of 5–25% smaller CCSN rates. The largest improvement is obtained for small jet energies and larger fraction of CCSN rates exhibiting such jets. The most stringent limit can be set for high Lorentz factors, while for small $\Gamma$ the constraints are weak. At 90% confidence level, a sub-population of SNe with typical values of $\Gamma \sim 10$ and $E_{\mathrm{jet}} = 3 \times 10^{51}$ erg does not exceed 4%.

---

[e]http://pan-starrs.ifa.hawaii.edu/public/
[f]An exception could be the delayed collapse of a neutron star to a black hole, recently also postulated to explain short radio bursts. Unlike in the original Blitzar model for short radio bursts, the magnetic shock wave would not accelerate electrons in almost empty space, but protons in a denser/dirtier environment.

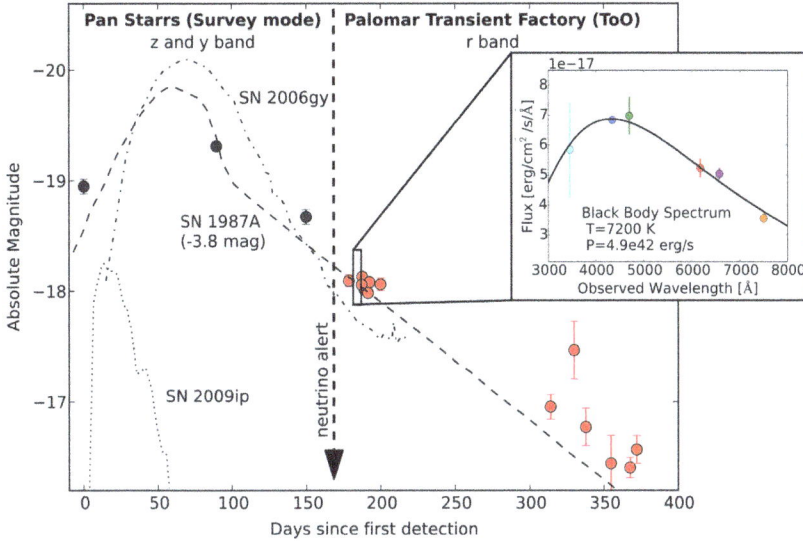

Fig. 2. Lightcurve of PTF12csy, a SN Type IIn detected by the Palomar Transient Facility as a result of the most significant alert by IceCube. The archival data before the neutrino alert came from the Pan-STARRS $3\pi$ survey and shows that the SN was at least 170 days old at the time of the neutrino alert. The insert shows the photometric data as a function of wavelength for a narrow range in time, fitted with a black body spectrum. The total bolometric energy released is estimated to be at least $10^{50}$ ergs (adapted from Ref. 34).

Since the energy spectrum from choked jet models is expected to be soft, it is worth extending a search for transient neutrino signal towards lower energies. This has been recently accomplished using IceCube/DeepCore data,[23] providing slightly weaker constraints on the SN parameters (jet energy and Lorenz factor and SN rate), however with generally fewer assumptions.

For longer emission times, e.g. as expected for SNe IIn where neutrinos are produced during the interaction of the ejecta and the CSM, it is also effective to search for neutrinos in coincidence with catalog/literature SNe. So far, one search attempted to identify gamma-rays from Fermi with SNe IIn from PTF and other sources.[36] As gamma-rays can be absorbed in the CSM, the upper limits cannot be directly translated to a prospective neutrino emission. It is expected, that a similar search performed for neutrinos/SN coincidences will lead to to constraints on neutrino emission at a level of $10^{48}$–$10^{49}$ ergs in the TeV to PeV band. That should allow testing the hypothesis that SNe IIn are the primary source behind the diffuse flux of high energy astrophysical neutrinos observed by IceCube.

The ability of IceCube to identify individual, high-energy astrophysical neutrinos provides extra motivation for a search for transient, electromagnetic counter parts. The sources of single neutrinos will on average be more distant compared to sources producing neutrino multiplets. Hence, the search will work best for rarer (to avoid source confusion) and bright objects (visible through large parts of the Universe).

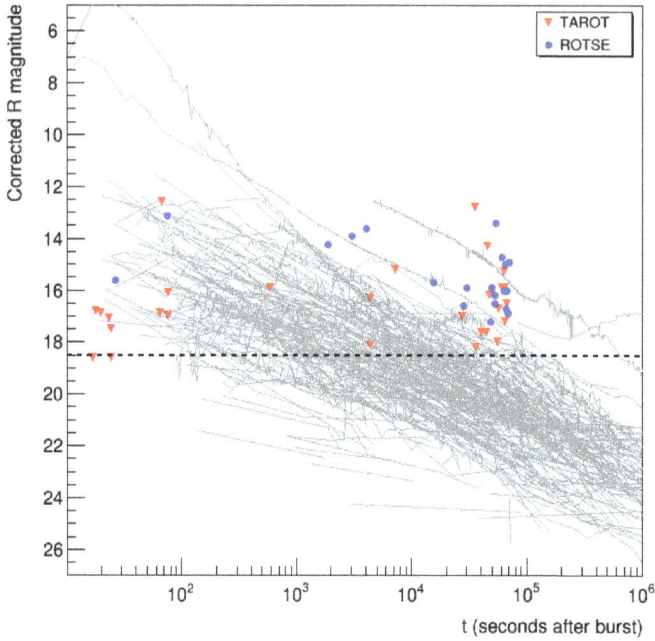

Fig. 3.   Search for optical afterglows in the follow-up of ANTARES alerts.[33] The triangles and circles correspond to upper limits from the TAROT and ROTSE follow-up observations, respectively, where the time corresponds to the first observation relative to the neutrino arrival time. The horizontal dashed line is the limiting magnitude of the telescopes. The gray lines correspond to afterglows of 301 independent GRBs.

A promising perspective is the identification of LLGRBs,[18] which could be detected through their electro-magnetic afterglow or associated Supernovae, requiring a high-cadence and deep survey. ANTARES has demonstrated the feasibility of the search for afterglows in the follow-up of single neutrino events[33] (see Fig. 3), while IceCube has also recently started to provide the public with real time information about detections of astrophysical neutrino candidates.

## 3. Low-energy (MeV) neutrinos

While the prospects of observing high-energy neutrinos from SNe remain uncertain until their first detection, the production of low-energy neutrinos in supernovae is well established through the observation SN1987A. Despite the fact that only few supernova neutrinos were detected from this unique event, a wealth of papers have been published in its wake reflecting the numerous and fundamental roles that neutrinos play in astrophysics as well as in particle physics (e.g. see Refs. 37, 38).

Due to the large sensor spacing and consequently high energy threshold, attempting to detect individual MeV neutrino events is not possible in current open water/ice neutrino detectors. However, due to the pristine environment of the Antarctic ice, IceCube's light sensors can be operated with very low noise rates, while detecting photons emitted hundreds of meters away.[4] Each IceCube sensor "surveys" an effective volume of about $600 \, \text{m}^3$ (for a neutrino interaction producing a 20 MeV positron), hence the complete IceCube array consisting of 5160 sensors has an effective target mass of $\sim 3$ Mton. While this is significantly larger than any other operating neutrino detector, the comparison has to be done with care as noise from individual sensors enters the significance.[4] In the fortunate case of a SN exploding close to the center of our galaxy, IceCube would measure fine details in the neutrino "lightcurve", imprinting the signature of the neutrino mass hierarchy, the observation of the accretion and cooling phases, an estimation of the progenitor mass from the shape of the neutrino light curve, the formation of a quark star or a black hole, as well as short term modulation due to turbulent phenomena and shocks produced during the explosion.[4] A core-collapse SN would be observed up to distances of 50 kpc at $10\sigma$, i.e. the distance of the Large Magelanic Cloud hosting SN1987A. Crude information about the spectrum (for instance the average neutrino energy, i.e. the temperature) would be obtained from the ratio of single to coincident photons, the later being more likely to come from more energetic neutrinos.[39]

While the Antarctic ice offers a specially pristine environment, the KM3NeT detector will also have a sensitivity to SNe beyond 20 kpc distances.[40] When a galactic SN appears, having multiple sites measure the neutrino burst can allow SN triangulation, hence allowing telescopes to point to the right direction, hours before the first photons emerge from the dense SN core.

However, with an expected rate of only 1–3 galactic SNe per century,[41,42] the chance for a detection during the lifetime of any neutrino experiment is not large. The situation will change drastically once neutrino detectors reach the sensitivity threshold to detect "mini-bursts" of neutrinos from supernovae in neighboring galaxies.[41–43] Not bound to our own galaxy, the rate of SN observations will depend only on the size of the detectors. An effective volume of $\sim 10$ Mtons is sufficient to detect SNe at a rate of $\sim 1-4$ per year — albeit most of them with less than ten individual neutrino events.[42,43] Despite the low number of detected neutrinos, these routine observations would provide a wealth of information and allow entirely new studies, such as determining the total SN rate in our local universe in a novel and less biased way. In addition, such routine observations could lead to the discovery of additional non-standard stellar explosions neutrino, such as failed or *dark* SNe (i.e. optically unobservable because of the subsequent collapse to a black hole),[44] merger of binary neutron stars[45] or the formation of quark stars[46] (see Refs. 42, 43 for a more extensive discussion of the science motivation).

Motivated by their scientific potential, several megaton scale neutrino detector concepts are being studied (e.g. DeepTITAND[42] and Hyper-Kamiokande[47]). An alternative study evaluated the potential of a $\sim 10$ Mton detector in the Antarctic ice shield, that builds on a continuation of the effort to reduce the energy threshold to a few GeV in the PINGU project.[48] The challenge is to reduce the energy threshold of the experiment by 2–3 orders of magnitude while controlling the background at a level required for the detection of supernovae. It was shown that by increasing the density of instrumentation while carefully controlling the noise level per sensor, this can in principle be achieved,[43] though the scale of such a project would be comparable to that of IceCube.

## 4. Conclusion

Over the next decade, the chance to observe a burst supernova in MeV neutrinos is about 10–30%, where the range reflects the current uncertainty in the galactic supernova rate. If lucky, we will be awarded with an unprecedented detailed monitor of the collapsing core of a supernova, making it possible to test a number of model predictions. The prospects of detecting higher energy neutrinos with current instrumentation are more uncertain. A number of models for high-energy neutrino production in stellar explosions exist, including SNe expanding into circumstellar medium, supernovae with choked or mildly relativistic jets as well as low luminosity gamma ray bursts. Identifying these objects is difficult due to their transient nature, however, dedicated multi-messenger searches that are being continuously improved could lead to a first discovery anytime soon. Not bound to our galaxy, there is the potential to observe extragalactic supernovae more routinely, thereby opening up a unique window into the otherwise hidden core of exploding stars.

## References

1. **Kamiokande-II** Collaboration, K. Hirata, *et al.*, "Observation of a neutrino burst from the supernova SN 1987a," *Phys. Rev. Lett.* **58** (1987) 1490–1493. [Erratum: *ibid.* **58** (1987) 727].
2. R.M. Bionta, *et al.*, "Observation of a neutrino burst in coincidence with supernova SN 1987a in the Large Magellanic Cloud," *Phys. Rev. Lett.* **58** (1987) 1494.
3. M. Ikeda, *et al.*, "Search for supernova neutrino bursts at super-kamiokande," *Astrophys. J.* **669**(1) (2007) 519.
4. **IceCube** Collaboration, R. Abbasi, *et al.*, "IceCube sensitivity for low-energy neutrinos from nearby supernovae," *Astron. Astrophys.* **535** (2011) A109, arXiv:1108.0171 [astro-ph.HE]. [Erratum: *ibid.* **563** (2014) C1].
5. **IceCube** Collaboration, M.G. Aartsen, *et al.*, "Searches for extended and point-like neutrino sources with four years of IceCube data," *Astrophys. J.* **796**(2) (2014) 109, arXiv:1406.6757 [astro-ph.HE].
6. A. Kappes, J. Hinton, C. Stegmann, and F.A. Aharonian, "Potential neutrino signals from galactic gamma-ray sources," *Astrophys. J.* **656** (2007) 870–896, arXiv:astro-ph/0607286 [astro-ph]. [Erratum: *ibid.* **661** (2007) 1348].
7. **KM3NeT** Collaboration, P. Sapienza, A. Trovato, and R. Coniglione, "KM3NeT and galactic point-like sources," *Nucl. Phys. Proc. Suppl.* **237–238** (2013) 246–249.
8. E. Waxman and A. Loeb, "TeV neutrinos and GeV photons from shock breakout in supernovae," *Phys. Rev. Lett.* **87** (2001) 071101, arXiv:astro-ph/0102317 [astro-ph].

9. T.J. Moriya, K. Maeda, F. Taddia, J. Sollerman, S.I. Blinnikov, and E.I. Sorokina, "Mass-loss histories of Type IIn supernova progenitors within decades before their explosion," *Mon. Not. Roy. Astron. Soc.* **439**(3) (2014) 2917–2926, arXiv:1401.4893 [astro-ph.SR].

10. K. Murase, T.A. Thompson, B.C. Lacki, and J.F. Beacom, "New class of high-energy transients from crashes of supernova ejecta with massive circumstellar material shells," *Phys. Rev.* **D84** (2011) 043003, arXiv:1012.2834 [astro-ph.HE].

11. B. Katz, N. Sapir, and E. Waxman, "X-rays, gamma-rays and neutrinos from collisoinless shocks in supernova wind breakouts," arXiv:1106.1898 [astro-ph.HE].

12. K. Kashiyama, K. Murase, S. Horiuchi, S. Gao, and P. Meszaros, "High energy neutrino and gamma ray transients from relativistic supernova shock breakouts," *Astrophys. J.* **769** (2013) L6, arXiv:1210.8147 [astro-ph.HE].

13. V.N. Zirakashvili and V.S. Ptuskin, "Type IIn supernovae as sources of high energy astrophysical neutrinos," arXiv:1510.08387 [astro-ph.HE].

14. P. Mészáros, "Gamma ray bursts as neutrino sources," in *Noutrino Astronomy: Current Status, Future Prospects* (World Scientific, 2017), arXiv:1511.01396 [astro-ph.HE].

15. E. Waxman and J.N. Bahcall, "High-energy neutrinos from cosmological gamma-ray burst fireballs," *Phys. Rev. Lett.* **78** (1997) 2292–2295, arXiv:astro-ph/9701231 [astro-ph].

16. **IceCube** Collaboration, R. Abbasi, *et al.*, "An absence of neutrinos associated with cosmic-ray acceleration in $\gamma$-ray bursts," *Nature* **484** (2012) 351–353, arXiv:1204.4219 [astro-ph.HE].

17. Z. Li, "Note on the normalization of predicted GRB neutrino flux," *Phys. Rev.* **D85** (2012) 027301, arXiv:1112.2240 [astro-ph.HE].

18. N. Senno, K. Murase, and P. Meszaros, "Choked jets and low-luminosity gamma-ray bursts as hidden neutrino sources," arXiv:1512.08513 [astro-ph.HE].

19. **IceCube** Collaboration, R. Abbasi, *et al.*, "Constraints on high-energy neutrino emission from SN 2008D," *Astron. Astrophys.* **527** (2011) A28, arXiv:1101.3942 [astro-ph.HE].

20. K. Murase and K. Ioka, "TeV–PeV neutrinos from low-power gamma-ray burst jets inside stars," *Phys. Rev. Lett.* **111**(12) (2013) 121102, arXiv:1306.2274 [astro-ph.HE].

21. **IceCube, ROTSE** Collaboration, R. Abbasi, *et al.*, "Searching for soft relativistic jets in core-collapse supernovae with the IceCube optical follow-up program," *Astron. Astrophys.* **539** (2012) A60, arXiv:1111.7030 [astro-ph.HE].

22. S. Ando and J.F. Beacom, "Revealing the supernova-gamma-ray burst connection with TeV neutrinos," *Phys. Rev. Lett.* **95** (2005) 061103, arXiv:astro-ph/0502521 [astro-ph].

23. **IceCube** Collaboration, M.G. Aartsen, *et al.*, "Search for transient astrophysical neutrino emission with IceCube-DeepCore," arXiv:1509.05029 [astro-ph.HE].

24. A. Venkatesan, M.C. Miller, and A.V. Olinto, "Constraints on the production of ultrahigh-energy cosmic rays by isolated neutron stars," *Astrophys. J.* **484** (1997) 323–328, arXiv:astro-ph/9612210 [astro-ph].

25. K. Fang, K. Kotera, and A.V. Olinto, "Newly-born pulsars as sources of ultrahigh energy cosmic rays," *Astrophys. J.* **750** (2012) 118, arXiv:1201.5197 [astro-ph.HE].

26. K. Fang, K. Kotera, K. Murase, and A.V. Olinto, "IceCube constraints on fast-spinning pulsars as high-energy neutrino sources," arXiv:1511.08518 [astro-ph.HE].

27. **IceCube** Collaboration, M.G. Aartsen, *et al.*, "Search for prompt neutrino emission from gamma-ray bursts with IceCube," *Astrophys. J.* **805**(1) (2015) L5, arXiv:1412.6510 [astro-ph.HE].

28. D.F. Cowen, A. Franckowiak, and M. Kowalski, "Estimating the explosion time of core-collapse supernovae from their optical light curves," *Astropart. Phys.* **33** (2010) 19–23, arXiv:0901.4877 [astro-ph.HE].

29. B.J. Shappee, *et al.*, "The man behind the curtain: X-rays drive the UV through NIR variability in the 2013 AGN outburst in NGC 2617," *Astrophys. J.* **788** (2014) 48, arXiv:1310.2241 [astro-ph.HE].

30. A. Rau, *et al.*, "Exploring the optical transient sky with the Palomar Transient Factory," *Publ. Astron. Soc. Pac.* **121** (2009) 1334–1351, arXiv:0906.5355 [astro-ph.CO].

31. E.A. Magnier, *et al.*, "The Pan-STARRS 1 Photometric Reference Ladder, Release 12.0," *Astrophys. J. Supp.* **205** (2013) 20, arXiv:1303.3634 [astro-ph.IM].

32. M. Kowalski and A. Mohr, "Detecting neutrino-transients with optical follow-up observations," *Astropart. Phys.* **27** (2007) 533–538, arXiv:astro-ph/0701618 [astro-ph].

33. **Zadko, TAROT, ROTSE, Swift, ANTARES** Collaboration, S. Adrian-Martinez *et al.*, "Optical and X-ray early follow-up of ANTARES neutrino alerts," arXiv:1508.01180 [astro-ph.HE].

34. **IceCube, Pan-STARRS1 Science Consortium, Swift, PTF** Collaboration, M. G. Aartsen *et al.*, "Detection of a Type IIn supernova in optical follow-up observations of IceCube neutrino events," *Astrophys. J.* **811**(1) (2015) 52, arXiv:1506.03115 [astro-ph.HE].

35. P.A. Evans, *et al.*, "Swift follow-up of IceCube triggers, and implications for the Advanced-LIGO era," *Mon. Not. Roy. Astron. Soc.* **448**(3) (2015) 2210–2223, arXiv:1501.04435 [astro-ph.HE].

36. **Fermi-LAT** Collaboration, M. Ackermann, *et al.*, "Search for early gamma-ray production in supernovae located in a dense circumstellar medium with the Fermi LAT," *Astrophys. J.* **807** (2015) 169, arXiv:1506.01647 [astro-ph.HE].

37. G.G. Raffelt, "Supernova neutrino observations: What can we learn?," *Nucl. Phys. Proc. Suppl.* **221** (2011) 218–229, arXiv:astro-ph/0701677 [astro-ph].

38. A. Dighe, "Physics potential of future supernova neutrino observations," *J. Phys. Conf. Ser.* **136** (2008) 022041, arXiv:0809.2977 [hep-ph].

39. M. Salathe, M. Ribordy, and L. Demirors, "Novel technique for supernova detection with IceCube," *Astropart. Phys.* **35** (2012) 485–494, arXiv:1106.1937 [astro-ph.IM].

40. **KM3NeT** Collaboration, A. Leisos, A.G. Tsirigotis, and S.E. Tzamarias, "A feasibility study for the detection of supernova explosions with an undersea neutrino telescope," arXiv:1201.5726 [astro-ph.IM].

41. S. Ando, J.F. Beacom, and H.Yüksel, "Detection of neutrinos from supernovae in nearby galaxies," *Phys. Rev. Lett.* **95**(17) (2005) 171101.

42. M.D. Kistler, H. Yüksel, S. Ando, J.F. Beacom, and Y. Suzuki, "Core-collapse astrophysics with a five-megaton neutrino detector," *Phys. Rev.* **D83** (2011) 123008, arXiv:astro-ph/0810.1959 [astro-ph].

43. S. Böser, M. Kowalski, L. Schulte, N.L. Strotjohann, and M. Voge, "Detecting extragalactic supernova neutrinos in the Antarctic ice," *Astropart. Phys.* **62** (2015) 54–65, arXiv:1304.2553 [astro-ph.IM].

44. L. Yang and C. Lunardini, "Revealing local failed supernovae with neutrino telescopes," (2011), arXiv:astro-ph/1103.4628 [astro-ph.CO].

45. Y. Sekiguchi, K. Kiuchi, K. Kyotoku, and M. Shibata, "Gravitational waves and neutrino emission from the merger of binary neutron stars," (2011), arXiv:gr-qc/1105.2125 [gr-qc].

46. B. Dasgupta, *et al.*, "Detecting the QCD phase transition in the next galactic supernova neutrino burst," *Phys. Rev.* **D81** (2010) 103005, arXiv:0912.2568 [astro-ph].

47. K. Abe, T. Abe, and H.E.A. Aihara, "Letter of intent: The Hyper-Kamiokande experiment — detector design and physics potential," (2011), arXiv:1109.3262 [hep-ex].
48. **IceCube PINGU** Collaboration, M.G. Aartsen, *et al.*, "Letter of intent: The precision iceCube next generation upgrade (PINGU)," arXiv:1401.2046 [physics. ins-det].

# Chapter 11

# The Quest for Dark Matter with Neutrino Telescopes

Carlos Pérez de los Heros

*Department of Physics and Astronomy,*
*Uppsala University, Uppsala, Sweden.*
*cph@physics.uu.se*

There should be no doubt by now that neutrino telescopes are competitive instruments when it comes to searches for dark matter. Their large detector volumes collect hundreds of neutrinos per day. They scrutinize the whole sky continuously, being sensitive to neutrino signals of all flavours from dark matter annihilations in nearby objects (Sun, Earth, Milky Way Center and Halo) as well as from far away galaxies or galaxy clusters, and over a wide energy range. In this review we summarize the analysis techniques and recent results on dark matter searches from the neutrino telescopes currently in operation.

## 1. Introduction

The need for a dark matter component in the universe is now overwhelming, but so far the indications arise from gravitational effects only: rotation curves of galaxies, gravitational lensing in clusters of galaxies or structure formation seeded by density fluctuations in the early universe, as derived from cosmic microwave background measurements (see e.g. Ref. 1). The fact is that dark matter must contribute to the energy budget of the universe approximately five times more than ordinary matter. Concrete evidence for any particular type of dark matter is, though, still lacking. Searches for dark matter are usually focused on scenarios where the candidates consist of stable relic particles whose present density is determined by the thermal history of the early universe. Such an approach is further justified by the fact that extensions of the Standard Model predict the existence of particles with the right interaction strength and quantum numbers required of a generic candidate for dark matter. The circle thus closes, and our theories of the smallest components of matter connect seamlessly with our understanding of the evolution of the universe at grand scales: a beautiful aspect which is hard to ignore.

In practice the particle dark matter paradigm just needs a stable (or sufficiently long-lived) massive particle with weak interactions, generically called a WIMP (Weakly Interacting Massive Particle). The mass and the couplings of WIMPs with baryonic matter are free parameters in the context of the astrophysical problem to be solved. These quantities are specified by the underlying particle physics theory and need to be determined experimentally. Good WIMP candidates arise in Super-symmetry:[2,3] from the neutralino in the Minimal Supersymmetric Standard Model (MSSM)[4,5] to the lightest particle in models with extra dimensions,[6,7] or models with R-parity violation where an unstable gravitino is the dark matter candidate. A feature of gravitino dark matter is that it would leave no signal in direct-detection experiments since the cross-section for the interaction between a gravitino and baryonic matter is suppressed by the Planck mass to the fourth power.[8,9]

However, the lack of evidence so far for supersymmetry from direct searches, first at LEP,[10] then at the Tevatron[11] and more recently at the LHC,[12] has restricted the allowed phase space of the theory and raised the supersymmetric particle mass scale to the $\mathcal{O}(\text{TeV})$ region. There are other flavours of supersymmetry, like the phenomenological MSSM (pMSSM)[13] or the Next-to-Minimal Supersymmetric Standard Model (NMSSM)[14] which can still accommodate WIMPs down to the few GeV region.

There are extensive reviews in the literature about the particle physics connection of the dark matter problem from the theoretical/phenomenological point of view.[15-19] This being a neutrino astronomy issue, we will concentrate on reviewing the latest experimental results from neutrino telescopes.

## 2. Indirect dark matter searches with neutrino telescopes

The riveting possibility of detecting dark matter with neutrino telescopes is based on the fact that it can be gravitationally trapped in the deep gravitational wells of heavy objects, like the Sun, the Earth and the halos of galaxies. Since dark matter candidates are electrically neutral they can be their own antiparticles, and subsequent pair-wise annihilation into Standard Model particles could lead to a detectable neutrino flux. This is a clear signal for a neutrino telescope: it is directional and has a different energy spectrum than the known atmospheric neutrino background flux. For such scenario to be viable, the WIMP must have some type of interaction with baryonic matter (for the capture in heavy objects to occur) and have a certain level of annihilation cross section (for a neutrino signal to be produced). Dark matter candidates might not be their own antiparticles. In this scenario, called "asymmetric dark matter",[20,21] the current dark matter halos would be populated with just dark matter (that is, not dark anti-matter) and no annihilation would then be taking place. This picture needs a mechanism to produce only dark-matter at the beginning of the universe, in a way similar to baryogenesis, and it would be bad news for indirect searches.[22]

The actual spectrum of neutrinos detectable at the Earth depends on the underlying particle physics model used to describe the WIMP, through the branching ratios to different final states.[23-25] Neutrinos arise from the annihilation of WIMPs into quarks, charged leptons or gauge bosons. Hadronization or decays of the annihilation products will produce a neutrino flux with an energy dependence determined by the annihilation channel. Note that highly energetic neutrinos (above a few TeV) produced from the products of annihilations of WIMPs inside the Sun will undergo neutral and charged current interactions in the dense solar interior on their way out, and the resulting outgoing flux is skewed to lower energies with respect to the original spectrum. That is not the case for searches from the center of the Earth or the galaxy, where the amount of material is not enough to distort the original annihilation spectrum.

## 3. Current experiments

There are currently three large-scale underwater/ice Cherenkov neutrino telescopes in operation, IceCube at the South Pole,[26] ANTARES off the coast of Toulon, France[27] and Baikal, in Lake Baikal, Russia.[28] These are open-volume detectors in the sense that the instrumentation is deployed into a huge volume of naturally occurring water or ice. Such an approach is the most cost-effective to instrument volumes of $\mathcal{O}(\mathrm{km}^3)$. At a lower scale, we have Super-Kamiokande[29] using 50 kT of ultra-pure water in a vessel located 1.000 m underground in the Mozumi mine in Kamioka. The detection method is similar for all these detectors: neutrinos are detected by the Cherenkov light emitted by secondary particles produced in a neutrino interaction inside or near the detector. The relative timing of the signals in the photo-multiplier tubes that surround the detector volume makes it possible to reconstruct the direction of the original neutrino, and the amount of light deposited is related to the neutrino energy. Also underground, but using scintillator instead of water, is the Baksan array,[30] situated in the North Caucasus at a depth of 850 meters of water equivalent. It consists of several planes of scintillators distributed in a four-storey 17 m×17 m×11 m cavern. The detector reconstructs upgoing muons measuring the time-of-flight through the detector planes. Baksan is the detector that has been running for the longest time, since 1977.

The main background of any analysis with neutrino telescopes is the overwhelming flux of muons produced in cosmic-ray interactions in the atmosphere, *atmospheric muons*. The same interactions produce a flux of neutrinos, *atmospheric neutrinos* which constitute an irreducible background to any search for new physics. Atmospheric muons can be filtered out by using the Earth as a filter, at the expense of reducing the sky coverage of the instrument. Full-sky coverage can be regained by defining a "veto region" in order to tag incoming tracks, most probably an atmospheric muon, and define "starting events", which must have been produced by a neutrino interaction inside the detector. This comes at the price of reducing the effective volume of the detector and having a somewhat different energy response for upgoing and downgoing events. For high enough neutrino energies ($\mathcal{O}(10)$ TeV), the possibility exists of rejecting atmospheric neutrinos when accompanied by a muon produced in the same air shower.[31]

## 4. Searches for dark matter from the Sun and Earth

WIMPs that may have accumulated gravitationally during the lifetime of the solar system in the center of the Sun or Earth can annihilate to produce a measurable neutrino flux.[32-41] While any other product of the annihilation will be absorbed, neutrinos will not, and a neutrino telescope can "look" inside the Sun or the Earth for a signal of such annihilations. The strength of the expected neutrino flux depends on several factors, not least on the inter-relationship between the capture rate of WIMPs, $\Gamma_C$, proportional to the WIMP–nucleon cross section, and the annihilation rate, $\Gamma_A$, proportional to the velocity averaged WIMP–WIMP annihilation cross section. WIMPs will in general have spin and can then interact with baryonic matter through a spin-dependent and a spin-independent coupling, which arise from axial and scalar terms in the Lagrangian, respectively.[42] Since the Sun is primarily a proton target (75% of H and 24% of He in mass)[43] the capture of WIMPs from the halo can be considered to be driven mainly via the spin-dependent scattering. Other, heavier, elements constitute less than 2% of the mass of the Sun, but can still play a role when considering spin-independent capture since the spin-independent cross section, $\sigma_{SI}$, is proportional to the square of the atomic mass number. Recent studies[44] have, however, pointed out the importance of heavier elements on the spin-dependent capture process in models with momentum-dependent interactions.

For the Earth the situation is rather different. The most abundant isotopes of the main components of the Earth's inner core, mantle and crust are spin 0 nuclei, $^{56}$Fe, $^{28}$Si and $^{16}$O.[45] Furthermore, the escape velocity from the center of the Earth is just 14.8 km/s. These values lie at the lower tail of the expected local WIMP velocity distribution, which is assumed to have a mean of the order of 220 km/s. Taken at face value, the Earth would appear to be very inefficient in trapping dark matter particles from the halo. But the composition of the Earth comes to the rescue, at least for WIMP masses which are resonant with the atomic mass of the main components of the Earth,[38] which favours the capture of relatively low-mass WIMPs ($m_\chi \lesssim 50$ GeV).

Indeed, the number of WIMP annihilations in the Sun or Earth, N, varies with time as $\dot{N} = \Gamma_C - \Gamma_A N^2/2$. An additional evaporation term from WIMP–nucleus scattering could be included, but it is negligible for WIMP masses above a few GeV.[33,46] Given the age of the Sun (4.5 Gyr), the estimated local dark matter density ($\rho_{\text{local}} \sim 0.4$ GeV/cm$^3$) and a weak-scale interaction between dark matter and baryons, many models predict that dark matter capture and annihilation in the Sun have reached equilibrium. Thus, annihilation is at its maximum possible value, $\Gamma_A = \Gamma_C/2$.

Experimentally, what a neutrino telescope measures are muons and particle showers from neutral and charged-current neutrino interactions inside or near the detector. We can relate the WIMP annihilation rate $\Gamma_A$ in the Earth or

Sun and the neutrino flux at the detector, $\Phi_\nu$, above a given energy threshold $E_{\text{thr}}$ as

$$\Phi_\nu(E_\nu) = \frac{dN_\nu}{dE_\nu dA dt} = \frac{\Gamma_A}{4\pi D^2} \int_{E_{\text{thr}}}^{\infty} dE'_\nu \epsilon(E'_\nu; E_\nu) \left(\frac{dN_\nu}{dE'_\nu}\right) \tag{1}$$

where $dN_\nu/dE'_\nu$ is the all-flavour neutrino flux at the center of the source and $\epsilon(E'_\nu; E_\nu)$ is a factor that takes into account the probability for a neutrino of energy $E'_\nu$ to loose energy on its way out of the Sun/Earth and reach the detector with an energy $E_\nu < E'_\nu$. This factor is not needed for the Earth case, but it becomes relevant for the dense interior of the Sun. Most of the experimental searches use the muon channel, since it gives better pointing and, in the end, dark matter searches from the Sun or Earth are really point-source searches. In that case an additional factor, $P^{i,\mu}_{\text{osc}}$, the oscillation probability of flavour $i$ to a muon neutrino, needs to be factored in.

In practice, the figure of merit of neutrino telescopes is the effective area, $A^{\text{eff}}_\nu$, the equivalent area with which it would detect a neutrino with 100% efficiency. The effective area is energy dependent and much smaller than the geometrical area of the detector, since it includes the neutrino–nucleon cross section and trigger and analysis efficiencies. It can only be calculated with the help of detailed detector Monte Carlo simulations and it is always given for a specific analysis and signal type. Equation (1) can be then translated into the more familiar form for the number of events expected at the detector from a neutrino flux $dN_\nu/dE_\nu dA dt$ produced at the source,

$$N_\nu = T_{\text{live}} \int_{E^{\text{thr}}_\nu}^{\infty} dE_\nu A^{\text{eff}}_\nu(E_\nu) \frac{dN_\nu}{dE_\nu dA dt} \tag{2}$$

where $T_{\text{live}}$ is the exposure time of the detector, and now the dependence on the annihilation rate at the source is incorporated in the calculation of the neutrino flux. Under the assumption that the capture rate is fully dominated either by the spin-dependent or spin-independent scattering, conservative limits can be extracted on either the spin-dependent or spin-independent WIMP–proton cross section from the limit on $\Gamma_A$.[47] Cross sections are useful quantities since they allow an easy comparison with the results of direct searches or predictions of a specific particle physics model. Such conversion introduces an additional systematic uncertainty in the calculation of the cross sections, due to the element composition of the Sun or Earth, the effect of planets on the capture of WIMPS from the halo[48] and nuclear form factors used in the capture calculations.[42,49–51]

### 4.1. *Current results*

Searches for dark matter accumulated in the center of the Sun have been carried out by ANTARES,[52,53] Baikal,[54] Baksan,[55] Super-K[56] and IceCube,[57–59] and a

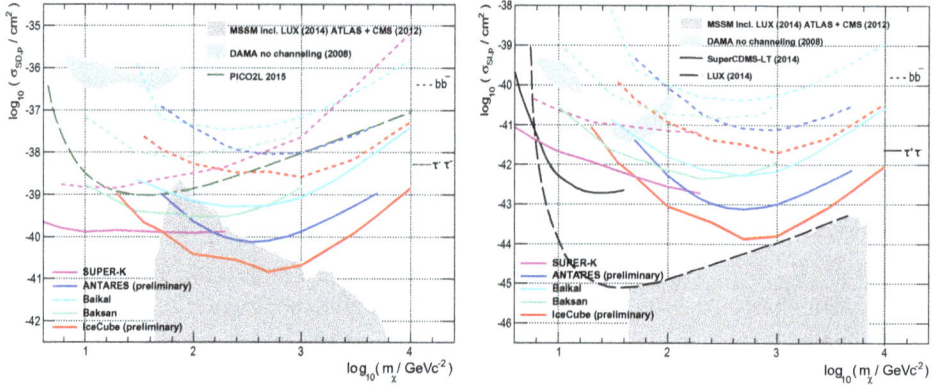

Fig. 1. Upper limits at 90% CL on the spin-dependent (left) and spin-independent (right) WIMP-proton cross section as a function of WIMP mass. Limits from IceCube,[60] Super-K,[56] ANTARES,[53] Baikal[54] and Baksan[55] are shown. Full lines refer to limits on the $\tau^+\tau^-$ annihilation channel and dashed lines to the $b\bar{b}$ channel. Direct search results from PICO[61] and LUX,[62] and tentative signal regions[63-65] (gray-shaded areas) are included for comparison. The brown-shaded region indicates the parameter space from a 25-parameter MSSM scan.[66]

summary of their results is shown in Fig. 1. Since the exact branching ratios of WIMP annihilation into different channels is model-dependent, experiments usually choose two annihilation channels which give extreme neutrino spectra to show their results. Annihilation into $b\bar{b}$ is chosen as a representative case producing a soft neutrino spectrum, and annihilation into $W^+W^-$ or $\tau^+\tau^-$ as a hard spectrum. Assuming a 100% branching ratio to each of these channels brackets the expected neutrino spectrum from any more realistic model with branching to more channels. The full and dashed curves in Fig. 1 illustrate this. Since large-volume neutrino telescopes are high-energy neutrino detectors, the sensitivity increases by more than an order of magnitude between the "soft" and "hard" spectra and, in both cases, it decreases rapidly with decreasing WIMP mass (softer neutrino spectra). The limits for the spin-dependent cross section are competitive with direct searches. Direct search experiments do not reach cross section values below $10^{-39}$ cm$^2$ at their best point, worsening rapidly away from it. IceCube or SuperK reach bounds at the $\sim 10^{-40}$–$10^{-41}$ cm$^2$ level, covering between the two experiments the WIMP mass range from a few GeV to 100 TeV.

The picture changes dramatically when we consider spin-independent limits. Here direct-search experiments have the advantage of dedicated spinless targets, and the limits from neutrino telescopes lie about three orders of magnitude above the best limit from Lux at a WIMP mass of about 50 GeV. The situation improves slightly for higher masses. But even though the limits of direct experiments worsen rapidly away from the resonance interaction with their target nucleus, the limits from neutrino telescopes on the spin-independent cross section lie above over the whole mass range.

## 5. Searches for dark matter from galaxies

### 5.1. *The halo issue*

The accepted scenario for the formation of cosmic structures assumes the formation of regions of increased dark matter density through gravitational collapse from primordial density fluctuations, which in turn attract atomic gas, seeding the formation of galaxies.[69] This scenario favours cold (or warm) dark matter over a relativistic species, since in the latter case the formation of large-scale structures would have been suppressed and we would not recover the observed universe. But, in order to predict the rate of annihilation of dark matter particles in galactic halos, the precise size and shape of the halo is of paramount importance. There is still some controversy on how dark halos evolve and what shape they have, which is reflected in the different parametrizations of the dark matter density around visible galaxies that are commonly used in the literature: the Navarro-Frenk-White (NFW) profile,[70] the Kravtsov profile,[71] the Moore profile[72] and the Burkert[73] profile being the most popular ones. The common feature of these profiles is a denser spherically symmetric region of dark matter in the center of the galaxy, with decreasing density as the radial distance to the center increases. Where they diverge is in the predicted shape of the central region. Simulations of galaxy formation and evolution are very time consuming and complex in nature, and have not been determinant in settling the issue. Profiles obtained from $N$-body simulations tend to predict a steep power-law type behaviour of the dark matter component in the inner parts of the halo, while profiles based on observational data (stellar velocity fields) tend to favour a constant dark matter density near the core of the galaxies. This is the core-cusp problem,[74] and it is an unresolved issue which affects the signal prediction from dark matter annihilations in neutrino telescopes. A general parametrization of the dark halo in galaxies can be found in.[75,76] Note that the shape of the dark halo can depend on the local characteristics of any given galaxy, like the size of the galaxy[77] or on its evolution history.[78,79]

The shape of the dark matter halo is important because the expected annihilation signal depends on the line-of-sight (l.o.s.) integral from the observation point (the Earth) to the source, and involves an integration over the square of the dark matter density. This is included in the so-called J-factor,[76,80] which is galaxy-dependent, and absorbs all the assumptions on the shape of the specific halo being considered. In the case of our galaxy, the expected signal from the galactic center using one halo parametrization or another can differ by orders of magnitude depending on the halo model used (see e.g. Fig. 1 in Ref. 76)

The differential neutrino flux from dark matter annihilation from a given galaxy, $d\phi_\nu/dE$, depends on the candidate dark matter mass, $m_{\mathrm{WIMP}}$, the neutrino energy spectrum, $dN_\nu/dE$, the thermally averaged product of the self-annihilation cross section, $\sigma_A$, times the WIMP velocity, $v$, and the J-factor, $J_\Psi$,

$$\frac{d\phi_\nu}{dE} = \frac{1}{2} \frac{\langle\sigma_A v\rangle}{4\pi m_{\mathrm{WIMP}}^2} J_\Psi \frac{dN_\nu}{dE} \qquad (3)$$

Since there is not yet a consensus on the distribution and shape of the dark halos in galaxies, neutrino telescopes usually present their results for a few benchmark halos. In this way the effect of different halo assumptions is factorized from other uncertainties, like detector systematics or the choice of the underlying particle physics model.

### 5.2. *Observable candidates*

The largest gravitational potential close to the solar system is the center of our own galaxy. Further away, we find dwarf galaxies: small, low-brightness galaxies orbiting the center of the Milky Way as remnants from our galaxy's formation process. A common feature of dwarf galaxies is that they have a low mass-to-light ratio, suggesting the presence of large amounts of dark matter.[81] Since these galaxies are small and simple in their structure, consisting of a small number of stars, they do not present any background from violent processes that could mask a signal from dark matter annihilation. The detection sensitivity can be further increased by stacking objects with similar characteristics. At cosmological scales, our closest galaxy, Andromeda, and galaxy clusters are other obvious candidates for dark matter detection. Andromeda is a spiral galaxy with a relatively well characterized dark matter halo,[82] while galaxy clusters are the largest gravitationally bound systems known and present an estimated 85% of dark matter in comparison with about 3% of luminous matter, the rest consisting of intracluster gas.[83]

### 5.3. *Current results*

Searches for dark matter from our own galaxy, dwarf galaxies, Andromeda and galaxy clusters have been carried our by IceCube[68,84–86] and ANTARES,[53,67] and are shown in Figs. 2 and 3. The limits obtained depend strongly on the studied galaxy, through the J-factor, and the assumed annihilation channel. All sources considered showed results compatible with the background-only hypothesis yielding limits on the velocity-averaged annihilation cross section at the level of $10^{-20}$ cm$^3$ s$^{-1}$–$10^{-23}$ cm$^3$ s$^{-1}$, depending on assumptions and WIMP mass.

## 6. The IceCube PeV events and dark matter

The recent discovery of a ultra high-energy astrophysical neutrino flux by IceCube[87–90] has triggered an intense theoretical activity trying to explain its possible origin. The still low statistics, 54 events detected in 1347 days of livetime with a background of atmospheric neutrinos and muons of about 21 events, and the relatively poor angular resolution of the cascade events, make it difficult to assign an

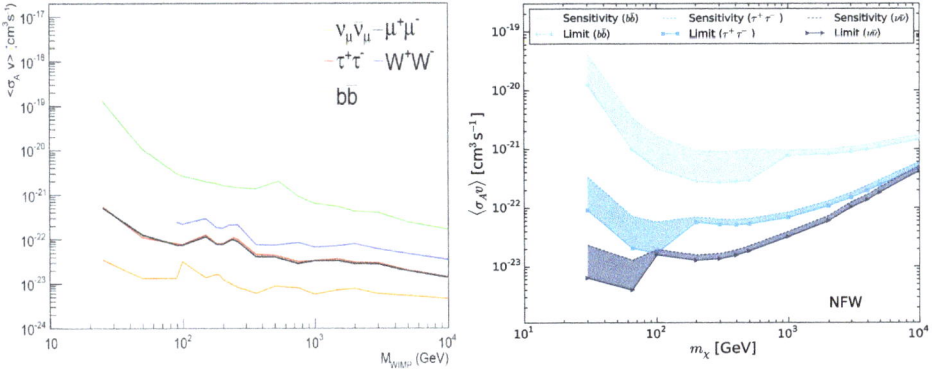

Fig. 2. Upper limits at 90% CL on the velocity-averaged WIMP annihilation cross section from ANTARES (left)[67] and IceCube (right).[68] The limits were obtained from analyses on the Galactic Center. The different curves show different annihilation channels, assuming a 100% branching ratio to each. The IceCube curves show the sensitivity (dashed lines) as well as the observed upper limits (solid lines). The shaded areas are to guide the eye between a sensitivity and its corresponding limit. All limits were obtained assuming an NFW-type halo profile.

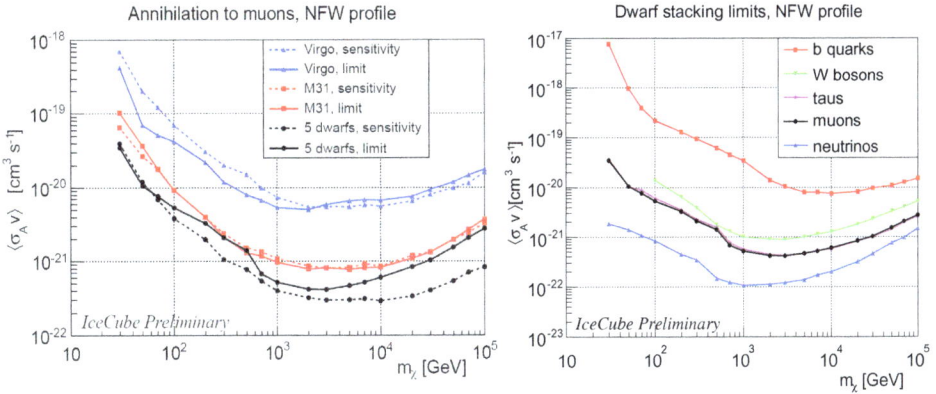

Fig. 3. Upper limits at 90% CL on the velocity-averaged WIMP annihilation cross section from IceCube. **Left:** For the Virgo cluster, the Andromeda galaxy and the stacking of five dwarf galaxies (Segue 1, Ursa Major II, Willman 1, Coma Berenices and Draco), assuming 100% annihilation to muons and a NFW profile. **Right:** For different benchmark annihilation channels for the stacking of the mentioned five dwarf galaxies. Figures taken from Ref. 86.

origin to the events. The arrival directions of the events are compatible with an isotropic flux, maybe with a small galactic component. Many proposed explanations are based on astrophysical processes, but there has been also a series of works pointing at the possible origin of the events as originating from heavy dark matter decay.[91–98] Since the rate of dark matter decay is proportional to the dark matter density, and not to the density squared as for the annihilation case, the resulting neutrino flux can easily accommodate both a galactic and a diffuse extra-galactic component of comparable strength, which is consistent with the distribution of the

IceCube events. In the case of annihilations, a galactic component could easily dominate due to the closeness of our galaxy. Another feature of dark matter decay is a monochromatic neutrino line at $m_{\mathrm{WIMP}}/2$, in models with a dominant branching ratio to the neutrino channel. However, and more realistically, other final states will also contribute to the final neutrino spectrum giving a lower-energy continuum from decays into quarks and charged leptons. This is also compatible with the energy distribution of the IceCube events (see Fig. 4) which can be interpreted as presenting a peak at around 2 PeV and a lower energy continuum, separated by a dip just below 1 PeV. These are the features of the IceCube results that have triggered the explanations based on decaying of heavy dark matter candidates (except for the analysis in Ref. 94, which concentrates on candidates in the $\mathcal{O}(100)$ TeV range). One can further argue that considering a heavy dark matter candidate is timely in view of the lack of evidence of new physics at the TeV scale from the LHC, which has put some strain on the vanilla WIMP paradigm with dark matter candidates on the $\mathcal{O}(100)$ GeV-TeV region.

The exact nature of the potential heavy dark matter sector is however completely open in view of the IceCube data, which is not constraining enough at this moment. The models proposed range from "neutrinophilic" models where the dark matter predominantly decays into a $\nu\bar{\nu}$ pair, e.g., Refs. 98 and 103, to boosted dark matter scenarios, e.g., Ref. 104 or rather model-independent analyses like in Ref. 94, or with minimal additions to the see-saw model in Ref. 95. Many authors assume the existence of an astrophysical power-law diffuse component in addition to the dark matter component. Such combination can compensate the slight tension between the IceCube measured flux and a pure power-law assumption, although such a tension can be mitigated by considering a power-law with a cutoff. In any case, the results

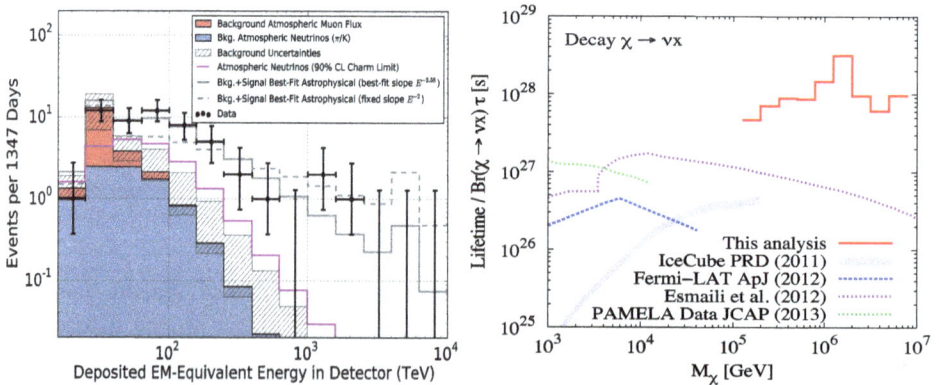

Fig. 4.   **Left:** Estimated energy deposited in the detector of the 4-year sample of ultra-high energetic IceCube events.[90] **Right:** Limit on the lifetime of a super-heavy dark matter candidate derived using the high-energy neutrino flux observed by IceCube (red line), compared to the previous experimental constraints from IceCube,[99] Fermi-LAT,[100] PAMELA[101] and derived limits from neutrino data.[102] Excluded are regions below the pictured lines. Figure from Ref. 95.

from the different analyses of the IceCube data tend to concur on a limit on the lifetime of a generic dark matter candidate at the level of $\geqslant \mathcal{O}(10^{27})$ s.

Decaying dark matter has also been proposed as an explanation of the observed $e^+$ excess in cosmic rays,[105–108] although a slight fine tuning of the models towards leptophilic dark matter candidates (decaying preferably to charged leptons rather than to quarks) seems necessary, since new-physics in the proton spectrum is strongly constrained by data from the same detectors.[109,110] The IceCube data can provide an escape route by confirming a neutrino annihilation channel while complying with current limits on annihilation to charged leptons from cosmic-ray detectors. But we must wait for more events in IceCube.

## 7. Outlook

Naturally, any model of dark matter producing a neutrino flux must be viewed in the grand scheme of dark matter searches and be consistent with limits from the $\gamma$ and cosmic-ray channels,[111–113] as well as with constraints from direct[62,64,114,115] and accelerator searches.[12] The near future will bring us more events from the neutrino telescopes in operation, as well as the next-generation, large-mass direct search experiments[116,117] and the directional detection efforts,[118–120] which will provide a quantitative jump in the dark matter search paradigm. Neutrino telescopes are also planning the next generation arrays, PINGU[121] and ORCA[122] on the low-energy side, and the high-energy extension of IceCube,[123] KM3NET[124] and GVD[125] at the high, energy end. There is both overlap and complementarity in the WIMP mass range and annihilation channels covered by all these experiments. It is just left to Nature to reveal what solution she has chosen as dark matter.

## Acknowledgments

I am indebted to M. Rameez, J. Zornoza, C. Tönnis and S. Demidov for kindly providing some of the data shown in Fig. 1.

## References

1. L. Bergström, Nonbaryonic dark matter: Observational evidence and detection methods, *Rept. Prog. Phys.* **63** (2000) 793.
2. H.P. Nilles, Supersymmetry, supergravity and particle physics, *Phys. Rept.* **110** (1984) 1–162.
3. S.P. Martin, A Supersymmetry primer, arXiv:hep-ph/9709356. (2011). [*Adv. Ser. Direct. High Energy Phys.* **18** (1998) 1].
4. A. Djouadi, *et al.*, The minimal supersymmetric standard model: Group summary report, arXiv:hep-ph/9901246. (1998).
5. M.C. Rodriguez, History of supersymmetric extensions of the Standard Model, *Int. J. Mod. Phys.* **A25** (2010) 1091–1121.
6. H.-C. Cheng, J.L. Feng, and K.T. Matchev, Kaluza–Klein dark matter, *Phys. Rev. Lett.* **89** (2002) 211301.

7. G. Bertone, G. Servant, and G. Sigl, Indirect detection of Kaluza–Klein dark matter, *Phys. Rev.* **D68** (2003) 044008.

8. L. Covi, M. Grefe, A. Ibarra, and D. Tran, Unstable gravitino dark matter and neutrino flux, *JCAP.* **0901** (2009) 029.

9. M. Grefe, Indirect searches for gravitino dark matter, *J. Phys. Conf. Ser.* **375** (2012) 012035.

10. A.C. Kraan. SUSY searches at LEP. In *Lake Louise Winter Institute 2005*, pp. 189–193, (2005).

11. D. Toback and L. Źlvković, Review of physics results from the Tevatron: Searches for new particles and interactions, *Int. J. Mod. Phys.* **A30** (06) (2015) 1541007.

12. A. Cakir Searches for beyond the Standard Model physics at the LHC: Run1 summary and Run2 prospects. In *13th Conference on Flavor Physics and CP Violation (FPCP 2015)* Nagoya, Japan, May 25–29, 2015, FPCP2015, p. 024, Nagoya, Japan. PoS.

13. C.F. Berger, J.S. Gainer, J.L. Hewett, and T.G. Rizzo, Supersymmetry without prejudice, *JHEP.* **02** (2009) 023.

14. U. Ellwanger, M. Rausch de Traubenberg, and C.A. Savoy, Particle spectrum in supersymmetric models with a gauge singlet, *Phys. Lett.* **B315** (1993) 331–337.

15. G. Jungman, M. Kamionkowski, and K. Griest, Supersymmetric dark matter, *Phys. Rept.* **267** (1996) 195–373.

16. J.L. Feng, K.T. Matchev, and F. Wilczek, Prospects for indirect detection of neutralino dark matter, *Phys. Rev.* **D63** (2001) 045024.

17. G. Bertone, D. Hooper, and J. Silk, Particle dark matter: Evidence, candidates and constraints, *Phys. Rept.* **405** (2005) 279–390.

18. J.L. Feng, Dark matter candidates from particle physics and methods of detection, *Ann. Rev. Astron. Astrophys.* **48** (2010) 495.

19. L. Bergström, Dark matter evidence, particle physics candidates and detection methods, *Annalen Phys.* **524** (2012) 479–496.

20. H. Davoudiasl and R.N. Mohapatra, On relating the genesis of cosmic baryons and dark matter, *New J. Phys.* **14** (2012) 095011.

21. K. Petraki and R.R. Volkas, Review of asymmetric dark matter, *Int. J. Mod. Phys.* **A28** (2013) 1330028.

22. M.L. Graesser, I.M. Shoemaker, and L. Vecchi, Asymmetric WIMP dark matter, *JHEP.* **1110** (2011) 110.

23. A. Bottino, N. Fornengo, G. Mignola, and L. Moscoso, Signals of neutralino dark matter from Earth and Sun, *Astropart.Phys.* **3** (1995) 65–76.

24. M. Cirelli, N. Fornengo, T. Montaruli, I.A. Sokalski, A. Strumia, *et al.*, Spectra of neutrinos from dark matter annihilations, *Nucl. Phys.* **B727** (2005) 99–138.

25. M. Blennow, J. Edsjö, and T. Ohlsson, Neutrinos from WIMP annihilations in the Sun including neutrino oscillations, *Nucl. Phys. Proc. Suppl.* **221** (2011) 37–38.

26. A. Achterberg, *et al.*, First year performance of the IceCube Neutrino Telescope, *Astropart. Phys.* **26** (2006) 155–173.

27. M. Ageron, *et al.*, ANTARES: The first undersea neutrino telescope, *Nucl. Instrum. Meth.* **A656** (2011) 11–38.

28. A. Avrorin, *et al.*, The Baikal neutrino experiment, *Nucl. Instrum. Meth.* **A626-627** (2011) S13–S18.

29. K. Abe, *et al.*, Calibration of the Super-Kamiokande detector, *Nucl. Instrum. Meth.* **A737** (2014) 253–272.

30. E. Alekseev, L. Alekseeva, V. Bakatanov, M. Boliev, A. Voevodsky, *et al.*, The Baksan underground scintillation telescope, *Phys. Part. Nucl.* **29** (1998) 254–256.

31. S. Schonert, T.K. Gaisser, E. Resconi, and O. Schulz, Vetoing atmospheric neutrinos in a high energy neutrino telescope, *Phys. Rev.* **D79** (2009) 043009.

32. W.H. Press and D.N. Spergel, Capture by the Sun of a galactic population of weakly interacting massive particles, *Astrophys. J.* **296** (1985) 679–684.

33. L.M. Krauss, M. Srednicki, and F. Wilczek, Solar system constraints and signatures for dark matter candidates, *Phys. Rev.* **D33** (1986) 2079–2083.

34. M. Srednicki, K.A. Olive, and J. Silk, High-energy neutrinos from the Sun and cold dark matter, *Nucl. Phys.* **B279** (1987) 804.

35. T. Gaisser, G. Steigman, and S. Tilav, Limits on cold dark matter candidates from deep underground detectors, *Phys. Rev.* **D34** (1986) 2206.

36. S. Ritz and D. Seckel, Detailed neutrino spectra from cold dark matter annihilations in the sun, *Nucl.Phys.* **B304** (1988) 877.

37. A. Gould, Direct and indirect capture of by the earth, *Astrophys. J.* **328** (1988) 919–939.

38. A. Gould, Resonant enhancements in WIMP capture by the Earth, *Astrophys. J.* **321** (1987) 571.

39. A. Gould, J.A. Frieman, and K. Freese, Probing the Earth with WIMPS, *Phys. Rev.* **D39** (1989) 1029.

40. L. Bergström, J. Edsjö, and P. Gondolo, Indirect detection of dark matter in km size neutrino telescopes, *Phys. Rev.* **D58** (1998) 103519.

41. J. Lundberg and J. Edsjö, WIMP diffusion in the solar system including solar depletion and its effect on earth capture rates, *Phys. Rev.* **D69** (2004) 123505.

42. J. Engel, S. Pittel, and P. Vogel, Nuclear physics of dark matter detection, *Int. J. Mod. Phys.* **E1** (1992) 1–37.

43. N. Grevesse and A.J. Sauval, Standard Solar Composition, *Space Sci. Rev.* **85** (1998) 161–174.

44. R. Catena and B. Schwabe, Form factors for dark matter capture by the Sun in effective theories, *JCAP.* **1504** 04 (2015) 042.

45. J.M. Herndon, The chemical composition of the interior shells of the Earth, *Proc. Roy. Soc. London. Series A* **372** (1748) (1980) 149.

46. K. Griest and D. Seckel, Cosmic asymmetry, neutrinos and the Sun, *Nucl. Phys.* **B283** (1987) 681.

47. G. Wikström and J. Edsjö, Limits on the WIMP–nucleon scattering cross-section from neutrino telescopes, *JCAP.* **0904** (2009) 009.

48. S. Sivertsson and J. Edsjö, WIMP diffusion in the solar system including solar WIMP–nucleon scattering, *Phys. Rev.* **D85** (2012) 123514.

49. A. Bottino, F. Donato, N. Fornengo, and S. Scopel, Implications for relic neutralinos of the theoretical uncertainties in the neutralino nucleon cross-section, *Astropart. Phys.* **13** (2000) 215–225.

50. J.R. Ellis, K.A. Olive, and C. Savage, Hadronic uncertainties in the elastic scattering of supersymmetric dark matter, *Phys. Rev.* **D77** (2008) 065026.

51. R. Ruiz de Austri and C. Pérez de los Heros, Impact of nucleon matrix element uncertainties on the interpretation of direct and indirect dark matter search results, *JCAP.* **1311** (2013) 049.

52. S. Adrian-Martinez, *et al.*, First results on dark matter annihilation in the Sun using the ANTARES neutrino telescope, *JCAP.* **1311** (2013) 032.

53. C. Tönnis, *et al.*, Overview of dark matter searches with ANTARES. In *Proc. of the 34th International Cosmic Ray Conference.* 30 July–6 August, 2015. The Hague, The Netherlands. *PoS(ICRC2015)*, p. 1207 (2015).

54. A. Avrorin, *et al.*, Search for neutrino emission from relic dark matter in the Sun with the Baikal NT200 detector, *Astropart. Phys.* **62** (2014) 12–20.

55. M. Boliev, S. Demidov, S. Mikheyev, and O. Suvorova, Search for muon signal from dark matter annihilations in the Sun with the Baksan Underground Scintillator Telescope for 24.12 years, *JCAP.* **1309** (2013) 019.

56. K. Choi, *et al.*, Search for neutrinos from annihilation of captured low-mass dark matter particles in the Sun by Super-Kamiokande, *Phys. Rev. Lett.* **114** (14) (2015) 141301.

57. M.G. Aartsen, *et al.*, Search for dark matter annihilations in the Sun using the completed IceCube neutrino telescope. In *Proc. of the 34th International Cosmic Ray Conference.* 30 July–6 August, 2015. The Hague, The Netherlands. *PoS(ICRC2015)*, (2015) p. 1209.

58. M.G. Aartsen, *et al.*, Improved methods for solar dark matter searches with the IceCube neutrino telescope. In *Proc. of the 34th International Cosmic Ray Conference.* 30 July–6 August, 2015. The Hague, The Netherlands. *PoS(ICRC2015)*, (2015) p. 1099.

59. M. Aartsen, *et al.*, Search for dark matter annihilations in the Sun with the 79-string IceCube detector, *Phys. Rev. Lett.* **110** (13) (2013) 131302.

60. T. DeYoung, *et al.*, Results from IceCube, In *Proc. of Very Large Volume Neutrino Telescope, VLnT2015,* 14–16 September 2015, Roma, Italy. To appear in *EPJ Web of Conferences.*

61. C. Amole, *et al.*, Dark matter search results from the PICO-2L $C_3F_8$ bubble chamber, *Phys. Rev. Lett.* **114** (23) (2015) 231302.

62. D. Akerib, *et al.*, First results from the LUX dark matter experiment at the Sanford Underground Research Facility, *Phys. Rev. Lett.* **112** (2014) 091303.

63. C. Savage, G. Gelmini, P. Gondolo, and K. Freese, Compatibility of DAMA/LIBRA dark matter detection with other searches, *JCAP.* **0904** (2009) 010.

64. C.E. Aalseth *et al.*, Search for an annual modulation in a p-type point contact germanium dark matter detector, *Phys. Rev. Lett.* **107** (2011) 141301.

65. R. Agnese, *et al.*, Silicon detector dark matter results from the final exposure of CDMS II, *Phys. Rev. Lett.* **111** (25) (2013) 251301.

66. H. Silverwood, P. Scott, M. Danninger, C. Savage, J. Edsjö, *et al.*, Sensitivity of IceCube-DeepCore to neutralino dark matter in the MSSM-25, *JCAP.* **1303** (2013) 027.

67. S. Adrian-Martinez, *et al.*, Search of dark matter annihilation in the galactic centre using the ANTARES neutrino telescope, arxiv:1505.04866. (2015).

68. M.G. Aartsen, *et al.*, Search for dark matter annihilation in the galactic center with IceCube-79, arxiv:1505.07259. (2015).

69. C. Knobel, An introduction into the theory of cosmological structure formation, arXiv:1208.5931. (2013).

70. J.F. Navarro, C.S. Frenk, and S.D. White, The structure of cold dark matter halos, *Astrophys. J.* **462** (1996) 563.

71. A.V. Kravtsov, A.A. Klypin, J.S. Bullock, and J.R. Primack, The cores of dark matter dominated galaxies: Theory versus observations, *Astrophys. J.* **502** (1998) 48.

72. B. Moore, T.R. Quinn, F. Governato, J. Stadel, and G. Lake, Cold collapse and the core catastrophe, *Mon. Not. Roy. Astron. Soc.* **310** (1999) 1147–1152.

73. A. Burkert, The structure of dark matter halos in dwarf galaxies, *Astrophys. J.* **171** (1996) 175.

74. W.J.G. de Blok, The core-cusp problem, *Adv. Astron.* **2010** (2010) 789293.

75. H. Zhao, Analytical dynamical models for double-power-law galactic nuclei, *Mon. Not. Roy. Astron. Soc.* **287** (1997) 525.

76. H. Yuksel, S. Horiuchi, J.F. Beacom, and S. Ando, Neutrino constraints on the dark matter total annihilation cross section, *Phys. Rev.* **D76** (2007) 123506.

77. M. Ricotti, Dependence of the inner DM profile on the halo mass, *Mon. Not. Roy. Astron. Soc.* **344** (2003) 1237.

78. A. Dekel and J. Silk, The origin of dwarf galaxies, cold dark matter, and biased galaxy formation, *Astrophys. J.* **303** (1986) 39–55.

79. G. Ogiya and M. Mori, The core-cusp problem in cold dark matter halos and supernova feedback: Effects of oscillation, *Astrophys. J.* **793** (2014) 46.

80. L. Bergström, P. Ullio, and J.H. Buckley, Observability of gamma-rays from dark matter neutralino annihilations in the Milky Way halo, *Astropart. Phys.* **9** (1998) 137–162.

81. M. Walker, Dark matter in the galactic dwarf spheroidal satellites, in *Planets, Stars and Stellar Systems: Vol. 5, Galactic Structures and Stellar Populations*, p. 1039. Springer, (2013).

82. A. Tamm, E. Tempel, P. Tenjes, O. Tihhonova, and T. Tuvikene, Stellar mass map and dark matter distribution in M31, *Astron. Astrophys.* **546** A4 (2012).

83. G.M. Voit, Tracing cosmic evolution with clusters of galaxies, *Rev. Mod. Phys.* **77** (2005) 207–258.

84. M.G. Aartsen, *et al.*, IceCube search for dark matter annihilation in nearby galaxies and galaxy clusters, *Phys. Rev.* **D88** (2013) 122001.

85. M.G. Aartsen, *et al.*, Multipole analysis of IceCube data to search for dark matter accumulated in the Galactic halo, *Eur. Phys. J.* **C75** (1) (2015) 20.

86. M.G. Aartsen, *et al.*, Searching for neutrinos from dark matter annihilations in (dwarf) galaxies and galaxy clusters with IceCube. In *Proc. of the 34th International Cosmic Ray Conference*. 30 July–6 August, 2015. The Hague, The Netherlands. PoS(ICRC2015), p. 1215, (2015).

87. M.G. Aartsen, *et al.*, First observation of PeV-energy neutrinos with IceCube, *Phys. Rev. Lett.* **111** (2013) 021103.

88. M.G. Aartsen, *et al.*, Evidence for high-energy extraterrestrial neutrinos at the IceCube detector, *Science.* **342** (2013) 1242856.

89. M.G. Aartsen, *et al.*, Observation of high-energy astrophysical neutrinos in three years of IceCube data, *Phys. Rev. Lett.* **113** (2014) 101101.

90. M.G. Aartsen, *et al.*, Observation of astrophysical neutrinos in four years of IceCube data. In *Proc. of the 34th International Cosmic Ray Conference*. 30 July–6 August, 2015. The Hague, The Netherlands. *PoS(ICRC2015)* p. 1081, (2015).

91. B. Feldstein, A. Kusenko, S. Matsumoto, and T.T. Yanagida, Neutrinos at IceCube from Heavy Decaying Dark Matter, *Phys. Rev.* **D88** (1) (2013) 015004.

92. A. Esmaili and P.D. Serpico, Are IceCube neutrinos unveiling PeV-scale decaying dark matter?, *JCAP.* **1311** (2013) 054.

93. Y. Bai, R. Lu, and J. Salvado, Geometric compatibility of IceCube TeV–PeV neutrino excess and its galactic dark matter origin, arXiv:1311.5864. (2013).

94. A. Bhattacharya, M.H. Reno, and I. Sarcevic, Reconciling neutrino flux from heavy dark matter decay and recent events at IceCube, *JHEP.* **06** (2014) 110.

95. C. Rott, K. Kohri, and S.C. Park, Superheavy dark matter and IceCube neutrino signals: Bounds on decaying dark matter, *Phys. Rev.* **D92** (2) (2015) 023529.

96. A. Esmaili, S.K. Kang, and P.D. Serpico, IceCube events and decaying dark matter: Hints and constraints, *JCAP.* **1412** (12) (2014) 054.

97. K. Murase, R. Laha, S. Ando, and M. Ahlers, Testing the dark matter scenario for PeV neutrinos observed in IceCube, *Phys. Rev. Lett.* **115** (7) (2015) 071301.

98. L.A. Anchordoqui, V. Barger, H. Goldberg, X. Huang, D. Marfatia, L.H.M. da Silva, and T.J. Weiler, IceCube neutrinos, decaying dark matter, and the Hubble constant, arXiv:1506.08788v2. (2015).

99. R. Abbasi, *et al.*, Search for dark matter from the galactic halo with the IceCube Neutrino Observatory, *Phys. Rev.* **D84** (2011) 022004.

100. M. Ackermann, *et al.*, Constraints on the galactic halo dark matter from Fermi-LAT diffuse measurements, *Astrophys. J.* **761** (2012) 91.

101. M. Cirelli and G. Giesen, Antiprotons from dark matter: Current constraints and future sensitivities, *JCAP.* **1304** (2013) 015.

102. A. Esmaili, A. Ibarra, and O.L.G. Peres, Probing the stability of superheavy dark matter particles with high-energy neutrinos, *JCAP.* **1211** (2012) 034.

103. S.M. Boucenna, M. Chianese, G. Mangano, G. Miele, S. Morisi, O. Pisanti, and E. Vitagliano, Decaying leptophilic dark matter at IceCube, arXiv:1507.01000. (2015).

104. J. Kopp, J. Liu, and X.-P. Wang, Boosted dark matter in IceCube and at the galactic center, *JHEP.* **04** (2015) 105.

105. O. Adriani, *et al.*, An anomalous positron abundance in cosmic rays with energies 1.5–100 GeV, *Nature.* **458** (2009) 607–609.

106. J. Chang, *et al.*, An excess of cosmic ray electrons at energies of 300–800 GeV, *Nature.* **456** (2008) 362–365.

107. M. Ackermann, *et al.*, Measurement of separate cosmic-ray electron and positron spectra with the Fermi Large Area Telescope, *Phys. Rev. Lett.* **108** (2012) 011103.

108. M. Aguilar, *et al.*, First result from the Alpha Magnetic Spectrometer on the International Space Station: Precision measurement of the positron fraction in primary cosmic rays of 0.5350 GeV, *Phys. Rev. Lett.* **110** (2013) 141102.

109. O. Adriani, *et al.*, PAMELA measurements of cosmic-ray proton and helium spectra, *Science.* **332** (2011) 69–72.

110. C. Consolandi, Primary cosmic ray proton flux measured by AMS-02. In *10th International Symposium on Cosmology and Particle Astrophysics (CosPA 2013)* Honolulu, Hawaii, USA, November 12–15, 2013, (2014).

111. M. Ackermann, *et al.*, Searching for dark matter annihilation from milky way dwarf spheroidal galaxies with six years of Fermi-LAT data, arXiv:1503.02641. (2015).

112. J. Aleksić, *et al.*, Optimized dark matter searches in deep observations of Segue 1 with MAGIC, *JCAP.* **1402** (2014) 008.

113. J. Grube, VERITAS limits on dark matter annihilation from dwarf galaxies, *AIP Conf. Proc.* **1505** (2012) 689–692.

114. E. Behnke, *et al.*, First dark matter search results from a 4-kg $CF_3I$ bubble chamber operated in a deep underground site, *Phys. Rev.* **D86** (5) (2012) 052001.

115. E. Aprile, *et al.*, Limits on spin-dependent WIMP–nucleon cross sections from 225 live days of XENON100 data, *Phys. Rev. Lett.* **111** (2) (2013) 021301.

116. E. Aprile, The XENON1T dark matter search experiment, *Springer Proc. Phys.* **C12-02-22** (2013) 93–96.

117. L. Baudis, DARWIN: Dark matter WIMP search with noble liquids, *J. Phys. Conf. Ser.* **375** (2012) 012028.

118. V.M. Gehman, A. Goldschmidt, D. Nygren, C.A.B. Oliveira, and J. Renner, A plan for directional dark matter sensitivity in high-pressure xenon detectors through the addition of wavelength shifting gaseous molecules, *JINST.* **8** (2013).

119. J.B.R. Battat, *et al.*, First background-free limit from a directional dark matter experiment: Results from a fully fiducialised DRIFT detector, *Phys. Dark Univ.* **9–10** (2014) 1–7.

120. A. Alexandrov, *et al.*, A novel approach to dark matter search based on nanometric emulsions, *JINST.* **9** (12) (2014) C12053.

121. M. Aartsen, *et al.*, Letter of Intent: The Precision IceCube Next Generation Upgrade (PINGU), arXiv:1401.2046. (2014).

122. U.F. Katz. The ORCA Option for KM3NeT. In *Proceedings of the 15th International Workshop on Neutrino Telescopes (Neutel 2013)*, (2014).

123. M.G. Aartsen, *et al.*, IceCube-Gen2: A vision for the future of neutrino astronomy in antarctica, arXiv:1412.5106. (2014).

124. R. Coniglione, The KM3NeT neutrino telescope, *J. Phys. Conf. Ser.* **632** (1) (2015) 012002.

125. A.D. Avrorin, *et al.*, The prototyping/early construction phase of the BAIKAL-GVD project, *Nucl. Instrum. Meth.* **A742** (2014) 82–88.

## Chapter 12

## KM3NeT: Astroparticle and Oscillations Research with Cosmics in the Abyss

M. de Jong

*Nikhef, Science Park 105*
*1098 XG Amsterdam, The Netherlands*
*mjg@nikhef.nl[*]*

KM3NeT[a] is a large research infrastructure, that will consist of a network of deep-sea neutrino telescopes in the Mediterranean Sea. The main objectives of KM3NeT are the discovery and subsequent observation of high-energy neutrino sources in the Universe (ARCA) and the measurement of the mass hierarchy of neutrinos (ORCA). A cost-effective technology for (very) large water Cherenkov detectors has been developed based on a new generation of low price 3-inch photo-multiplier tubes. Following the successful deployment and operation of two prototypes, the construction of the KM3NeT research infrastructure has started. The improvements in the technology and the prospects of the different phases of the implementation of KM3NeT are summarised.

## 1. Introduction

The main objectives of KM3NeTv are the discovery and subsequent observation of high-energy neutrino sources in the Universe (ARCA) and the measurement of the mass hierarchy of neutrinos (ORCA).[1] The successful construction and operation of the ANTARES detector (Mediterranean Sea) has demonstrated the feasibility of deep-sea neutrino telescopes.[2] Furthermore, the transparency of the deep waters, the size of the detector and the geographical location make KM3NeT an ideal instrument to study sources of high-energy neutrinos in the galaxy. The neutrino signal reported by IceCube supports the hypothesis that the conditions in some astrophysical sources are conducive to high-energy neutrino production.[5] The next challenge is to identify the sources of cosmic neutrinos. The measurement of the mass hierarchy of neutrinos (KM3NeT/ORCA) is separately documented in this book.

---

[*]also at Leiden University, the Netherlands.
[a]multi-<u>KM</u>$^3$ <u>N</u>eutrino <u>T</u>elescope

## 2. Technology

The detection principle is based on the detection of the Cherenkov light induced
by relativistic charged particles emerging from an interaction of a neutrino in the
vicinity of the detector. The angular resolution is primarily determined by the lever
arm between the light sensors and the measurement precision of their positions and
the arrival time of the Cherenkov light. Of the different types of neutrino interactions, the charged current interaction of a muon neutrino yields the best angular
resolution because the muon that emerges from such an interaction has the longest
range. The optical properties of the water can be summarised by the probability
density function (PDF) of the arrival time of the Cherenkov light. As an example,
the PDF corresponding to a muon with an energy of 1 TeV is shown in Fig. 1.

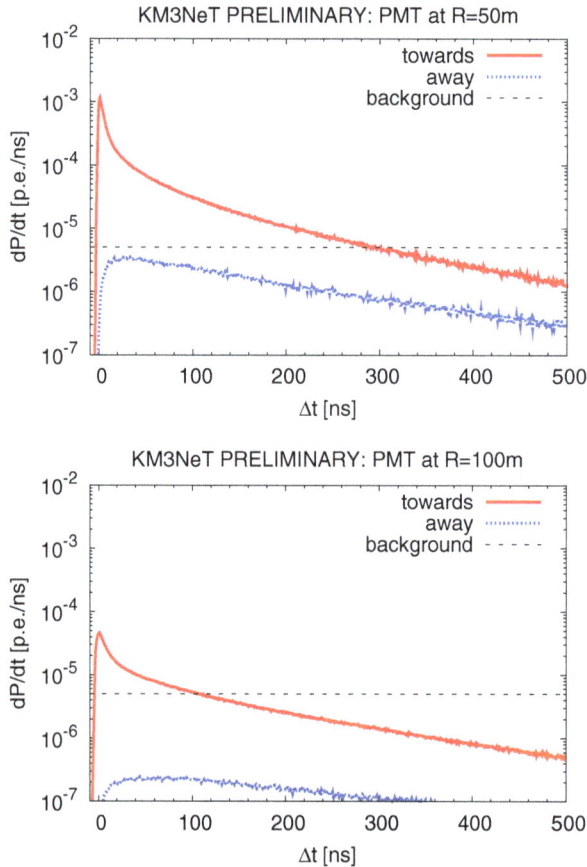

Fig. 1.   The probability density function (PDF) of the arrival time of light from a 1 TeV muon on
a 3-inch PMT. The top (bottom) figure is for a minimal distance of approach of 50 (100) m. The
red (blue) curve corresponds to a PMT with the front (back) facing the muon trajectory and the
dashed horizontal line to the random background due to Potassium decays (and bioluminescence).
The time is offset to the Cherenkov hypothesis.

As can be seen from Fig. 1, the signal-to-noise ratio is high ($> 100$) at typical distances of approach around 50 m, the FWHM small ($< 5$ ns) and the pointing capability evident. The light transmission properties of deep-sea water combined with the presently feasible position (10 cm) and time (1 ns) resolution of the PMTs make it possible to reconstruct the direction of high-energy muons with an accuracy of about 0.1°. A new generation of 3-inch photo-multiplier tubes (PMTs) has been developed for KM3NeT. These PMTs combine good timing (RMS less than 2 ns), relatively high quantum efficiency (around 30%) and low price (per unit photo-cathode area less than 10-inch PMTs). A time synchronisation system based on White Rabbit[6] provides sub-nanosecond precision on the measured time of single photo-electrons, while the position and orientation of the PMTs is measured to a few centimetres and few degrees, respectively. The PMTs and the readout electronics are installed within pressure-resistant glass spheres, so called digital optical modules (DOMs). The DOMs are distributed in space along flexible strings, of which one end is fixed to the sea floor and the other end is held almost vertical by a buoy. The concept of strings is modular by design. The construction and operation of the research infrastructure thus allows for a phased and distributed implementation. A collection of strings forms a single KM3NeT building block.

A **building block** comprises a large number of strings which are connected to junction boxes via interlink cables running along the seabed. Each string hosts a fixed number of optical modules which contain the PMTs. A main electro-optical cable connects the deep sea infrastructure to the shore station. The depth/distance of the infrastructure from shore are 2500 m/50 km, 3500 m/100 km, 4500 m/30 km, for the Toulon, Capo Passero and Pylos sites, respectively. It is foreseen that each site will host at least one building block. The site off shore from Toulon will host the ORCA detector and the site off shore from Capo Passero will host the ARCA detector. The detection efficiency of the ARCA detector has been studied as a function of *i)* the number of strings, *ii)* the number of optical modules per string, *iii)* the horizontal spacing between strings and *iv)* the vertical spacing between optical modules for different absorption lengths of the water. In this, the efficiency is defined as the number of detected events due to an assumed flux of neutrinos from RXJ1713.[8] The optimal spacing is found to be 90 m horizontally and 36 m vertically. The number of events as a function of the number of strings and number of optical modules is shown in Fig. 2. In this, the various detector configurations consistently comprise a total of 12,320 optical modules. As can be seen from Fig. 2, the normalised detection efficiency increases up to a point where it flattens out. The smallest detector with optimal efficiency corresponds to about 115 strings and 18 optical modules per string. This configuration is referred to as a building block.

A **string** is about 700 m in length and hosts 18 optical modules spaced 36 m apart. It comprises two thin parallel ropes which hold the optical modules in place. Attached to the ropes is the vertical electro-optical cable ("backbone" cable), an oil-filled plastic tube which contains the electrical wires and optical fibres used

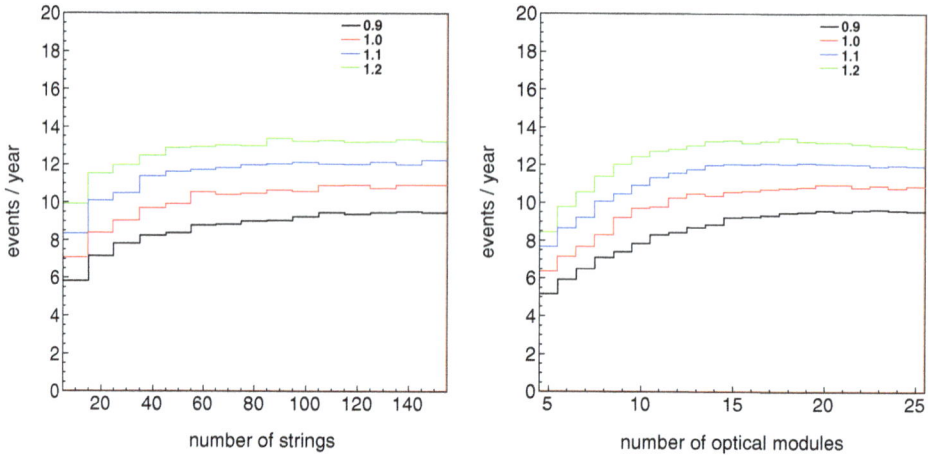

Fig. 2.    The number of events per year due to an assumed spectrum of neutrinos from RXJ1713 as a function of the number of strings and the number of optical modules per string. The neutrino spectrum is taken from Ref.8. The colour coding refers to a scaling factor applied to the absorption length.

for the power and data transmission. A surface boat will be used to deploy multiple strings in a single cruise. The strings are initially deployed on the seabed coiled around a spherical frame (so-called launcher vehicle). A remotely operated vehicle (ROV) is used to deploy and connect interlink cables from the base of a string to the junction box. The ROV also triggers the autonomous unfurling of a string after deployment.

A **digital optical module** is a transparent, pressure resistant, 17-inch glass sphere containing a total of 31 3-inch PMTs and their associated readout electronics. This design offers a number of improvements compared to previous designs based on a single large area PMT, most notably: Larger photo-cathode area, digital photon counting, directional information, wider field of view and reduced ageing effects. A position calibration device (acoustic piezo sensor) and a time calibration device (nano-beacon) are also housed inside each glass sphere. The readout electronics feature low power consumption (7 W), high-bandwidth (Gb/s) data transmission using dense wavelength division multiplexing (DWDM), time over threshold measurement of each PMT signal and precision time synchronisation via the White Rabbit protocol. The optical module also incorporates an acoustic sensor and compass used for position and orientation calibration. An exploded view and a photo of the optical module are shown in Fig. 3.

A **shore station** is a (small) building which houses the equipment for GPS, power and real-time computing and a remotely operated control room. It provides a high-bandwidth internet connection to the central data repository. A crew of 2–3 persons is adequate for computer, network and power maintenance and security.

Fig. 3. Exploded view of the KM3NeT optical module (left) consisting of a 17-inch pressure resistant glass sphere housing 31 3-inch PMTs and a real KM3NeT optical module (right). The small white dot in the middle is an acoustic piezo sensor which is used for position calibration.

A **prototype** of the KM3NeT optical module with 31 3-inch PMTs was mounted on the so-called instrumentation line of ANTARES and deployed in April 2013. An analysis of the first data confirms the specifications of the system.[7] In May 2014, a prototype string housing three optical modules was deployed offshore from Portopalo di Capo Passero, Italy. Evidence of correlated signals in the three optical modules from atmospheric muons has been found (see Fig. 4).

The **readout** of the KM3NeT detector is based on the *"All-data-to-shore"* concept which has been pioneered by ANTARES. In this, all analog signals from the PMTs that pass a preset threshold (typically 0.3 p.e.) are digitised and all digital data are sent to shore where they are processed in real time. The physics events are filtered from the background using designated software. To maintain all available information for the offline analyses, each event will contain a snapshot of all data from the detector during the event. Different filters can be applied to the data simultaneously. The building blocks are separately read out.

The **data** contain the time of the leading edge and the time over threshold of every analogue pulse, commonly referred to as a hit. Each hit corresponds to 6 bytes of data (1 B for PMT address, 4 B for time and 1 B for time over threshold). The least significant bit of the time information corresponds to 1 ns. The optical background due to decays of $^{40}$K and bioluminescence amounts to typically 6 kHz for a single PMT. The total data rate then amounts to about 25 Gb/s per building

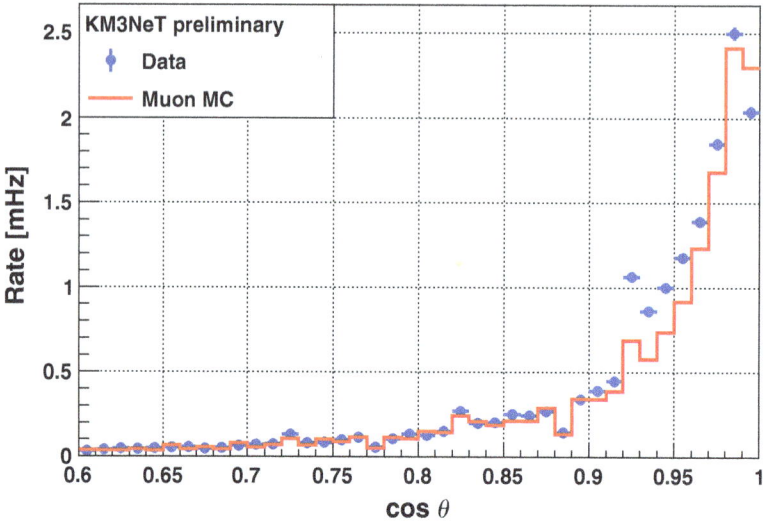

Fig. 4. Distribution of the reconstructed zenith angle. The data have been obtained from the so-called pre-production model of a detection unit (PPM-DU). The red curve corresponds to a Monte Carlo (MC) simulation of the detector response to atmospheric muons.

block. A reduction of the data rate by a factor of about $10^5$ is required to store the filtered data on disk. In addition to physics data, summary data containing the singles rates of all PMTs in the detector are stored with a sampling frequency of 1 Hz. This information can be used in the simulations and the reconstruction to take into account the actual status and background conditions of the detector.

The **data filter** runs on a farm of commodity PCs. For the detection of muons and showers, the time-position correlations that are used to filter the data follow from causality. In the following, the level-zero filter (L0) refers to the threshold for the analogue pulses which is applied off shore. All other filtering is applied on shore. The level-one filter (L1) refers to a coincidence of two (or more) L0 hits from different PMTs in the same optical module within a fixed time window. The scattering of light in deep-sea water is such that the time window can be very small. A typical value is $\Delta T = 10$ ns, as can be seen from Fig. 1. The estimated L1 rate per optical module is then about 1,000 Hz of which about 600 Hz is due to genuine coincidences from $^{40}$K decays. The remaining part arises from random coincidences which can be reduced by a factor of two by making use of the known orientations of the PMTs. This is referred to as level-two filter (L2). A general solution to trigger an event consists of a scan of the sky combined with a directional filter. In the directional filter, the direction of the muon is assumed. For each direction, an intersection of a cylinder with the 3D array of optical modules can be considered. The diameter of this cylinder (i.e. road width) corresponds to the maximal distance traveled by the light. It can safely be set to few times the absorption length without a significant loss of the signal. The number of PMTs to be considered is then reduced by a factor of 100 or more, depending on the assumed direction. Furthermore, the time window that follows from causality is reduced by a similar factor. (Only the transverse distance between the PMTs should be taken into account as a correction can be made for the propagation time of the muon). This improves the signal-to-noise ratio (S/N) of an L1 hit by a factor of (at least) $10^4$ compared to the general causality relation. With a requirement of five (or more) L1 hits, this filter shows a very small contribution of random coincidences. The field of view of the directional filter is about 10°. So, a set of 200 directions is sufficient to cover the full sky. By design, this trigger can be applied to any detector configuration. Furthermore, the minimum number of L1 hits to trigger an event can be lowered for a limited number of directions. A set of astrophysical sources can thus be tracked continuously with a lower detection threshold for each source. For showers, the maximal 3D-distance between PMTs can be applied without consideration of the direction of the shower. Hence, a similar improvement of the S/N ratio can be obtained. Alternative signals with different time-position correlations, such as slow monopoles, can be searched for in parallel. It is obvious but worth noting that the number of computers and the speed of the algorithms determine the performance of the system and hence the physics output of KM3NeT for the same price.

## 3. Prospects

The first observation by IceCube of extra-terrestrial neutrinos included a significant contribution of high-energy shower events.[5] These events are due to neutral current interactions of neutrinos or charged current interactions of electron neutrinos or tau neutrinos (in which the tau does not decay into a muon). Given the excellent angular resolution of the KM3NeT neutrino telescope for charged current interactions of muon neutrinos, the question then arises how well these events can be reconstructed. The energy resolution and the angular resolution of charged current electron neutrinos are shown in Fig. 5.

As can be seen from Fig. 5, the energy resolution is about 10 % and the angular resolution is about 2° (the angular resolution is defined as the median space angle between the reconstructed direction and the true neutrino direction). These results make it possible to do astronomy with any neutrino flavour.

The construction and operation of the KM3NeT research infrastructure allows for a phased implementation. Recently, the KM3NeT collaboration started the first construction phase (phase-1): 31 strings equipped with 558 optical modules will be assembled and deployed at the French and Italian sites. The resulting arrays will be different in size, the setup at the Italian site being significantly larger and providing the equivalent of about 20% of the instrumented volume of the IceCube detector. The strings to be deployed at the Italian site will be configured for high-energy neutrino detection (ARCA) and the strings to be deployed at the French site for low-energy neutrino detection (ORCA). In the next phase, KM3NeT 2.0, one ORCA and two ARCA building blocks will be realised. The ORCA and ARCA detectors

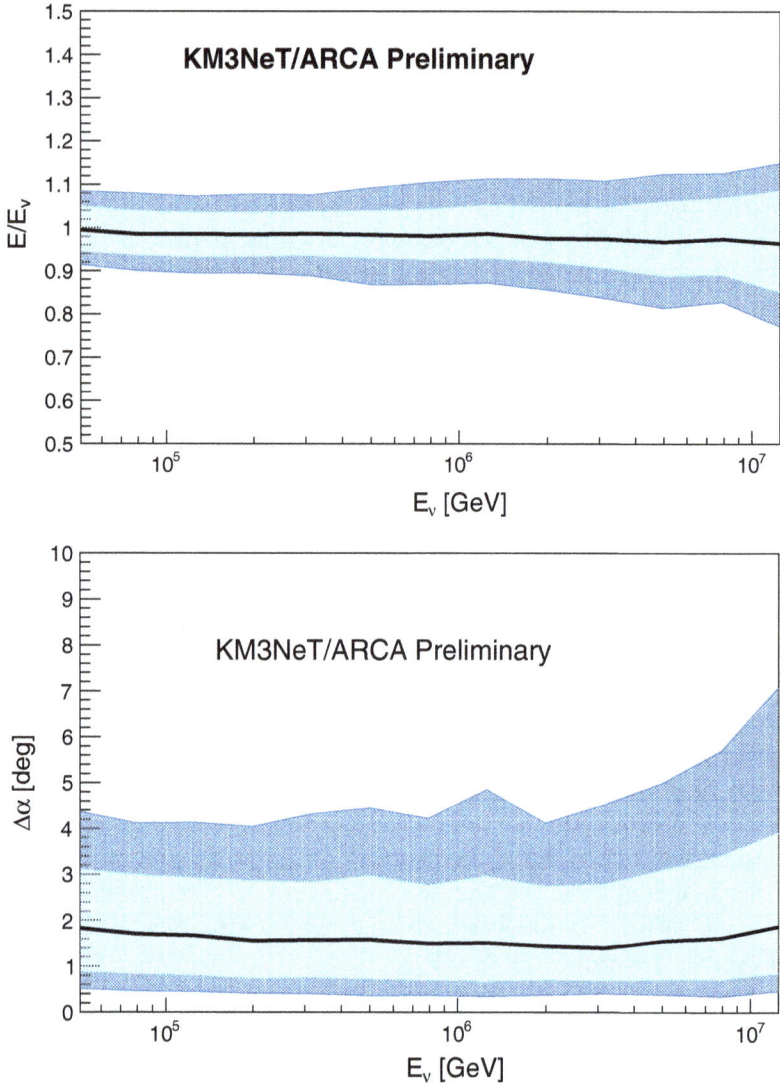

Fig. 5.   The energy resolution (top) and the angular resolution (bottom) of charged current electron neutrinos as a function of the neutrino energy.

will be built with the same technology. The dimensions of the ORCA detector are reduced by a factor of about four compared to those of the ARCA detector, to realise a lower detection threshold for the atmospheric neutrinos needed for the determination of the neutrino mass hierarchy. The ultimate goal is to fully develop the KM3NeT research infrastructure to comprise a distributed installation at the three foreseen sites (Italy, France and Greece), with almost 700 strings equipped with 12,400 optical modules in total (phase-3). The phase-2 detector will allow

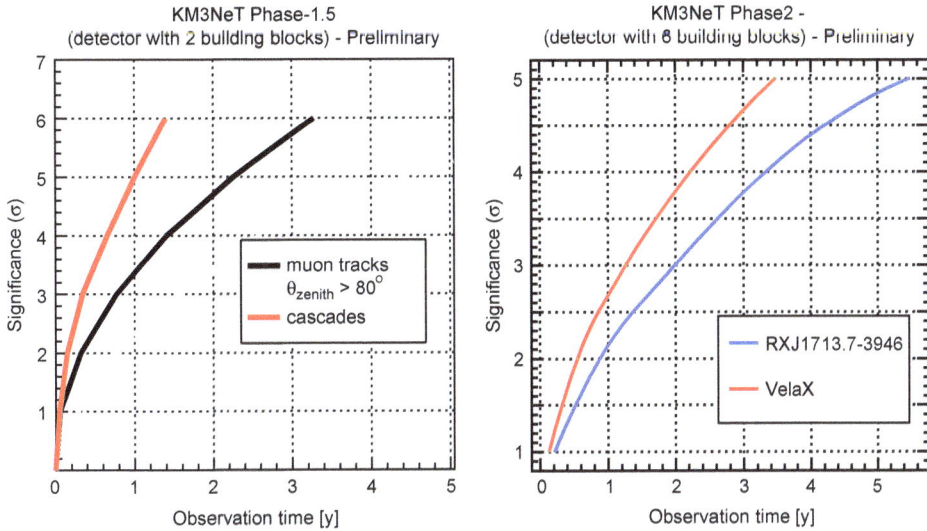

Fig. 6.    The significance ($\sigma$) as a function of the observation time (years) for a measurement of the signal reported by IceCube[5] using ARCA (left); and the assumed neutrino signals from the Supernova Remnant RXJ1713[8] and the Pulsar Wind Nebula Vela X[9] using KM3NeT phase-3 (right).

for an independent measurement of the IceCube signal with different methodology, improved resolution, complementary field of view and larger detection volume. An overview of the prospectives of phase-2 and phase-3 is shown in Fig. 6.

## References

1. S. Adrián-Martínez, *et al.*, [KM3NeT Collaboration], *J. Phys. G: Nucl. Part. Phys.* **43** (2016) 084001.
2. M. Ageron, *et al.*, [ANTARES Collaboration], *Nucl. Instrum. Meth. A* **656** (2011) 11.
3. www.km3net.org
4. icecube.wisc.edu
5. M.G. Aartsen, *et al.*, [IceCube Collaboration], *Science* **342** (2013) 1242856.
6. http://www.ohwr.org/projects/white-rabbit
7. S. Adrián-Martínez, *et al.*, [KM3NeT Collaboration], *Eur. Phys. J. C* **76** (2016) 54.
8. S.R. Kelner, F.A. Aharonian and V.V. Bugayov, *Phys. Rev. D* **74** (2006) 034018. [Erratum: *ibid. D* **79** (2009) 039901]
9. F.L. Villante and F. Vissani, *Phys. Rev. D* **78** (2008) 103007.

# Chapter 13

# A Next-Generation IceCube: The *IceCube-Gen2* Facility for High-Energy Neutrino Astronomy

Erik Blaufuss

*University of Maryland, College Park, MD*

Albrecht Karle

*University of Wisconsin-Madison, Madison, WI*

Given the recent observations of an astrophysical flux of neutrinos by the IceCube Neutrino Observatory, design work for a next-generation Antarctic neutrino observatory is underway. *IceCube-Gen2* is envisioned to include an instrumented array of an approximately 10 km$^3$ volume of clear glacial ice at the South Pole, able to deliver substantially larger astrophysical neutrino samples of all neutrino flavors. This detector would support a rich physics program, including searches for point sources, a detailed spectral and flavor characterization of the astrophysical neutrinos, searches for cosmogenic neutrinos, studies of cosmic rays, and searches for signatures of beyond-the-standard-model neutrino physics. In this paper, we highlight the scientific case for *IceCube-Gen2*, the design considerations and constraints, the development work toward a new class of optical sensors, and the increase in sensitivity to astrophysical neutrinos from the addition of an extensive surface detector that would identify and reject atmospheric backgrounds originating from the Southern Hemisphere. The *IceCube-Gen2* facility, including a larger detector and other components in combination with the existing IceCube Neutrino Observatory, would be the flagship experiment of the new field of neutrino astronomy.

## 1. Introduction

High-energy neutrinos are unique in their ability to probe the extreme universe. They reach the Earth from the edge of the universe without absorption or deflection by magnetic fields. They can escape from accelerators in the deepest regions of the universe where the highest energy cosmic rays are created. While these properties give them advantages over other astrophysical messengers such as photons and charged particles, their weak interactions make neutrinos very difficult to detect.

Detectors searching for astrophysical neutrinos are constructed by instrumenting large volumes of natural water or ice to detect Cherenkov emission from the charged particles produced when neutrinos interact with matter in or near the detector. The deep ice of the Antarctic glacier is host to IceCube,[1] the first kilometer-scale neutrino observatory, which has recently reported an observed astrophysical neutrino signal.[2–4,10] Proposed next-generation deep-water-based detectors include KM3NeT[11] in the Mediterranean Sea and GVD[12] in Lake Baikal, both of which provide a view of the sky that is complementary to IceCube's.

For an underground detector such as IceCube, the primary background is downward-directed muons created in cosmic-ray interactions in the Earth's atmosphere. These atmospheric muons trigger IceCube at a rate of $\sim$3 kHz. Atmospheric neutrinos are detected at a rate of $\sim$300 per day, have a mean energy of $\sim$1 TeV and form an isotropic background in searches for astrophysical neutrinos. At energies in excess of $\sim$100 TeV, the flux of atmospheric neutrinos is small, and upward-directed events of higher energy are likely to be of astrophysical origin.

A robust method to identify neutrino events is to distinguish events that start inside the detector from those that enter from outside, which applies to neutrino events originating from the entire sky. A search using this veto technique has been successfully applied to four years of data from the IceCube detector.[2,21] The 54 events found from this search had deposited energies ranging from 30 TeV to 2 PeV, clearly reflecting a significant observation of astrophysical neutrinos. An independent analysis of the spectrum of upgoing muons arising from neutrinos passing through the Earth has confirmed the existence of the astrophysical component.[4,5] The measured flux for this analysis, with results from other select IceCube searches for astrophysical neutrinos, as well as model predictions are shown in Fig. 1.

These modest sample sizes of astrophysical neutrinos represent the "first light" in the field of high-energy neutrino astronomy. Given present statistics, these

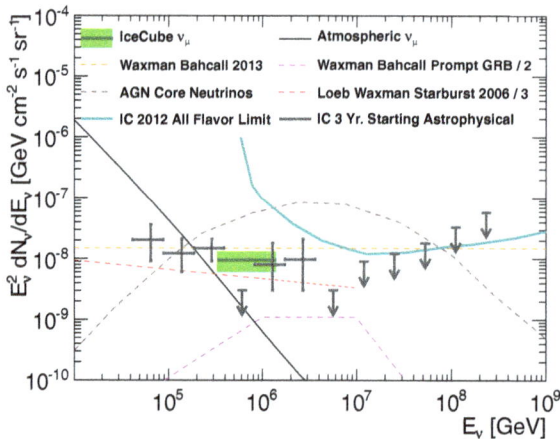

Fig. 1. Comparison of best-fit results for an $E^{-2}$ spectrum from selected IceCube measurements, including the two-year diffuse track search,[4] the three-year starting track search,[2] and limits from the extremely high energy all-flavor search,[13] as well as several model predictions.[14,16–20] All fluxes are given per neutrino flavor.

astrophysical neutrino samples appear consistent with an isotropic, diffuse flux of neutrinos equally distributed among the three neutrino flavors. Additionally, all searches for individual galactic and extragalactic source candidates[22,23] have so far only resulted in upper limits.

The effectiveness of IceCube as a tool for neutrino astronomy over the next decade is constrained by the limited number of astrophysical neutrinos that can be measured in the cubic-kilometer array. Here, we present a vision for the next-generation IceCube neutrino observatory, the *IceCube-Gen2* high-energy array, which is an expanded array of light-sensing modules that instrument a $\sim$10 km$^3$ volume for the detection of high-energy neutrinos.[24] With its unprecedented sensitivity and improved angular resolution, this instrument would explore extreme energies (PeV-scale) and would collect high-statistics samples of astrophysical neutrinos of all flavors, enabling detailed spectral studies, significant point source detections and new discoveries.

## 2. Science motivation

The observation of astrophysical neutrinos opens a unique window through which to view the universe. At the highest energies, above 100 TeV, neutrinos are the only astronomical messengers that are not absorbed or deflected en route from their astrophysical source, and they serve as powerful probes of the high-energy nonthermal universe. As exciting as the initial discovery has been, it has raised as many questions as it has answered. Studies that address these questions include searches for the origin of these events; a full characterization of the flux, spectrum and flavor composition of the neutrinos; searches for neutrinos arising from the GZK process; searches for signals in coordination with other observatories and messenger particles; and improved measurements of the cosmic-ray flux that drives our primary backgrounds. A detector with a significant increase in instrumented volume would be able to explore these questions.

With no indication of a steady point source in the current data, identification of a time-independent source by IceCube becomes challenging due to the observed background rates. Searches for point sources, therefore, become increasingly sensitive to rare transient sources, such as GRBs[17] or other flaring sources, where observation of a multiplet of events in a short time can yield a significant detection. Sensitivity gains to transient events would grow quickly in a larger detector. A next-generation neutrino observatory with five times the point-source sensitivity of IceCube and otherwise similar detector performance is predicted to have an increased sensitivity to transient source densities and rates by about a factor of $(5^3)$.[25]

While spectral and flavor composition of the observed astrophysical neutrino events are consistent with a single power law and equal distribution among all neutrino flavors, better measurement of these quantities from an increased sample size would make great strides in probing the interiors of the cosmic accelerators that generate these events. The neutrino production mechanisms (*p-p* and *p-γ* interactions, or neutron decay) and details of magnetic fields in the source region all leave their imprint[24,26] on the observed event sample. Additionally, galactic and extragalactic components could be resolved independently.

The highest energy cosmic rays are known to exist beyond energies of $10^{20}$ eV. At these energies, cosmic rays will interact with cosmic microwave background photons and produce neutrinos. These cosmogenic neutrinos[27] have energies in excess of 100 PeV, and either an observation or a stringent upper limit on their flux would provide constraints on the yet unknown composition of cosmic rays at the highest energies. We refer to other chapters in this volume for more detailed discussions of astrophysical sources.[6,7]

Multimessenger astronomy, the combined observations of cosmic rays, neutrinos, photons of all wavelengths, and, in the near future, gravitational waves, represents a powerful opportunity to decipher the physical processes that govern the nonthermal universe. Neutrinos play a central role in multimessenger astronomy, as they are an unambiguous signature for the acceleration and interaction of protons and nuclei. IceCube already has longstanding coordinated observation programs with several instruments.[8,28–30] With its improved sensitivity, $IceCube$-$Gen2$ would be a unique instrument to complement the next generation of wide-field optical telescopes, providing the possibility of searching large portions of the sky for neutrinos from large classes of supernova.[31]

Given the harder observed spectrum, astrophysical neutrinos will begin to dominate the irreducible background from atmospheric neutrinos and muons at high energies. Figure 2 highlights this transition from atmospheric neutrinos to astrophysical neutrinos at a zenith angle of 135°, showing both the neutrino spectra from atmospheric neutrinos and astrophysical neutrinos as well as the flux of muons

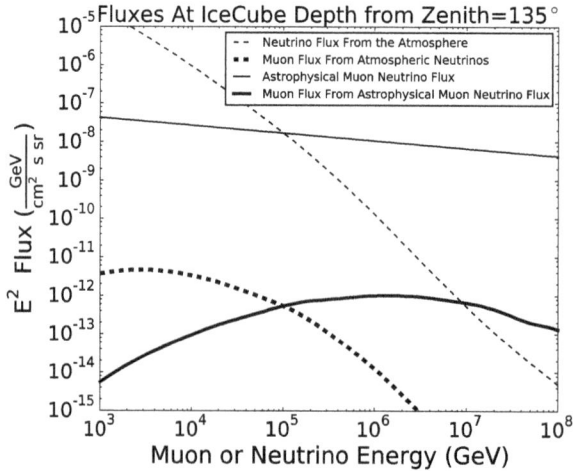

Fig. 2. Upgoing neutrino fluxes and resulting muon fluxes at a deep in-ice detector from atmospheric[15] and astrophysical[4] neutrinos at a representative zenith angle of 135° (declination +45°). Neutrino and muon fluxes begin to dominate the observed signal around energies of ~50 TeV.

resulting from their interactions in or near a deep detector. Both neutrinos and the resulting muons from astrophysical neutrinos begin to dominate above energies of ∼50 TeV, and point to the key energy regime for the *IceCube-Gen2* detector.

## 3. Design of a next-generation high-energy array

The design goals of *IceCube-Gen2* include a substantial increase in sensitivity in several channels compared to IceCube. Detection of muons arising from neutrino interactions remain the work horse for point source searches; therefore, increasing the muon effective area is crucial. Detection of these tracks, arising from neutrino interactions in or near the detector, would scale with the detector cross-sectional area. Good angular resolution is important. However, at high energies, the atmospheric background is greatly reduced and multimessenger and transient searches will play a more important role. Searches for electron, tau, or neutral current neutrino interactions depends on observing an electromagnetic or hadronic cascade resulting from the interaction of the neutrino with nucleons inside the instrumented region, and therefore scale with detector volume. High statistics in the cascade channel — $\nu_e, \nu_\tau$, and neutral current events — will allow a precise measurement of the energy spectrum for different regions in the sky and a precise favor ratio assessment.

*IceCube-Gen2* would build upon the existing IceCube detector infrastructure. By taking advantage of the very long absorption lengths found in the glacial ice at the South Pole, significantly larger string separation distances[35] can be used for the additional instrumentation, achieving a much larger instrument with a number of strings comparable to the existing IceCube detector. The larger detector would target neutrino energies above ∼50 TeV with high efficiency. Rate, angular reconstruction performance, energy resolution, and background rejection efficiency for the detection of these astrophysical events are key metrics in the design process underway within the collaboration.

The most important aspect for the design of a larger instrument in the glacial ice is the optical properties of the ice, in particular, the absorption length for Cherenkov photons. Typical absorption lengths are between 50 m and 200 m in the upper half of the current detector, and often exceed 200 m in the lower half. Although the optical properties vary with the layered structure of the ice, the average absorption and scattering lengths dictate the distance the strings of sensors can be spaced apart without impacting the uniform response of the detector. Early studies indicate that spacings of ∼240–300 m still maintain high efficiency for detecting astrophysical neutrinos.

The optical properties of the glacial ice prevent us from using optical modules at depths much shallower than the current instrumented range used by IceCube (with depths between 1450 m and 2450 m). Measurements of the depth dependence of the absorptivity of the Antarctic ice, shown in Fig. 3, suggest to increase the instrumented depth of the strings by ∼250 m leading to an increase in the geometric area for horizontal track events and therefore a 25% increase in effective area

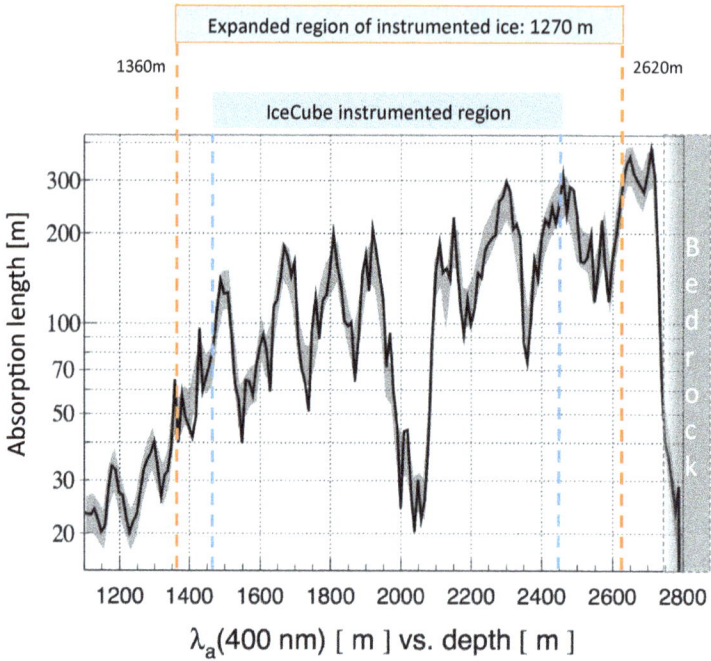

Fig. 3.   Absorption length in the glacial ice versus depth.[35] Note the layer of high dust concentration starting at about 2000 m depth. The ice above and below that layer is very clear. The current instrumented depth range used in IceCube and an extended string length, adding about 260 m to each string, are indicated. Note that not all simulations shown in this report have been performed with the extended string length.

for such events. The same increase is gained for the contained volume for cascade events.

To investigate the sensitivity of a larger detector, several benchmark geometries are being evaluated in software simulations, shown in Fig. 4. These benchmark geometries use a nonregular grid pattern to avoid symmetries that deteriorate acceptance and resolution for muon tracks and veto efficiencies. They are compared to the IceCube detector in its completed 86-string configuration and are used to scale sensitivities to a $10 \, \text{km}^3$ instrument. As the detector volume grows in these geometries, the exposed area increases and reaches up to $\sim 10 \, \text{km}^2$, substantially larger than the IceCube area, as summarized in Table 1.

For point source searches, which rely on throughgoing or starting muon tracks, the sensitivity increases with the projected cross-sectional area relative to source direction. At the energies of interest for astrophysical neutrino searches, these muons have significant ranges where their energy loss in the instrumented volume is large enough to separate them from background events. For example, a 10 PeV (100 TeV) muon will travel on average 6.3 km (5.9 km) before its energy drops to 10% of its initial energy. A time-independent point source search sensitivity will scale approximately with the square root of the increase in cross-sectional area and, to the extent

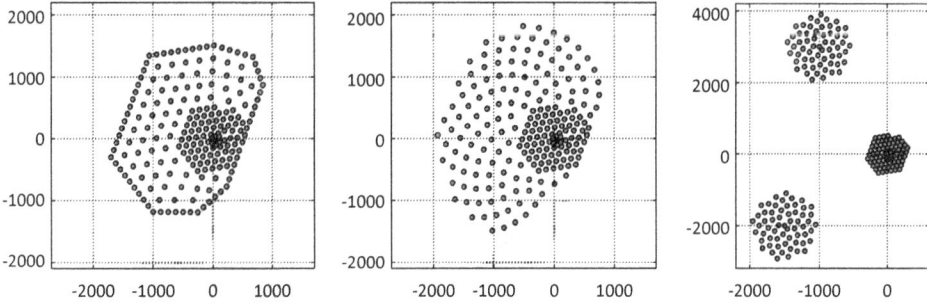

Fig. 4. Example benchmark detector string layouts under study. Each expands on IceCube by adding 120 strings constrained to the South Pole "Dark Sector" (shaded in light green). For the central panel, uniform string spacing of ∼240 m is shown, with an instrumented volume of 8.0 km³. The left panel presents a string layout with a denser edge weighting, where edge strings are spaced at ∼125 m, while interior strings are spaced at ∼240 m, with an instrumented volume of 6.2 km³. The right panel presents an option with multiple clusters of strings, with a total instrumented volume of 8.4 km³.

Table 1. Table of benchmark geometry volumes, effective detector cross-sectional areas sensitive to throughgoing muons, and comments on the relative merits of each design.

| Detector | Instr. volume | Cross section at 45° | Comments |
|---|---|---|---|
| IceCube 86 strings | 1.1 km³ | 1.7 km² | — IceCube as built |
| Sunflower 240 m | 8.9 km³ | 8.4 km² | — Large volume and cross-sectional area<br>— Veto efficiency reduced at lower energies |
| Edge-weighted 240 m | 7.2 km³ | 6.7 km² | — Large volume and cross-sectional area<br>— Veto efficiency maintained at lower energies |
| Three-cluster | 8.4 km³ | 11 km² | — Large volume and optimized cross-sectional area<br>— Reduced volume for vetoed and contained analyses |

that it is still background dominated, linearly with the improvement in angular resolution. In scenarios where the atmospheric backgrounds are negligible (e.g., short transients or searches for sources of very high energy neutrinos), sensitivities are expected to scale nearly linearly with cross-sectional area.

Electron or tau neutrinos generally interact with nucleons in the ice via deep-inelastic scattering processes, but at $E_\nu \sim 6.3\,\text{PeV}$ the resonant formation of an on-shell $W^-$-boson, the so-called Glashow resonance,[36] enhances the the cross section for electron antineutrinos. The resonance would be observable mostly as a peak in the cascade energy spectrum and serves as a key tag for neutrino flavor in a larger detector. Benchmark detectors show increased rates proportional to the volume gains, with about a factor of 10 in gains in observable Glashow resonance events. Given the similar event signatures for tau neutrinos, the same event rate gains in the larger detector are also indicated for tau neutrino events.[24]

## 4. Optical sensors

The design of the optical sensors will use as a reference the IceCube optical sensors, which are based on high quantum efficiency 10-inch Hamamatsu PMTs. An array of 120 strings with 80 sensors each would require 9600 sensors for the high-energy strings alone. R&D efforts are underway for alternate designs that consider differences in cost, directional sensitivity, and signal readout. The following designs are being investigated, listed in order of maturity:

- Baseline IceCube style: P-DOM[37]
  Photodetector: $1 \times 10$-inch PMT (Hamamatsu R7081-02)
  Minimum PMT cathode area (datasheet): 380 cm$^2$
  Readout: waveform digitization 250 Msamples/s; 14 bit
  Housing diameter: 33 cm
- KM3Net-inspired: M-DOM[38]
  Photodetector: $24 \times 3$-inch PMT (candidate: Hamamatsu R12199)
  Minimum PMT cathode area (datasheet): 977 cm$^2$
  Readout: multithreshold time-over threshold for each channel
  Housing diameter: 35.6 cm
- Dual PMT in a elliptical glass housing (D-Egg)[39]
  Photodetector: $2 \times 8$-inch PMT (candidate: Hamamatsu R5912)
  Minimum PMT cathode area (datasheet): 567 cm$^2$
  Readout: TBD
  Housing diameter: 30.5 cm
- Wavelength-shifting optical module[40]
  Photodetector: Wavelength shifting material coupled to a light-guiding tube, which is connected to a small photodetector; average photon effective area comparable the above.
  Housing diameter: 11 cm, and 26 cm for the case of assembly with three sensors and cable

Figure 5 shows images of these sensors, a photograph for the existing IceCube DOM and graphical illustrations for the other design concepts.

One of the advantages of the M-DOMs is the omnidirectional sensitivity and the directional information obtained with 24 pixels. The D-Egg is a pragmatic and interesting compromise that provides 50% more coverage than the IceCube DOM and fits in a housing of slightly smaller diameter.

The cost-per-unit of effective photon area of a sensor is driven not only by the cost of the sensor but to a significant degree by the cost for other parts such as electronics and pressure housing. In the case of IceCube, the PMT accounted for

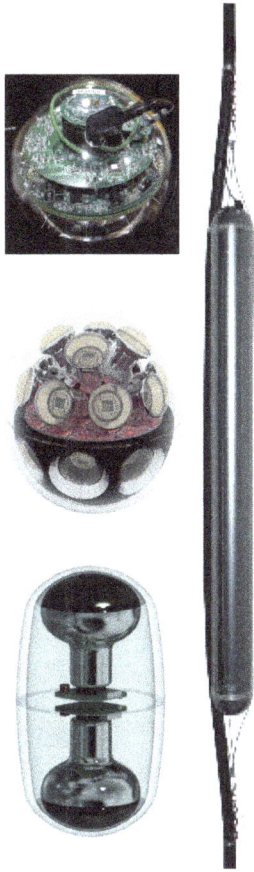

Fig. 5. The four types of sensors discussed are shown from top left: IceCube optical sensor, KM3NeT-inspired mDOM, D-Egg, and WOM.

only about one-quarter of the cost of a shipped DOM. To that extent, it can be cost-efficient to use sensors with greater photocathode area. This is especially the case for a detector that is optimized for high energies, as considered here, where photons travel larger distances.

The concept of a wavelength-shifting optical module is to take advantage of the extremely UV-rich Cherenkov signal, especially in the glacial ice that is transparent down to a wavelength of ∼200 nm. At first glance, this does not seems obvious since the tube must be immersed in a dense medium. However, simulations indicate that such an approach could be very cost effective. The time resolution would be on the order of fewer than 10 nsec for single photons. However, for the high-energy signals, where photons tend to travel larger distances, this should still be effective.

The diameter of the sensors has impact on the required diameter of the bore hole. For IceCube, the sensors as assembled for deployment had an outer diameter of 43 cm, and the accommodating bore hole was drilled to a diameter of about

60 cm. Drilling a single hole required about 32 hours. Smaller assemblies would allow for a smaller diameter borehole. Based on past drilling simulations and data, an instrument with a diameter reduced by 2 inches would allow a savings of 15% of drill time and fuel consumption. Such a difference could easily result in one less season required for construction. An instrument with less than 25 cm diameter, such as the WOM, could shave off 40% of the fuel needs compared to the reference 13-inch P-DOM resulting in more than $10M of savings.

In summary, while a straightforward basic extrapolation for an IceCube-like P-DOM can easily be budgeted and planned for, there is still opportunity for new ideas, and more R&D might lead to both substantial cost savings per deployed photon detection effective area as well as savings of up to 40% in drill time and fuel.

## 5. Expanded surface detectors

IceCube's surface array, IceTop,[41] has proved to be a very valuable component of the detector, generating critical measurements of the primary cosmic ray spectrum from 1 PeV to beyond 1 EeV.[32] Accordingly, the Gen2 high-energy array would include a surface detector near the top of each deployed string. With a spacing of ~250 m, such a surface array would provide a high-resolution measurement of the primary spectrum from 10 PeV to above 1 EeV. With the larger aspect ratio of the *IceCube-Gen2* high-energy array, the acceptance for coincident events seen by both the surface array and the deep array increases by a factor of 40, from 0.26 km²sr to ~10 km²sr. Such a detector would allow unprecedented measurement of the evolution of the primary composition in the region where a transition from galactic to extragalactic cosmic rays is predicted.[42]

A surface detector also opens up the possibility of vetoing the background of cosmic ray muons and even atmospheric neutrinos. The method is simple. Atmospheric muons and neutrinos are produced in air showers from cosmic rays. If the air shower can be detected, it can be used to tag a primary muon bundle in the deep detector and identify it as background. A large surface detector could veto the high flux of punch-through muons present at a depth of two km. The method applies also to atmospheric neutrinos, opening up the possibility of a background-free sky up to a certain zenith angle and above an energy of 100 to a few hundred TeV. This atmospheric neutrino self-veto[44,45] applies both muons and atmospheric neutrinos. Studies with IceCube and IceTop[43] suggest that above an energy of 200 to 300 TeV, background can be suppressed to a level smaller than the astrophysical flux observed with IceCube. The energy threshold would obviously depend on the density of the surface veto detector.

One possibility for the array would be several thousand scintillators distributed over a large area of over 50 km². Strategies are being studied to extend the surface array beyond the footprint of the high-energy array.[46,47] While the cost of such an array would not be insignificant, it offers the prospect of doubling the rate of cosmic neutrinos above a few 100 TeV. At energies more than 100 TeV, IceCube records about 10 muon neutrinos in the Northern Hemisphere (upgoing). About 13 tracks expected from the Southern Hemisphere are not visible because of background. A surface veto of up to a 5 km radius would increase the number of muons with energy >100 TeV (300 TeV) from cosmic neutrinos by 75% (100%). These numbers suggest that a surface veto array would be a powerful component of a Gen2 detector facility.

## 6. Drilling and logistics

IceCube demonstrated the ability to deploy 86 strings on time and on budget in a hostile environment. Drilling at the South Pole is the most formidable challenge for engineering and logistics support. The enhanced hot water drill[48] developed for IceCube is capable of drilling to 2500 m depth within about 30 hours. Up to 20 holes have been drilled in a single Antarctic summer. For *IceCube-Gen2*, preliminary design studies have been performed for an optimal drill strategy and a reduced logistical impact. A future upgrade would require substantial demands on logistics and support. While IceCube has demonstrated a very successful operation, it is still worthwhile to question the challenges to logistical support for creating a larger detector. Detailed strategies have been developed to ensure such a project would not compete with other science experiments for logistical support at the South Pole, which are summarized as follows:

- Mobile drill: The drill is designed to be mobile in order to most efficiently drill holes over an area on the order of 10 km². While for IceCube there was a seasonal drill camp, for *IceCube-Gen2* five major drill subsystems would be situated on large sleds that can be easily moved during a given season. The design would be optimized for less maintenance.
- Design for minimal logistical impact: During the initial phase and over the course of IceCube construction, the hot water drilling technique was refined, and detailed simulations were developed that accurately described the drill data.[49] Based on these data, it is clear that narrower holes allow for a faster drill time. Given that and the fact that optical sensor designs include potentially smaller diameters, as discussed above, a savings of up to 40% of fuel cost and drill time per hole appears feasible.
- Population at the South Pole station: An obvious concern for the construction of *IceCube-Gen2* is the required population at the South Pole. The South Pole station is designed for a summer population of about 160 people. Approximately 50 people would be required for *IceCube-Gen2* construction efforts, using IceCube as a guide. A simple solution to avoid adverse impacts on other science projects would be to construct and run a separate summer camp where IceCube personnel could stay. Such a summer camp would be funded on project cost and not compete with other program resources.

- Transport of equipment: For IceCube, a total of nine million pounds of cargo was transported to the South Pole by more than 300 LC130 aircraft missions. For *IceCube-Gen2*, alternate means of logistics support are being considered. All major cargo would be transported by air-ride cargo sleds (ARCS). Such sleds are in active development already for high-payload traverses in Antarctic and Greenland. No planes would be needed to transport fuel, sensors or cables. There would be zero impact to other science projects in Antarctica. Indeed, the *IceCube-Gen2* investments in logistics solutions would be of long-term benefit to the US Antarctic Program. Another variable could be the duration of the drill season. A regular summer season set a constraint to 8 weeks of active drilling. Experts consider an increase of the duration by use of other aircraft as feasible, allowing for a possible increase of the drill season to drill more than 30 holes.

Only personnel and small equipment would be required to be flown in by aircraft, a minor task compared to the case for IceCube where everything was transported by plane, even while the South Pole Telescope was being built and the South Pole station had not yet been completed. Overall, we conclude that the logistics needs for *IceCube-Gen2* would not compete with the needs of other science projects.

## 7. Outlook

With the detection of a clear astrophysical neutrino signal, IceCube has observed the "first light" in the field of high-energy neutrino astronomy. However, detailed spectral studies and searches for specific source locations in this signal remain a challenge with the event samples available from the current IceCube instrument.

Design studies for an enhanced instrument, the *IceCube-Gen2* high-energy array, are well underway. Current investigations suggest that they would result in an instrumented volume approaching $10 \, \text{km}^3$ and would lead to significantly larger neutrino detection rates, across all neutrino flavors and detection channels. Figure 6 shows a schematic of the envisioned detector.

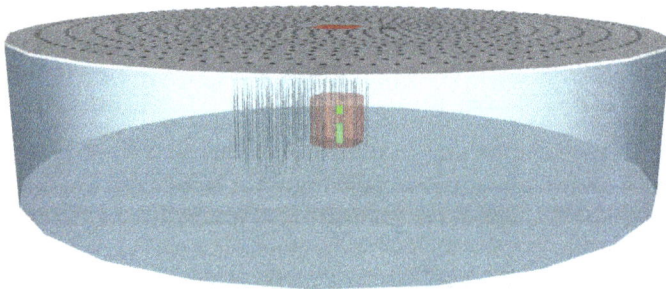

Fig. 6. Artist's representation of the *IceCube-Gen2* detector, including an $8.0 \, \text{km}^3$ volume array with an extended surface veto array. The current IceCube dectector components are indicated in red; not shown are PINGU or a possible radio detector.

Identifiable tracks created by astrophysical muon neutrinos are expected to increase by a factor of about 5, while events from electron, tau, and neutral-current neutrino interactions would increase by a factor of about 8, depending on the final string layout. The addition of a surface veto component would greatly increase the sensitivity to tracks from the Southern Hemisphere, yielding roughly another factor of 2 increase in track signals.

We have not discussed here the possible augmentation of such a neutrino facility with a large radio detector. A review of the radio detection method is given in elsewhere in this volume.[9] A detector like ARA[53] may serve as an ideal complement to an optical detector to achieve the large increase at energies above 30 PeV.

The next step in neutrino astronomy is clear. A complete set of design requirements needs to be developed that is based on the science goals. It combines the robust systems for drilling and detector instrumentation demonstrated by the IceCube detector with an optimized geometrical sensor arrangement that maximizes sensitivity to astrophysical neutrinos. Once in operation, the *IceCube-Gen2* detector, would truly be the flagship enterprise for the emerging field of neutrino astronomy.

## References

1. J. Ahrens, *et al.*, "Sensitivity of the IceCube detector to astrophysical sources of high energy muon neutrinos," *Astropart. Phys.* **20** (2004) 507–532.
2. M.G. Aartsen, *et al.*, "Observation of high-energy astrophysical neutrinos in three years of IceCube," *Phys. Rev. Lett.* **113** (2014) 101101.
3. M.G. Aartsen, *et al.*, "A combined maximum-likelihood analysis of the high-energy astrophysical neutrino flux measured with IceCube," *Astrophys. J.* **809** (2015) 98.
4. M.G. Aarsten, *et al.*, "Evidence for astrophysical muon neutrinos from the northern sky with IceCube," *Phys. Rev. Lett.* **115** (2015) 081102.
5. Ch. Wiebusch, "Observations of diffuse fluxes of cosmic neutrinos," chapter in this volume.
6. E. Waxman, "The origin of IceCubes neutrinos: Cosmic ray accelerators embedded in star forming calorimeters," chapter in this volume.
7. M. Ahlers, "Galactic neutrino sources," chapter in this volume.
8. M. Kowalski, "Neutrinos from core-collapse supernovae," chapter in this volume.
9. A. Connolly and A. Vieregg, "Radio detection of high-energy neutrinos," chapter in this volume.
10. IceCube Coll., "High energy astrophysical neutrino flux characteristics for neutrino-induced cascades using IC79 and IC86-string IceCube configurations," *Proceedings, 34th International Cosmic Ray Conference (PoS ICRC 2015)*, p. 1109.
11. A. Margiotta, "Status of the KM3NeT project," *JINST* **9** (2014) C04020.
12. A.D. Avrorin, *et al.*, "The prototyping/early construction phase of the BAIKAL-GVD project," *Nucl. Instrum. Meth.* **A742** (2014) 82.
13. M.G. Aartsen, *et al.*, "Probing the origin of cosmic rays with extremely high energy neutrinos using the IceCube Observatory," *Phys. Rev. D* **88** (2013) 112008.
14. M. Honda, T. Kajita, K. Kasahara and S. Midorikawa, "New calculation of the atmospheric neutrino flux in a three-dimensional scheme," *Phys. Rev. D* **70** (2004) 043008.

15. M. Honda, T. Kajita, K. Kasahara, S. Midorikawa, and T. Sanuki, "Calculation of atmospheric neutrino flux using the interaction model calibrated with atmospheric muon data," *Phys. Rev. D* **75** (2007) 043006.

16. R. Enberg, M. Reno and I. Sarcevic, "Prompt neutrino fluxes from atmospheric charm," *Phys. Rev. D* **78** (2008) 043005.

17. E. Waxman and J. Bahcall, "High energy neutrinos from cosmological gamma-ray burst fireballs," *Phys. Rev. Lett.* **78** (1997) 2292.

18. E. Waxman, "IceCube's Neutrinos: The beginning of extra-galactic neutrino astrophysics?" (2013) [arxiv:1312.0558].

19. F.W. Stecker, "Note on high-energy neutrinos from active galactic nuclei cores," *Phys. Rev. D* **72** (2005) 107301.

20. A. Loeb and E. Waxman, "The cumulative background of high energy neutrinos from starburst galaxies," *J. Cosmol. Astropart. Phys.* **5** (2006) 003.

21. IceCube Coll., "Observation of astrophysical neutrinos in four years of IceCube data," *Proceedings, 34th International Cosmic Ray Conference (PoS ICRC 2015)*, p. 1081.

22. M.G. Aartsen, *et al.*, "Search for time-independent neutrino emission from astrophysical sources with 3 yr of IceCube data," *Astrophys. J.* **779** (2013) 132.

23. S. Adrian-Martinez, *et al.*, "Searches for point-like and extended neutrino sources close to the galactic center using the ANTARES neutrino telescope," *Astrophys. J.* **786** (2014) L5.

24. M.G. Aartsen, *et al.*, "IceCube-Gen2: A vision for the future of neutrino astronomy in antarctica," [arxiv:1412.5106].

25. M. Ahlers and F. Halzen, "Pinpointing extragalactic neutrino sources in light of recent IceCube observations," [arxiv:1406.2160].

26. T. Kashti and E. Waxman, "Astrophysical neutrinos: Flavor ratios depend on energy," *Phys. Rev. Lett.* **95** (2005) 181101.

27. V. Berezinsky and G. Zatsepin, "Cosmic rays at ultra high energies (neutrino?)," *Phys. Lett.* **B28** (1969) 423.

28. IceCube Coll., "Neutrino-triggered target-of-opportunity programs in IceCube," *Proceedings, 34th International Cosmic Ray Conference (PoS ICRC 2015)*, p. 1052.

29. The IceCube, Pierre Auger and Telescope Array collaborations, "Search for correlations between the arrival direction of IceCube neutrino events and ultrahigh-energy cosmic rays detected by the Pierre Auger Observatory and the Telescope Array," *J. Cosmol. Astropart. Phys.* **2016** (2016) 037.

30. M. Santander, *et al.*, "Searching for TeV gamma-ray emission associated with IceCube high-energy neutrinos using VERITAS," *Proceedings, 34th International Cosmic Ray Conference (PoS ICRC 2015)*, p. 785.

31. S. Ando, J. Beacom and H. Yuksel, "Detection of neutrinos from supernovae in nearby galaxies," *Phys. Rev. Lett.* **95** (2005) 171101.

32. M.G. Aartsen, *et al.*, "Measurement of the cosmic ray energy spectrum with IceTop-73," *Phys. Rev. D* **88** (2013) 042004.

33. M.G. Aartsen, *et al.*, "Search for galactic PeV gamma rays with the IceCube Neutrino Observatory," *Phys. Rev. D* **87** (2013) 062002.

34. R. Abbasi, *et al.*, "Observation of an anisotropy in the galactic cosmic ray arrival direction at 400 TeV with IceCube," *Astrophys. J.* **746** 33.

35. M.G. Aartsen, *et al.*, "Measurement of South Pole ice transparency with the IceCube LED calibration system," *Nucl. Instr. Meth.* **A711** (2013) 73.

36. S.L. Glashow, "Resonant scattering of antineutrinos," *Phys. Rev.* **118** (1960) 316.

37. IceCube Coll., "Generation 2 IceCube Digital Optical Module and DAQ," *Proceedings, 34th International Cosmic Ray Conference (PoS ICRC 2015)*, p. 1148.

38. IceCube Coll., "Multi-PMT optical modules for IceCube-Gen2," *Proceedings, 34th International Cosmic Ray Conference (PoS ICRC 2015)*, p. 1147.

39. IceCube Coll., "A dual-PMT optical module (D-Egg) for IceCube-Gen2," *Proceedings, 34th International Cosmic Ray Conference (PoS ICRC 2015)*, p. 1137.

40. D. Hebecker, *et al.*, "Progress on the development of a wavelength-shifting optical module," *Proceedings, 34th International Cosmic Ray Conference (PoS ICRC 2015)*, p. 1134.

41. R. Abbasi, *et al.* [IceCube Collaboration], "IceTop: The surface component of IceCube," *Nucl. Instrum. Meth. A* **700** (2013) 188, doi:10.1016/j.nima.2012.10.067 [arXiv:1207.6326 [astro-ph.IM]].

42. IceCube Coll., "Cosmic ray science potential for an extended surface array at the IceCube observatory," *Proceedings, 34th International Cosmic Ray Conference (PoS ICRC 2015)*, p. 694.

43. IceCube Coll., "IceTop as Veto for IceCube," *Proceedings, 34th International Cosmic Ray Conference (PoS ICRC 2015)*, p. 1086.

44. S. Schönert and T.K. Gaisser, E. Resconi and O. Schulz, "Vetoing atmospheric neutrinos in a high energy neutrino telescope," *Phys. Rev.* **D79** (2009) 043009.

45. T.K. Gaisser, K. Jero, A. Karle, J. van Santen, "Generalized self-veto probability for atmospheric neutrinos," *Phys. Rev.* **D90** (2014) 023009.

46. IceCube Coll., "Simulation studies for a surface veto array to identify astrophysical neutrinos at the South Pole," *Proceedings, 34th International Cosmic Ray Conference (PoS ICRC 2015)*, p. 1070.

47. IceCube Coll., "Motivations and techniques of a surface detector to veto air showers for neutrino astronomy with IceCube at the southern sky," *Proceedings, 34th International Cosmic Ray Conference (PoS ICRC 2015)*, p. 1156.

48. T. Benson, *et al.* IceCube Enhanced Hot Water Drill functional description, *Annals of Glaciology* **55** (2014) 105–114, doi: 10.3189/2014AoG68A032.

49. L. Greenler, T. Benson, J. Cherwinka, A. Elcheikh, F. Feyzi, A. Karle and R. Paulos, Modeling hole size, lifetime and fuel consumption in hot-water ice drilling, *Annals of Glaciology*, **55** (2014) 115–123, doi: 10.3189/2014AoG68A033.

50. IceCube Coll., "Status of the PINGU Detector" *Proceedings, 34th International Cosmic Ray Conference (PoS ICRC 2015)*, p. 1174.

51. R. Abbasi, *et al.*, "Calibration and characterization of the IceCube photomultiplier tube," *Nucl. Instrum. Meth.* **A618** (2010) 139.

52. R. Abbasi, *et al.*, "The IceCube data acquisition system: Signal capture, digitization, and timestamping," *Nucl. Instrum. Meth.* **A601** (2009) 294.

53. P. Allison, *et al.* [ARA Collaboration], Performance of two Askaryan Radio Array stations and first results in the search for ultra-high energy neutrinos, (2015), arXiv:1507.08991, submitted to *Phys. Rev. D*.

# Chapter 14

# Neutrino Physics with Very Large Volume Neutrino Telescopes

Tyce DeYoung

*Department of Physics and Astronomy, Michigan State University,*
*567 Wilson Road, East Lansing, MI 48824, USA*
*deyoung@pa.msu.edu*

Antoine Kouchner

*Laboratoire AstroParticule et Cosmologie (APC)*
*Université Paris Diderot, CNRS/IN2P3, CEA/IRFU, Observatoire de Paris,*
*Sorbonne Paris Cité, 75205 Paris, France,*
*kouchner@apc.univ-paris7.fr*

Although designed to observe very high energy neutrinos from astrophysical sources, existing neutrino telescopes have also demonstrated their potential for exploring neutrino physics through measurements of oscillation parameters via disappearance of atmospheric muon neutrinos. As the specialized techniques required to reconstruct neutrino events accurately at energies of tens of GeV have been refined, the results from the current generation of detectors have begun to approach the precision of dedicated long-baseline oscillation experiments. Dedicated components focusing on low-energy neutrinos are planned for both KM3NeT (ORCA) and IceCube-Gen2 (PINGU), which would greatly enhance these measurements and also permit determination of the ordering of the neutrino mass eigenstates through observation of matter effects on neutrino oscillations.

## 1. Atmospheric neutrinos

The vast majority of the neutrinos observed in neutrino telescopes are not produced directly in astrophysical sources, but rather through the decay of mesons produced in cosmic ray interactions with the Earth's atmosphere. These "atmospheric" neutrinos constitute a background (in astronomical terms, a foreground) to searches for astrophysical neutrinos, but they have also provided a useful tool for validating the sensitivity of the telescopes[1-3] and calibrating the detectors' response. The very large volumes of neutrino telescopes implies that correspondingly large data sets of atmospheric neutrinos are recorded — in IceCube, some half a million atmospheric neutrino events have been identified to date — which provides the opportunity for a variety of studies of fundamental physics.

At the relevant energies, the flux of cosmic rays incident on the atmosphere is essentially isotropic; modulation by solar magnetic field affects cosmic rays below approximately $10\,\text{GeV}$[4] and anisotropies of unknown origin observed at higher energies[5,6] are well below the percent level. To first order, the spectrum of cosmic rays falls as $dN/dE \sim E^{-2.65}$ in this energy range, although variations in the spectrum are observed at different energies for the various nuclear species that make up the cosmic rays.[7–10]

The vast majority of the atmospheric neutrinos are produced through the "conventional" mechanism: the decay chain of charged pions produced when a cosmic ray primary such as a proton interacts with an atmospheric nucleus

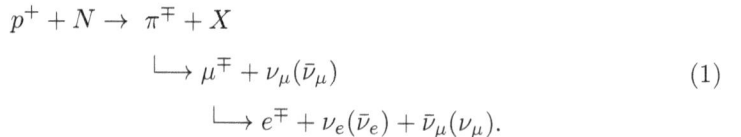

$$p^+ + N \rightarrow \pi^\mp + X$$
$$\quad\quad \hookrightarrow \mu^\mp + \nu_\mu(\bar{\nu}_\mu) \tag{1}$$
$$\quad\quad\quad \hookrightarrow e^\mp + \nu_e(\bar{\nu}_e) + \bar{\nu}_\mu(\nu_\mu).$$

Above $E_\nu \sim 100\,\text{GeV}$ a similar chain for kaons is the dominant process. The need for the secondary mesons and muons to decay before reaching the Earth's surface steepens the neutrino energy spectrum by one power relative to the primary cosmic ray spectrum. It also introduces an anisotropy in the flux, as the longer path lengths available for meson decay in atmosphere near the horizon increase the neutrino flux relative to the vertical. The geomagnetic field also affects neutrino production at energies below about $10\,\text{GeV}$, and asymmetries due to the local field configuration are included in the atmospheric flux model used for each detector. At much higher energies, short-lived charmed mesons are also relevant, but these "prompt" neutrinos make up only a negligible fraction of the overall atmospheric neutrino data samples.

As a result, atmospheric neutrinos provide a relatively well understood flux of neutrinos observable with high statistics from energies of a few GeV up to tens of TeV. Current measurements of the spectra of muon and electron neutrinos (and anti-neutrinos, which are generally not distinguished) are shown in Fig. 1. The measurements agree well with newer measurements by Super-Kamiokande[11] and with theoretical calculations by Honda *et al.*,[12] adjusted for the latest cosmic ray composition measurements following the H3a model of Gaisser.[13] The contribution of the prompt $\nu_\mu$ flux as calculated by Enberg *et al.*[14] is visible above $100\,\text{TeV}$. Lower energy measurements by the Fréjus experiment[15] are shown for comparison.

At first approximation, uncertainties in the atmospheric neutrino flux can be modeled as affecting the normalization and spectral index. This approach has been followed for DeepCore and ANTARES analyses, where the flux uncertainties are not dominant systematics. A more detailed treatment taking into account hadron

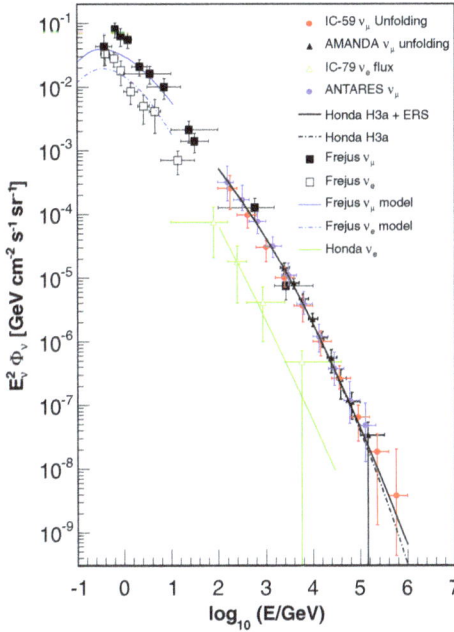

Fig. 1.  Spectra of atmospheric $\nu_\mu + \bar{\nu}_\mu$ and $\nu_e + \bar{\nu}_e$ measured by IceCube and ANTARES, compared to theoretical calculations (see text for details).

production uncertainties[16] parametrizes uncertainties in the $\nu/\bar{\nu}$ ratio, the $\nu_e/\nu_\mu$ ratio, and the angular dependence as functions of energy. These ratios are generally known to better precision than the $\sim 15\%$ uncertainty on the overall normalization, but the energy- and angle-dependent effects can be more problematic for neutrino physics measurement. This treatment is being incorporated in performance estimates for the more advanced ORCA and PINGU detectors.

## 2. Neutrino oscillations in vacuum

Neutrino oscillations arise from the fact that the weak interaction eigenstates in which neutrinos are produced are not identical to the mass eigenstates which propagate through space. The different mass eigenstates correspond to different kinetic energies and thus different wavelengths, so that interference effects modify the probabilities of detecting the various flavor eigenstates in a periodic manner as the neutrino travels. In a simplified two-neutrino model, which is a good approximation over the range of energies and propagation baselines accessible to ANTARES and DeepCore, the probability that a $\nu_\mu$ produced in the atmosphere with energy $E_\nu$ will remain a $\nu_\mu$ after traveling a distance $L$ is

$$P_{\mu\mu} \approx 1 - \sin^2 2\theta_{23} \sin^2 \left[1.27\Delta m_{32}^2 L/E_\nu\right], \tag{2}$$

where $\theta_{23}$ is the mixing angle and $\Delta m_{32}^2 = m_3^2 - m_2^2$ the mass splitting (in eV$^2$) between the eigenstates, and $L$ and $E_\nu$ are measured in km and GeV respectively.

Between production and detection, atmospheric neutrinos travel over baselines ranging from tens of kilometers for vertically down-going neutrinos up to 12,700 km for neutrinos traveling upward along the Earth's diameter. For the latter trajectory, complete disappearance of the $\nu_\mu$ flux occurs at approximately 24 GeV. As a result, the large atmospheric neutrino data sets accessible to neutrino telescopes permit the periodic oscillation pattern to be mapped out as a two-dimensional function of energy and zenith angle, referred to as an *oscillogram*.

By measuring behavior dependent on $L/E_\nu$ as $L$ and $E_\nu$ vary independently, the effects of oscillations can be distinguished from detector- or flux-related systematic uncertainties which would also affect the observed rates of neutrinos; *i.e.*, the complexity of the oscillatory behavior described in Eq. (2) enables it to be disentangled from other effects. The strategy for making competitive oscillation measurements with neutrino telescopes is thus to exploit the statistical power of these detectors to compensate for their less precise calibrations, by simultaneously constraining systematic uncertainties with the same data set used to extract the oscillation parameters.

Several theories of new physics beyond the Standard Model with energy- and baseline-dependent effects can also be constrained with this data, including additional "sterile" neutrino flavors,[17–19] non-standard neutrino interactions,[20] violations of Lorentz invariance and other signatures of quantum gravity.[21]

## 3.  First measurements

Measurements of atmospheric neutrino oscillations require observation of events at energies well below 100 GeV, lower than those for which neutrino telescopes were originally designed. At higher energies, muon neutrinos generally interact outside the instrumented volume, and the resulting muons are observed as through-going tracks whose energy can be estimated from their brightness. Contained events are predominantly cascades produced by $\nu_e$, $\nu_\tau$, or $\nu_x$ neutral current (NC) interactions.

Below 100 GeV, muon track lengths are shorter than the characteristic size of the detectors, and so most $\nu_\mu$ events are partially or fully contained within the detector, with both the muon track and the hadronic cascade at the $\nu N$ interaction vertex visible. As a result, discrimination of $\nu_\mu$ CC events ("tracks") from other types of events (generically referred to as "cascades") becomes challenging as the track length becomes shorter than the sensor spacing in the detector. Muons from these events are approximately minimum-ionizing particles, and their energy is estimated from the muon track's length rather than its brightness. Specialized analysis techniques are thus required for these events.

Initial measurements of oscillations with neutrino telescopes relied on selection of up-going muon tracks with high quality reconstructions to reject the atmospheric muon background, similar to traditional astronomical analyses, with a subsequent selection of tracks which appear to have finite range within the detector.

The first successful analysis, by ANTARES[22] used the hyperbolic hit time pattern produced when unscattered photons from the Cherenkov cone intersect a string of photosensors to estimate neutrino direction, with the energy provided by the estimated track length. This approach does not provide azimuthal information unless such patterns are observed on multiple strings, but for oscillation studies only the zenith angle is required. Selecting events with very high quality track reconstructions strongly reduced events other than $\nu_\mu$ CC. This selection also preferred $\nu_\mu$ events with low inelasticity, so that the muon range alone provided a good energy estimator. The distribution of reconstructed $L/E_\nu$ for these events showed evidence for neutrino oscillations at the 97.9% confidence level, with values for $\theta_{23}$ and $\Delta m^2_{\text{atm}}$ consistent with measurements by K2K,[23] MINOS,[24] and Super-Kamiokande,[25] as shown in Fig. 2.

A first IceCube analysis[26] used a modified version of the astrophysical point source analysis, comparing the angular distributions of events with signatures of finite muon tracks (low energy) to those with apparently through-going tracks (high energy). This analysis likewise established the observation of oscillations

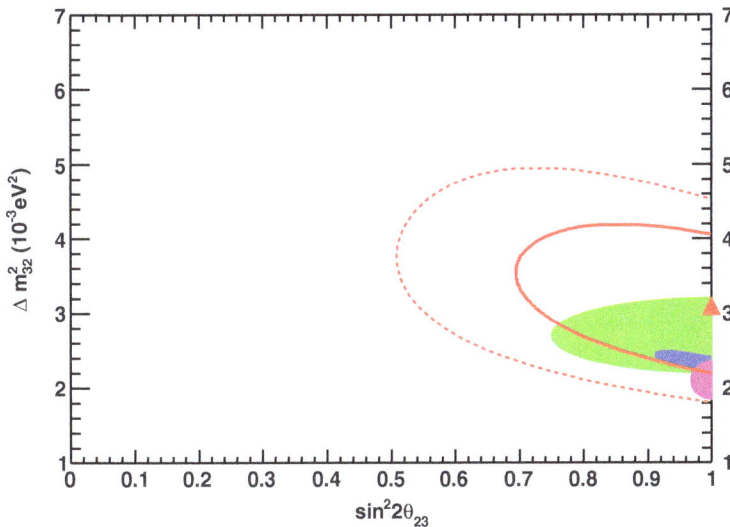

Fig. 2.   68% and 90% C.L. contours (solid and dashed red lines) of the ANTARES oscillation measurement. The best fit point is indicated by the triangle. The solid filled regions show contemporary 68% C.L. results from K2K (green), MINOS (blue) and Super-Kamiokande (magenta) for comparison.

and extracted measurements of the atmospheric parameters consistent with those measured by other experiments, but still with substantially larger uncertainties.

More recent IceCube analyses use more efficient event selection methods that reject particles entering the fiducial volume from outside. With DeepCore, the outer IceCube detector is used as an active veto, tagging events with activity in the veto region prior to the DeepCore event, which minimizes selection bias in the resulting neutrino sample. A second-generation analysis from IceCube[27] combined these techniques with a modified version of the specialized ANTARES low-energy reconstruction to analyze a three-year data set. This improved both the event selection efficiency and the angular resolution of the analysis, and likelihood-based estimators of muon track length and the energy deposition at the $\nu N$ vertex provided energy information. For the first time, the oscillation parameters were measured with a precision approaching that of Super-K,[28] and within a factor of two of contemporary measurements from long-baseline neutrino beam experiments,[29,30] as shown in Fig. 3.

In parallel, maximum-likelihood reconstructions tailored for low energy neutrinos have been developed, providing substantial improvements in detector resolution. The fact that both the $\nu N$ vertex cascade and the primary lepton are visible means that the hypothesis space for reconstruction is ten-dimensional: $(x, y, z, t, \theta_c, \phi_c, E_c, \theta_\ell, \phi_\ell, E_\ell)$, where $c$ and $\ell$ refer to the cascade and lepton, respectively. Full consideration of the particles produced at the hadronic vertex, rather than collecting them into a single "cascade," would expand the dimensionality further. Computationally, a full maximum-likelihood treatment is thus difficult —

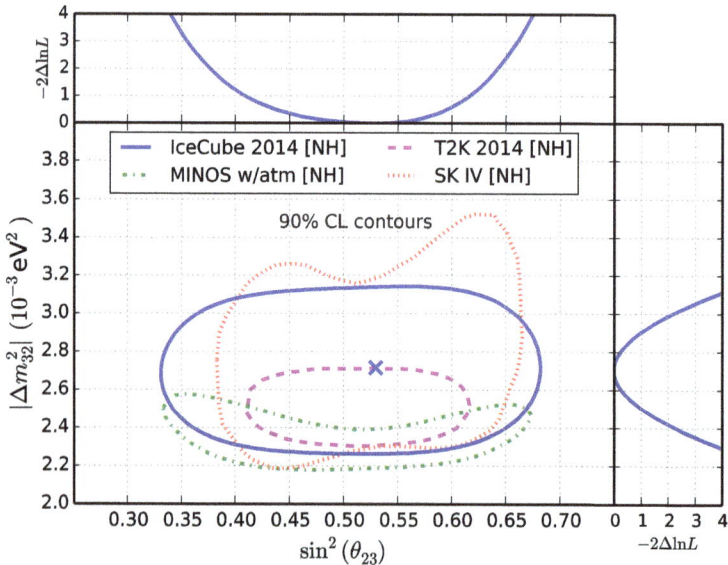

Fig. 3.   Current measurement of the oscillation parameters by IceCube, compared to 2014 results from T2K, MINOS, and Super-Kamiokande.

particularly in IceCube where Cherenkov photon transport through the ice is complex and spatially non-uniform — and simplifications must be made. At present, these algorithms ignore the transverse momentum of the outgoing particles and model the cascade and leading lepton as collinear, which fundamentally limits the reconstruction of the neutrino direction. Nevertheless, these reconstruction methods, along with refinements of the event selection methods discussed above, permit the identification of contained low energy neutrino event samples with efficiency more than an order of magnitude higher than in the original studies. Although these analyses are still being finalized, Monte Carlo data challenges indicate that these improvements will provide significantly improved precision for measurements of atmospheric mixing parameters, comparable to the leading long-baseline measurements shown in Fig. 3.

## 4. Matter effects and the mass ordering

The model of two neutrino species mixing in vacuum shown in Eq. (2) is a good approximation to oscillations at energies above roughly 15 GeV. As lower energy thresholds are achieved and larger sets of neutrino events are recorded, the effects of both mixing with the third neutrino eigenstate and matter on neutrino propagation become visible. Matter effects are particularly interesting, because they enable determination of the ordering of the mass eigenstates. At present, it is unknown whether the mass of the third eigenstate is greater or less than that of the first two, a question with significant implications for theories of the origin of neutrino mass. This question is known as the *mass hierarchy* or *mass ordering*; the case of a heavier third eigenstate is referred to as the 'normal' ordering, while a lighter third eigenstate is called the "inverted" ordering.

The presence of electrons in the Earth with number density $n_e$ contributes a term $V = \sqrt{2}G_F n_e$ to the Hamiltonian for electron-type neutrinos due to coherent forward charged-current scattering, a phenomenon known as the Mikheyev–Smirnov–Wolfenstein (MSW) effect.[31,32] Diagonalizing the Hamiltonian including this extra term leads to "effective" mass eigenstates different from the vacuum masses, an effect analogous to a refractive index for light. This in turn produces an effective mixing angle $\theta_{13}^m$ of

$$\sin 2\theta_{13}^m \simeq \frac{\Delta m_{31}^2}{\Delta^m} \sin 2\theta_{13}, \tag{3}$$

where

$$\Delta^m \simeq \sqrt{(\Delta m_{31}^2 \cos 2\theta_{13} \mp 2\sqrt{2}G_F n_e E_\nu)^2 + (\Delta m_{31}^2 \sin 2\theta_{13})^2} \tag{4}$$

is the mass splitting between the effective eigenstates and the plus and minus signs refer to antineutrinos and neutrinos, respectively. Notably, this leads to resonant enhancement of the mixing when

$$|\Delta m_{31}^2| \cos 2\theta_{13} = 2\sqrt{2}G_F n_e E_\nu, \tag{5}$$

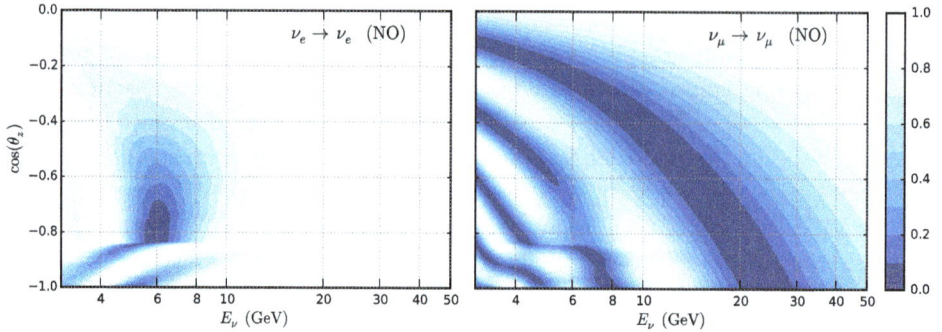

Fig. 4.   Oscillograms of survival probabilities of electron (left) and muon (right) neutrinos assuming the normal mass ordering.[35] Vacuum oscillations produce smooth curving bands in the muon disappearance oscillogram, while matter effects visibly distort this pattern below 10 GeV, and induce electron neutrino disappearance in the same energy range. The horizontal discontinuity around $\cos(\theta_z) = -0.85$ reflects the core-mantle boundary.

but with the enhancement affecting neutrinos if the mass ordering is normal $(\Delta m_{31}^2 > 0)$ and antineutrinos if the mass ordering is inverted $(\Delta m_{31}^2 < 0)$. Although the resonance condition of Eq. (5) is well defined, the varying density of the Earth and the finite, energy-dependent oscillation length lead to a complicated distortion of the oscillation probabilities, as shown in Fig. 4. This figure also includes other effects which arise from the interplay of oscillation lengths with sharp density transitions within the Earth such as the core-mantle boundary, which are known as *parametric enhancements*.[33,34]

The existence of mass ordering-dependent distortions of the oscillation probabilities at energies slightly lower than are accessible with ANTARES and DeepCore has led to the development of proposals to build next-generation very large volume neutrino detectors focused on neutrino physics measurements: PINGU at the South Pole and ORCA in the Mediterranean.

## 5. PINGU and ORCA

ORCA and PINGU are conceptually similar: large volume Cherenkov neutrino detectors with sensor densities substantially higher than those in existing neutrino telescopes, though still much lower than in underground neutrino detectors such as Super-K. Both designs have fiducial masses of approximately 5 MTon, instrumented by several thousand optical modules — similar to the number of sensors used to cover a GTon volume in the IceCube or KM3NeT designs.

The PINGU design closely follows that of IceCube DeepCore, with an additional 40 strings each consisting of 96 Digital Optical Modules (DOMs) containing 10″ Hamamatsu super-bialkali PMTs. The string spacing is approximately 22 m with a DOM spacing of 3 m on each string, leading to a cylindrical instrumented volume roughly 75 m in radius and 300 m high. The ORCA design uses KM3NeT's multi-PMT DOMs, which each contain 31 smaller 3″ PMTs arranged around a 17″ sphere,[36] offering improved photon counting capabilities with respect to the ANTARES optical modules.[37] The ORCA benchmark design contains 115 strings of 18 multi-PMT DOMs, with a 20 m string spacing and a 6 m DOM spacing, although the optimization of the geometry now indicates a preferred spacing of 9 m, endorsed by the collaboration. The overall photocathode coverage of the two designs is similar; ORCA has half as many DOMs as PINGU, but each ORCA DOM provides three times the photocathode area. In water, relatively few Cherenkov photons are scattered, so amplitude information can be extracted in ORCA from the time over threshold; the PINGU electronics will record the full waveform of the PMT trace with a planned sampling rate of 250 MHz in order to cope with the irregular arrival times of scattered photons.

The costs of each project are also comparable, although a detailed comparison is difficult due to the different accounting systems used in the US and Europe. ORCA's cost is estimated to be approximately €40 M, equivalent to $46 M at current exchange rates, and is dominated by the cost of detector hardware. The cost of PINGU hardware is estimated to be $48 M, with a further $23 M in expected logistical costs associated with deployment of the detector. A first 'demonstrator' phase of ORCA, consisting of 6–7 strings, has already been funded and is expected to be deployed by the end of 2016. In an optimistic scenario the full detector could be operational by 2020. PINGU would be constructed as the first part of the IceCube-Gen2 facility, with deployment anticipated over three austral summer seasons once the hot water drill equipment is available at the South Pole station.

While similar in many respects, there are two noteworthy differences between PINGU and ORCA. First, PINGU would be sited within the IceCube detector, which could be used as an active veto for atmospheric muons. ORCA simulations show that the muon background can be controlled by selecting well-reconstructed up-going contained neutrino events, and the present performance estimates for both ORCA and PINGU make use of only up-going events. However, the IceCube veto layers would provide a data-driven model of the residual muon backgrounds for comparison to Monte Carlo simulations of these events, and potentially also permit constraints on systematic uncertainties using the unoscillated down-going neutrino flux if the backgrounds can be controlled with sufficient precision.

Second, attenuation of Cherenkov photons in ORCA will be dominated by the ∼60 m absorption length (for blue light),[38] with relatively little scattering. In contrast, attenuation in the Antarctic ice cap is dominated by scattering, with effective scattering lengths of 40–50 m at PINGU's depths and absorption lengths of around 200 m.[39] In both detectors, the attenuation lengths are several times the inter-string spacing, but the high level of scattering in PINGU is more difficult to deal with computationally, especially given the spatial variation in the optical properties of the medium. As a result, the angular resolution and low-energy muon track identification performance achieved in ORCA simulations are superior to PINGU's. In addition, accounting for the broader Cherenkov emission profile from hadronic showers in ORCA reconstruction provides an estimate of the inelasticity of the event. In addition to identifying NC events, the inelasticity provides some separation of neutrinos from antineutrinos, with potentially significant enhancement of the determination of the mass ordering.[40] Efforts are underway to incorporate these techniques in IceCube, but are significantly complicated by the non-uniformity of the optical medium. Continued algorithmic development is ongoing for both detectors, and the performance of both detectors is likely to improve over present projections.

## 6. Projected performance

The primary scientific goals of ORCA and PINGU are the determination of the neutrino mass ordering via observation of the matter-induced distortions discussed in Sec. 4, and improved measurements of the mixing between the second and third eigenstates ("atmospheric" mixing), via both muon neutrino disappearance and tau neutrino appearance. With expected data sets of tens of thousands of neutrino events per year, the limiting factor in these measurements will be the degree to which systematic uncertainties can be constrained from the data and their effects disentangled from the parameters of interest.

Highly detailed Monte Carlo simulations of both detectors have been produced, based on the extensive experience operating the DeepCore and ANTARES detectors. The reconstruction algorithms developed for PINGU and ORCA are also inspired by DeepCore and ANTARES algorithms; the projected performance achieved with the benchmark detectors and reconstruction algorithms described above is shown in Figs. 5 and 6. Similar to what is described in Sec. 3, the PINGU

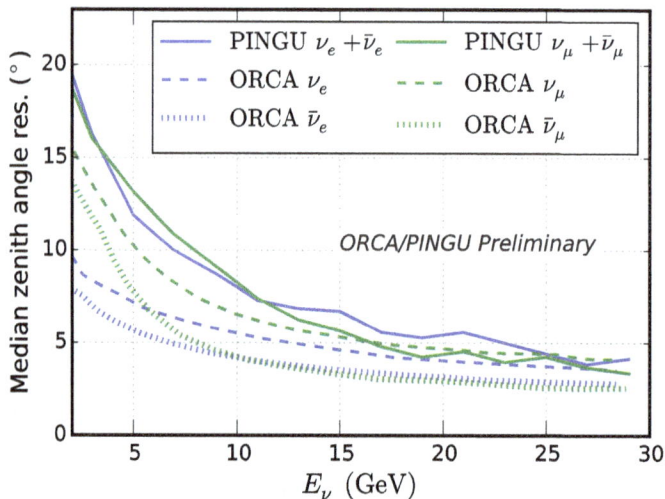

Fig. 5.   Median zenith resolutions.

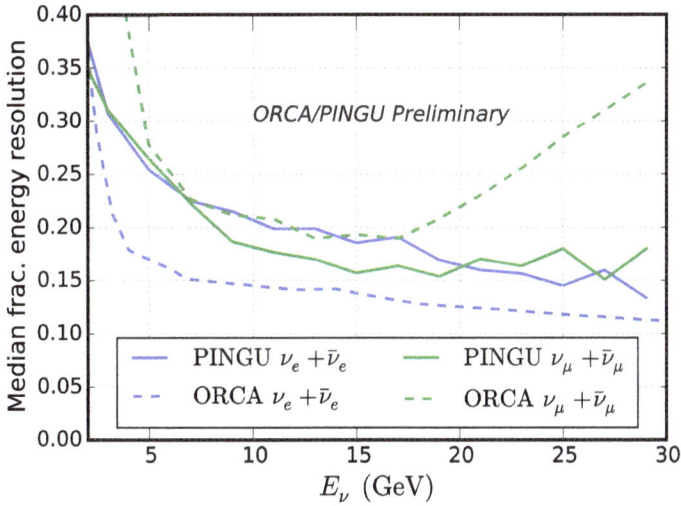

Fig. 6.   Median energy resolutions.

strategy uses a simultaneous fit of eight parameters for all event types, providing similar resolutions for track-like and cascade-like topologies, and robust energy estimates. This approach is currently implemented with an assumption that the incoming neutrino, charged lepton and hadronic system are collinear; the misalignment between neutrino and outgoing particles is included in the angular resolution.

On the ORCA side, two distinct reconstructions have been developed to search for the charged leptons in $\nu_e$ and $\nu_\mu$ CC interactions. Here again the assumption is that the neutrino direction is that of the reconstructed lepton. The reconstruction is performed in several steps. First the interaction vertex is identified, with an accuracy better than 1 m. Then the direction of the charged lepton is determined. In the muon channel, the energy estimate mostly relies on the observed path length inside the detector. This limits the resolution above 20 GeV as the muons escape the instrumented volume, as can be seen in Fig. 6. (For PINGU, this limitation arises at higher energies as the muon range is also measured with the surrounding IceCube DOMs.) In the electron channel, all the neutrino energy is deposited inside the detector and the resolution continues to improve with energy. The ORCA reconstruction for the electron channel also fits the inelasticity of the event by studying the hit angle distribution with respect to shower direction: low inelasticity event have a better defined peak at the Cherenkov angle. In the muon channel, some sensitivity to the inelasticity is also achieved by comparing light arrival times to the hypotheses of spherical emission (high inelasticity, dominated by the hadronic shower at the vertex) or according to a Cherenkov cone (low inelasticity, dominated by the muon). The use of these techniques to statistically distinguish neutrinos and antineutrinos, and consequently further improve the discrimination of the neutrino mass ordering, is currently being studied.

One of the most important ingredients for measuring the atmospheric mixing parameters and discriminating the mass ordering is the ability to properly separate the topology of tracks and cascades. This is often referred to as particle identification (PID). It is achieved in both ORCA and PINGU by means of multivariate algorithms relying on the quality of the reconstruction under each topology hypothesis. The final classification probability therefore depends on the details of the trigger and reconstruction algorithms. The performances currently achieved yield, above 10 GeV, a correct classification in ~80% or more of cases for both PINGU and ORCA. A lower energies, where clean muon tracks are more difficult to observe, the classification rate slowly decreases towards a random 50% value.

A third category of events is included in the ORCA PID chain, which relates to atmospheric muons. Those are primarily rejected by selecting up-going reconstructed events. In addition, simulations of the proposed ORCA detector indicate that the long scattering lengths in seawater mean that light will be reliably observed from elements of tracks outside the fiducial volume. As a consequence, the reconstructed interaction vertex is preferentially found outside the instrumented volume for atmospheric muons. This strongly suppresses the muon background, while preserving most atmospheric neutrinos interacting in the detector. Extra information, such as hit time residuals, topological variables, etc., is added in a Random Decision Forest to achieve a final contamination at an adjustable level of contamination of order of few percent. The remaining atmospheric muons equally fall into the track and cascade categories. Similar strategies are being developed in PINGU, but most of the rejection of atmospheric muons will come from the use of IceCube/DeepCore as an external veto. This tagging technique is already in use for DeepCore, for which the muon contamination in up-going neutrino samples is between ~1% and ~7% depending on the analysis. This approach can also be used to minimize selection bias in the resulting neutrino sample, and potentially to study the unoscillated down-going neutrino flux to better constrain systematic uncertainties.

### 6.1. *Determining the neutrino mass ordering*

Two statistical methods are used to estimate the significance with which the mass ordering can be determined by ORCA and PINGU. The first uses ensembles of pseudo-data drawn from parametric descriptions of the detectors effective mass, PID efficiency, and energy and zenith resolutions. These are in turn based on full Monte Carlo simulations of the detectors response to individual neutrino events and include non-Gaussian tails, biases due to invisible particles produced in hadronic showers, etc. A binned likelihood method with Gaussian penalty terms encoding prior knowledge of the nuisance parameters is used to fit the oscillation parameters for each pseudo-data set. The log-likelihood ratio (LLR) of this fit is then calculated with respect to the best-fitting hypothesis with the opposite mass ordering, and the expected significance is determined by integrating the tail of the LLR distribution for the opposite ordering beyond the median LLR obtained under the correct ordering. The opposite-ordering hypothesis not only reverses the sign of $\Delta m^2_{\text{atm}}$, but also alters the values of the other oscillation parameters (e.g. $\theta_{23}$) to most closely match the expected oscillogram for the injected parameters. A second, faster method is used to facilitate inclusion of larger numbers of systematic uncertainties and evaluation of performance for a wider range of values of the atmospheric mixing parameters. This method uses the same parametric descriptions of the detector response, but instead of generating large ensembles of pseudo-data relies on the "Asimov" assumption that the most likely values of the detector observables under a given hypothesis will provide the median result. The results of the two methods have been compared for selected values of the oscillation parameters and agree very well.

In studies using the full pseudo-data set method, both PINGU and ORCA include the effects of uncertainties in the $\nu/\bar{\nu}$ ratio, the $\nu_\mu/\nu_e$ ratio, and both scalar and linearly energy-dependent factors multiplying the overall event rate (which could be due to errors in effective area, atmospheric neutrino flux, or the neutrino-nucleon cross section $\sigma_{\nu N}$). ORCA studies include an additional uncertainty on the NC/CC ratio, and PINGU an additional term representing possible errors in the energy calibration (assumed to be a proportional factor). Using the faster $\Delta\chi^2$ method enables PINGU studies where the linear uncertainties in event rate are replaced by the full suite of neutrino-nucleon interaction model systematics implemented in GENIE[41] and the parametrization of atmospheric neutrino uncertainties described by Barr *et al.*[16] The neutrino-nucleon effects reduce the expected significance slightly for short exposure times, but have negligible effect for exposures of 3 years or more as other systematics become dominant. The flux uncertainties reduce the expected four-year significance by as much as $0.2\sigma$ if no constraints from down-going neutrinos are imposed.

The median significances with which the mass ordering can be determined by ORCA and PINGU with three and four years of data, respectively, are shown in Fig. 7, for both the NO and the IO cases, as a function of the true value of $\theta_{23}$. Both detectors are projected to be able to determine the mass ordering with approximately 3–4 years of data, although it should be noted that these projections represent the median expectation and the actual data could provide more or less significance. Larger values of $\theta_{23}$ increase the magnitude of the matter effects and thus the difference between the two orderings. However, if the true ordering is inverted, degeneracies between the octant of $\theta_{23}$ and the ordering largely cancel this effect, leading to much less variation in the expected significance. Consistent results from investigations by both PINGU and ORCA suggest that the sensitivity also depends somewhat on the true value of $\delta_{\text{CP}}$, with reductions of up to $0.5\sigma$ at 3 years possible; a fixed value of $\delta_{\text{CP}} = 0$ was assumed in Fig. 7 for both detectors.

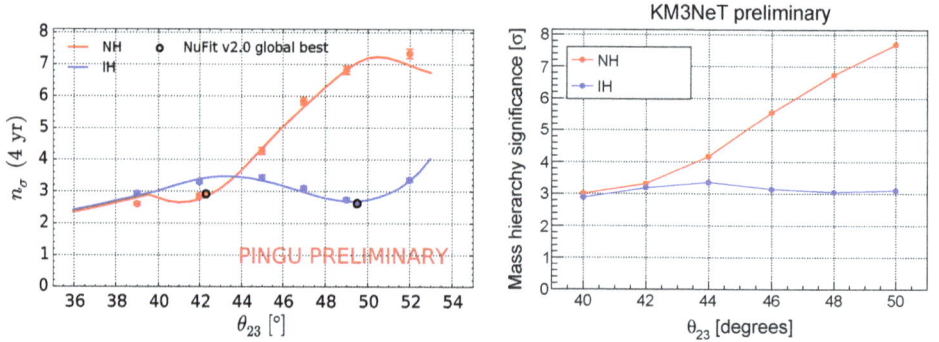

Fig. 7.   Estimated significance with which PINGU (left) and ORCA (right) can determine the mass ordering as a function of $\theta_{23}$. A four-year data set is assumed for PINGU and a three-year set for ORCA. For PINGU, the dots indicate the results from ensembles of pseudo-experiments while the solid line shows the prediction of a faster method (see text for details). For the normal ordering, larger values of $\theta_{23}$ increase the significance, while if the true ordering is inverted, degeneracies between the octant and the ordering largely annihilate these improvements.

## 6.2.   Measurements of atmospheric neutrino oscillations

The improved detector resolutions possible with ORCA and PINGU will also permit better measurements of the atmospheric neutrino mixing parameters $\theta_{23}$ and $\Delta m^2_{\mathrm{atm}}$. Statistical analyses similar to those used to estimate the sensitivity to the mass ordering are used to estimate the precision with which these parameters can be measured, shown in Fig. 8. The expected 68% and 90% confidence regions from four years of PINGU data are calculated assuming the normal ordering, for a mass splitting of $\Delta m^2_{32} = 2.38 \times 10^{-3}$ eV$^2$ and three possible mixing angles: $\sin^2 \theta_{23} = 0.386$, from the 2012 fit of Fogli et al.,[42] the current best fit of $\sin^2 \theta_{23} = 0.452$,[43] and maximal mixing. The expected three-year ORCA 68% confidence intervals are also calculated assuming normal ordering, for injected values of $\Delta m^2_{32} = 2.45 \times 10^{-3}$ eV$^2$ and three test values of $\sin^2 \theta_{23}$: 0.58, 0.50, and 0.42. The measurement of $\Delta m^2_{32}$ depends strongly on accurate calibration of the energy of neutrino events, and an ad hoc uncertainty on the reconstructed neutrino energy scale is included in both studies as a systematic. These preliminary studies show that this uncertainty can be constrained at least to the level of a few percent from the data itself. Current measurements by MINOS[44] and T2K[45] and projected sensitivities of T2K[45] and NOvA[46,47] in 2020 are also shown in Fig. 8 (right) for comparison.

In addition to measurements of $\nu_\mu$ disappearance, the size and relatively high energy range at which ORCA and PINGU observe the neutrino flux permits uniquely precise measurements of the rate of $\nu_\tau$ appearance. In the standard neutrino oscillation paradigm, the main channel through which $\nu_\mu$ disappear is by oscillation into $\nu_\tau$. However, at the energies probed by long-baseline experiments,

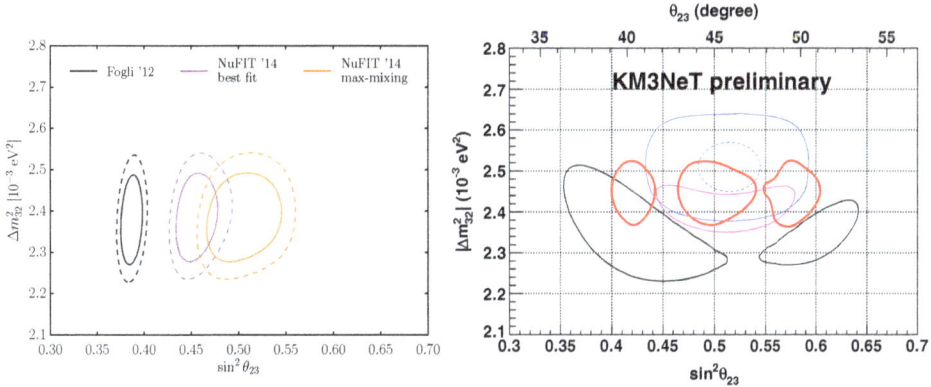

Fig. 8. Estimated precision with which PINGU (left) and ORCA (right) will determine atmospheric mixing parameters. On the left, 68% (solid) and 90% (dashed) confidence regions are shown for three possible true sets of parameters. On the right, 68% regions for ORCA (red) are shown for three sets of injected parameters, in comparison to reported 68% regions from MINOS (black) and T2K (blue), and the projected 2020 precision of NOvA (magenta) and T2K (dashed blue).

the CC cross section for these $\nu_\tau$ is strongly suppressed by the mass of the $\tau$ lepton, which limits the precision possible from long-baseline measurements and leaves room for the possibility of new physics, including the existence of sterile neutrinos which do not couple to the $Z$ boson or non-standard $\nu_\tau$ interactions.[50]

While $\nu_\tau$ CC events cannot be distinguished individually by PINGU or ORCA, the rate of $\nu_\tau$ appearance can be observed on a statistical basis as an excess of cascade-like events over the baseline from atmospheric NC and $\nu_e$ CC events. (The intrinsic $\nu_\tau$ atmospheric flux is essentially zero.) This excess cascade component will follow the oscillatory $L/E$ dependence of Eq. (2) multiplied by the rising $\nu_\tau$ CC cross section, leading to primarily vertically up-going events and a distinct spectral feature. By contrast, the NC and $\nu_e$ CC events follow a power-law spectrum and are most numerous at the horizon. As a result, the $\nu_\tau$ component can be separated effectively from the other cascade flux components and measured with high precision, as shown in Fig. 9.

## 7. Outlook

Neutrino telescopes have proven to be useful tools for measuring neutrino properties, taking advantage of the abundant flux of atmospheric neutrinos. The first measurement from ANTARES followed by more accurate studies with IceCube/DeepCore have demonstrated the potential of this type of detectors for precise measurements in the neutrino sector. While the latest results from IceCube/DeepCore already provide competitive results in the measurement of $\theta_{23}$ and $\Delta m^2_{\text{atm}}$, further improvement is expected in the coming years. However, the current generation of neutrino telescopes is only marginally sensitive to the matter effects that can reveal the neutrino mass ordering through subdominant effects in the flavor conversion rates. Such kind

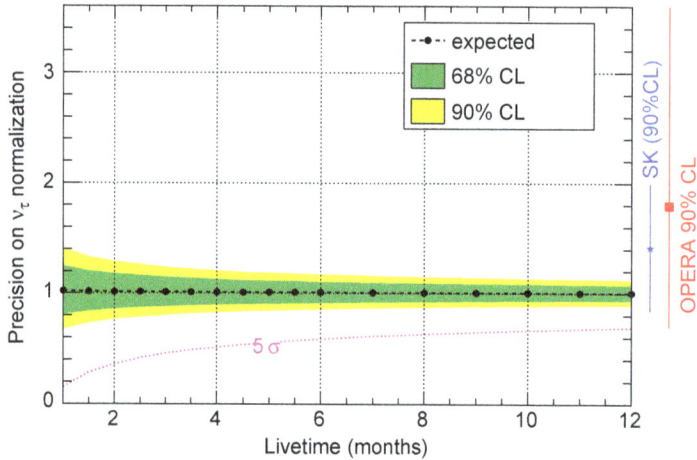

Fig. 9. Expected precision with which the rate of tau neutrino appearance can be measured in PINGU. The green and yellow bands show the median 68% and 90% uncertainties on the measured rate, assuming the rate is that predicted by standard oscillation theory. The bars at right show the 90% uncertainties of current measurements[48,49] for comparison.

of studies indeed require a denser PMT network than currently available, in order to lower the energy threshold down to the GeV range. This is the main goal of the planned ORCA and PINGU detectors being developed in the framework of the KM3NeT and IceCube-Gen2 collaborations, respectively. The performance of these detectors has been assessed relying on full Monte Carlo simulations and incorporating the main identified sources of systematics. Although the PINGU and ORCA designs differ in several aspects, their expected performance appears to be comparable. These studies provide independent and robust indications that the neutrino mass ordering can be established with a median expected significance of at least $3\sigma$ (depending on the true value of $\theta_{23}$) after 3 to 4 years of data taking. Funding permitting, this could be achieved as early as 2023. These detectors would also provide greatly improved measurements of atmospheric mixing parameters in a higher energy range and through different channels than will be probed by current and planned long baseline neutrino beam experiments, providing considerable potential for discovery of new physics.

## References

1. E. Andrés, *et al.*, *Nature* **410** (2001) 441.
2. R. Abbasi, *et al.* [IceCube Collaboration], *Phys. Rev. D* **83** (2011) 012001.
3. S. Adrián-Martínez, *et al.* [ANTARES Collaboration], *Eur. Phys. J. C* **73**(10) (2013) 2606.

4. K.A. Olive, *et al.* [Particle Data Group Collaboration], *Chin. Phys. C* **38** (2014) 090001.
5. M. Amenomori, *et al.* [Tibet AS Gamma Collaboration], *Astrophys. J.* **626** (2005) L29.
6. A.A. Abdo, *et al.*, *Phys. Rev. Lett.* **101** (2008) 221101.
7. H.S. Ahn, *et al.*, *Astrophys. J.* **707** (2009) 593.
8. H.S. Ahn, *et al.*, *Astrophys. J.* **714** (2010) L89.
9. A.D. Panov, *et al.*, *Bull. Russ. Acad. Sci. Phys.* **73** (2009) 564.
10. O. Adriani, *et al.* [PAMELA Collaboration], *Science* **332** (2011) 69.
11. E. Richard, *et al.* [Super-Kamiokande Collaboration], arXiv:1510.08127 [hep-ex].
12. M. Honda, T. Kajita, K. Kasahara, S. Midorikawa and T. Sanuki, *Phys. Rev. D* **75** (2007) 043006.
13. T.K. Gaisser, *Astropart. Phys.* **35** (2012) 801.
14. R. Enberg, M.H. Reno and I. Sarcevic, *Phys. Rev. D* **78** (2008) 043005.
15. K. Daum, *et al.* [Fréjus Collaboration], *Z. Phys. C* **66** (1995) 417.
16. G.D. Barr, T.K. Gaisser, S. Robbins and T. Stanev, *Phys. Rev. D* **74** (2006) 094009.
17. V. Barger, Y. Gao and D. Marfatia, *Phys. Rev. D* **85** (2012) 011302.
18. S. Razzaque and A.Y. Smirnov, *Phys. Rev. D* **85** (2012) 093010.
19. A. Esmaili and A.Y. Smirnov, *JHEP* **1312** (2013) 014.
20. A. Esmaili and A.Y. Smirnov, *JHEP* **1306** (2013) 026.
21. R. Abbasi, *et al.*, [IceCube Collaboration], *Phys. Rev. D* **82** (2010) 112003
22. S. Adrián-Martínez, *et al.*, [ANTARES Collaboration], *Phys. Lett. B* **714** (2012) 224.
23. M.H. Ahn, *et al.*, [K2K Collaboration], *Phys. Rev. D* **74** (2006) 072003.
24. P. Adamson, *et al.*, [MINOS Collaboration], *Phys. Rev. Lett.* **106** (2011) 181801.
25. K. Abe, *et al.*, [Super-Kamiokande Collaboration], *Phys. Rev. Lett.* **107** (2011) 241801.
26. M.G. Aartsen, *et al.*, [IceCube Collaboration], *Phys. Rev. Lett.* **111**(8) (2013) 081801.
27. M.G. Aartsen, *et al.*, [IceCube Collaboration], *Phys. Rev. D* **91**(7) (2015) 072004.
28. A. Himmel, [Super-Kamiokande Collaboration], *AIP Conf. Proc.* **1604** (2014) 345.
29. P. Adamson, *et al.*, [MINOS Collaboration], *Phys. Rev. Lett.* **110**(25) (2013) 251801.
30. K. Abe, *et al.*, [T2K Collaboration], *Phys. Rev. Lett.* **112**(18) (2014) 181801.
31. L. Wolfenstein, *Phys. Rev. D* **17** (1978) 2369.
32. S.P. Mikheev and A.Y. Smirnov, *Sov. J. Nucl. Phys.* **42** (1985) 913 [*Yad. Fiz.* **42** (1985) 1441].
33. S. Choubey and P. Roy, *Phys. Rev. D* **73** (2006) 013006.
34. M. Blennow and A.Y. Smirnov, *Adv. High Energy Phys.* **2013** (2013) 972485.
35. J.P. Yáñez and A. Kouchner, arXiv:1509.08404, to appear in the special issue "Neutrino Masses and Oscillations" of *Adv. High Energy Phys.*
36. S. Adrián-Martínez, *et al.*, [KM3NeT Collaboration], *Eur. Phys. J. C* **74**(9) (2014) 3056.
37. P. Amram, *et al.*, [ANTARES Collaboration], *Nucl. Instrum. Meth. A* **484** (2002) 369.
38. J.A. Aguilar, *et al.*, [ANTARES Collaboration], *Astropart. Phys.* **23** (2005) 131.
39. M.G. Aartsen, *et al.*, [IceCube Collaboration], *Nucl. Instrum. Meth. A* **711** (2013) 73.
40. M. Ribordy and A.Y. Smirnov, *Phys. Rev. D* **87**(11) (2013) 113007.
41. C. Andreopoulos, C. Barry, S. Dytman, H. Gallagher, T. Golan, R. Hatcher, G. Perdue and J. Yarba, arXiv:1510.05494 [hep-ph].
42. G.L. Fogli, E. Lisi, A. Marrone, D. Montanino, A. Palazzo and A.M. Rotunno, *Phys. Rev. D* **86** (2012) 013012.
43. M.C. Gonzalez-Garcia, M. Maltoni and T. Schwetz, *JHEP* **1411** (2014) 052.
44. P. Adamson, *et al.*, [MINOS Collaboration], *Phys. Rev. Lett.* **112** (2014) 191801.

45. K. Abe, *et al.*, [T2K Collaboration], *Phys. Rev. D* **91**(7) (2015) 072010.
46. M.D. Baird, Ph.D. Thesis, Indiana University (2015), FERMILAB-THESIS-2015-24.
47. S.M. Lein, Ph.D. Thesis, University of Minnesota (2015), FERMILAB-THESIS-2015-21.
48. N. Agafonova, *et al.*, [OPERA Collaboration], *Phys. Rev. Lett.* **115** (2015) 121802.
49. K. Abe, *et al.*, [Super-Kamiokande Collaboration], *Phys. Rev. Lett.* **110** (2013) 181802.
50. S. Parke and M. Ross-Lonergan, arXiv:1508.05095 [hep-ph].

# Chapter 15

# Radio Detection of High Energy Neutrinos

Amy Connolly

*Ohio State University, Department of Physics and CCAPP,*
*Columbus, OH 43210 USA*

Abigail G. Vieregg

*University of Chicago, Department of Physics, Enrico Fermi Institute, Kavli*
*Institute for Cosmological Physics, Chicago, IL 60637 USA*

## 1. Introduction

Ultra-high energy (UHE) neutrino astrophysics sits at the crossroads of particle physics, astronomy, and astrophysics.[a] Through neutrino astrophysics, we can uniquely explore the structure and evolution of the universe at the highest energies at cosmic distances and test our understanding of particle physics at energies greater than those available at particle colliders.

The detection of UHE neutrinos would shed light on the nature of the astrophysical sources that produce the highest energy particles in the universe. Astrophysical sources almost certainly produce UHE neutrinos in hadronic processes. Also, neutrinos above $10^{17}$ eV should be produced through the GZK effect,[1,2] where extragalactic cosmic rays above $10^{19.5}$ eV interact with the cosmic microwave background within tens of Mpc of their source. It was first pointed out by Berezinsky and Zatsepin[3,4] that this cosmogenic neutrino flux, sometimes called "BZ" neutrinos, could be observable. These neutrinos would point close to the cosmic ray production site both because of the close proximity of the interaction to the source and because the decay products are Lorentz boosted along the line of sight. The latter

---

[a]Here, we loosely define the UHE regime to be the energy range near $10^{18} - 10^{21}$ eV. A few radio experiments have sensitivities that reach to lower energies, while others have energy thresholds at higher energies.

effect constrains the direction the most strongly (1 MeV/1 EeV = $10^{-12}$ whereas 100 Mpc/1 Gpc = 0.1). Since cosmic rays do not follow straight paths in magnetic fields and get attenuated above the GZK threshold, and high energy photons ($E > 10^{14}$ eV) are attenuated by the cosmic infrared background, neutrinos offer the unique ability to point to the location of the highest energy cosmic accelerators in the sky.

UHE neutrino measurements will also have important implications for high energy particle physics and determining neutrino properties. A sample of UHE neutrinos would allow for a measurement of the $\nu$-p cross section[5,6] and a direct test of weak interaction couplings at center-of-mass (CM) energies beyond the Large Hadron Collider (a $10^{18}$ eV neutrino collides with a proton at rest at 45 TeV CM). A strong constraint on the UHE neutrino flux can even shed light on models of Lorentz invariance violation.[7,8]

IceCube has recently discovered neutrinos with energies up to a few $10^{15}$ eV that are likely produced by astrophysical sources directly.[9,10] Therefore, there is now a pressing motivation to measure the energy spectrum above $10^{15}$ eV with improved sensitivity, to provide insight into the origin of the seemingly cosmic events and the particle acceleration mechanisms that give rise to them and to determine the high-energy extent of the spectrum.

Figure 1 shows the current best limits on the high energy diffuse neutrino flux compared with a variety of GZK production models. Beyond the astrophysical neutrino events measured up to ~$10^{15}$ eV, IceCube sets the best limits on the high energy neutrino flux up to $10^{17.5}$ eV, and now shares the claim for the

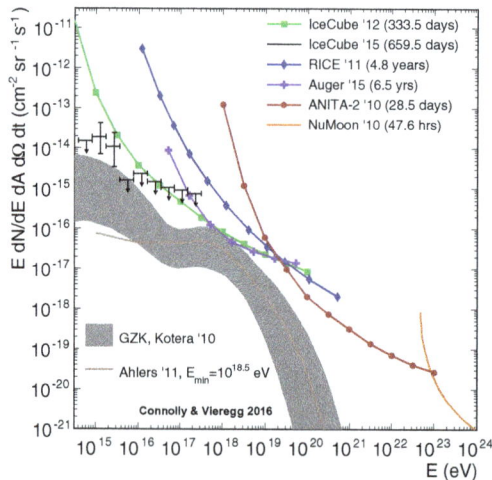

Fig. 1.   The current most competitive experimental constraints on the all-flavor diffuse flux of the highest energy neutrinos compared to representative model predictions.[11,12] Limits are from IceCube, Auger, RICE, ANITA and NuMoon.[13–18] Also shown is the astrophysical neutrino flux measured by IceCube.[17]

best constraints in the region up to $10^{19.5}$ eV with the Pierre Auger Observatory (Auger).[13,17] The current best limit on the flux of neutrinos above $10^{19.5}$ eV comes from the Antarctic Impulsive Transient Antenna (ANITA) experiment.[15,16]

Note that on the vertical axis of Fig. 1 is the differential flux $dN/dE$ multiplied by one power of $E$; the product is proportional to $dN/d\log_{10} E$. On this plot, an experiment that increases its sensitivity in an energy-independent way, for example by increasing its live time, will have its limits move only downward. An experiment that decreases its energy threshold with no other change will have its constraints move only to the left on this plot.

Despite the competitive limits currently imposed by optical detection experiments, it would be prohibitively costly to build a detector utilizing the optical signature that would be sensitive to the full range of predicted possible cosmogenic neutrino populations above $10^{17}$ eV, due to the detector spacings set by the absorption and scattering lengths of optical light in ice.[19] IceCube-Gen2 is a proposed IceCube expansion to 100–300 km$^2$ scale for the detection of neutrinos above $10^{13.5}$ eV, focusing on energies above which the atmospheric neutrino background is not overwhelmingly dominant, and with the discovery potential for BZ neutrinos.[20] However, to achieve sensitivity to the full range of BZ neutrino models, we must instead turn to a different detection mechanism that allows us to instrument larger volumes for comparable cost. Neutrino telescopes that utilize the radio detection technique search for the coherent, impulsive radio signals that are emitted by electromagnetic particle cascades induced by neutrinos interacting with a dielectric. Radio UHE neutrino detection requires a volume $\mathcal{O}(100)$ km$^3$ of dielectric material, which limits the detection medium to be naturally-occurring, that allows radio signals to pass through without significant attenuation over lengths $\mathcal{O}(1)$ km. Current and proposed experimental efforts in this field monitor or plan to monitor immense volumes of glacial ice, whose radio attenuation properties have been directly measured at multiple locations in Antarctica and Greenland, and have the desired clarity.[21–26] We note that although IceCube-Gen2 is nominally on an expansion of the array of optical sensors, a radio array component may be considered for enhancement of sensitivity in the energy range $10^{16}$–$10^{20}$ eV.[20]

Even some of the most pessimistic models predicting BZ neutrino fluxes are within reach of planned experiments using the radio technique. These more pessimistic models tend to have heavy cosmic-ray composition or a weak dependence of source densities on redshift, within constraints set by other measurements. Recent measurements with the Auger and the Telescope Array disfavor a significant iron fraction in the cosmic-ray composition at energies up to and even exceeding $10^{19.5}$ eV, which in turn favors higher fluxes of BZ neutrinos than if the highest energy cosmic rays were pure iron.[27,28] An experiment that has a factor of ~50 improvement over the best sensitivity currently achieved by IceCube near $10^{18}$–$10^{19}$ eV, or reduces the energy threshold with an ANITA-level sensitivity by about a factor of ~50 would reach these pessimistic neutrino flux expectations.

For most models, such an experiment would observe enough events to make an important impact in our understanding of UHE astrophysics and particle physics utilizing the highest energy observable particles in the universe. We note, however, that even these pessimistic models can be evaded through alternate explanations for the cosmic ray data, or more exotic scenarios.[7,29,30]

## 2. Motivation for the radio technique

### 2.1. *Askaryan effect*

In 1962, physicist Gurgen Askaryan predicted that an energetic electromagnetic cascade in a dense dielectric medium should produce observable, coherent electromagnetic radiation.[31] When an energetic charged particle or photon produces an electromagnetic cascade in a dielectric medium, the shower will acquire a ∼20% negative charge excess. This comes about primarily through Compton scattering of electrons in the medium, but also from the annihilation of positrons in the shower with electrons in the medium. If the charge excess is moving at a velocity greater than the phase velocity of light in the medium, the medium will emit Cherenkov radiation. The radiation is emitted most strongly at the angle given by $\cos\theta = 1/n$ with respect to the shower axis, and the power radiated at optical wavelengths is proportional to the shower energy. In ice, a common medium for detection of neutrino interactions, $n = 1.78$ at radio frequencies, giving $\theta = 57°$. For wavelengths longer than the size of the shower along the transverse dimension (perpendicular to the shower axis), the signal is emitted coherently and the electric field strength is proportional to the shower energy. The Cherenkov radiation is coherent at wavelengths longer than the Molière radius in a dense medium ($\sim 10$ cm), corresponding to frequencies $\lesssim 1$ GHz.

The Askaryan effect was first observed in a beam test at SLAC National Accelerator Laboratory in 2001.[32] Using a target of silica sand to produce showers induced by brehmsstrahlung photons from a 28.5 GeV electron beam and a series of broadband, microwave horn antennas with bandwidths covering the range 0.3 to 6 GHz, the beam test showed the first experimental evidence of the coherence effect predicted by Askaryan in the 1960's. The results of the beam test are shown in Fig. 2.[32] Figure 2a shows the measured shower profile as a function of distance along the shower axis, with an inset of the time-domain electric field measured near shower maximum. The measured shower profile (diamonds) matches the prediction well, and the signal is broadband and impulsive. Figure 2b shows the measured electric field strength as a function of shower energy, controlled by changing the number of photons that initiated the shower. The linear dependence of the electric field on the energy of the shower is evidence for coherence. Figure 2c shows the measured electric field strength as a function of frequency compared to results from a Monte Carlo simulation.

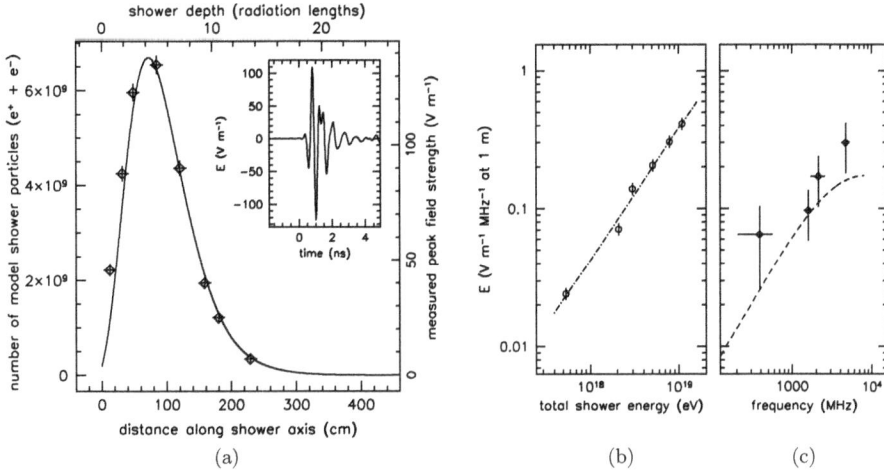

Fig. 2. The first observation of the Askaryan effect, from Ref. 32. (a) The field strengths closely follow the expected shower profile along the shower axis. The time-dependent electric field measured near shower maximum is shown in the inset. (b) The measured electric field plotted against shower energy. The dashed line shows a least-squares fit to the data, consistent with complete coherence. (c) The measured electric field spectrum compared to a semi-empirical parametrization is shown as a dashed line.

The observation of the Askaryan effect was confirmed later in rock salt and again in ice at separate subsequent beam tests at SLAC in 2004 and 2006, respectively.[33,34] The observation of the Askaryan effect in ice is of particular note, since many experimental efforts monitor large volumes of naturally occurring ice. Although the lunar regolith (sand) continues to be used as a target for neutrino searches[35,36] and rock salt has been investigated as a potential target medium,[37] most efforts use ice as a target due to its remarkable naturally-occurring volume, radio clarity, and uniformity.

## 2.2. *A radio clear dielectric for a detector*

To determine the suitability of different naturally-occurring dielectric media for the detection of radio emission from highly energetic neutrinos, many measurements have been made of the dielectric attenuation at radio frequencies at different sites around the world. The field attenuation length $L_\alpha$ is defined to be the distance over which the distance-corrected electric field $(E(r))$ drops by a factor of $1 - 1/e$. At a distance $r$:

$$rE(r) = E(0)e^{-r/L_\alpha}.$$ (1)

The following equation given in Ref. 26 is useful for relating the field attenuation length reported in particle astrophysics to the dielectric loss per distance $N_L$

reported by geophysicists:

$$N_L[\text{dB}/\text{km}] = 8686.0 \langle L_\alpha[\text{m}] \rangle. \tag{2}$$

Askaryan himself proposed dense media such as ice, salt, and even the lunar regolith for the detection of cosmic ray showers.[31] Ice was later proposed as a promising, cost-effective medium for the detection of showers produced by high energy neutrino interactions as well.[38] Several measurements have confirmed that there are naturally-occurring ice sheets that exhibit the dielectric properties necessary to make an excellent medium for neutrino detection. Measurements in naturally-occurring rock salt have found more modest attenuation lengths, and the lunar regolith has also been used as a target of observation for neutrino interactions above $\sim 10^{21}$ eV.

### 2.2.1. *Attenuation length of glacial ice*

One of the most promising sites for a large radio detector is in the ice near the South Pole, which sits atop $\sim 2.8$ km of radio-clear glacial ice. The measurement of the radio attenuation length in ice at the South Pole with the smallest uncertainties and largest horizontal component[23,24] gives a depth-averaged field attenuation length $\langle L_\alpha \rangle$ at 300 MHz of $1600^{+255}_{-120}$ m over the top 1500 m of ice,[24] the depth over which a surface or sub-surface experiment would be the most sensitive to neutrino interactions. This result is derived from a comparison between the amplitude of a broadband signal transmitted from an antenna deployed on an IceCube string 2500 m below the surface and the amplitude of the same signal received at an antenna buried at about 30 m below the surface, 2000 m away horizontally.

A site with similar properties is found at the peak of the the high plateau in Greenland, at Summit Station, which sits on top of 3 km of glacial ice. The attenuation length of the ice at Summit Station was measured by bouncing an impulsive radio signal off of the bottom of the glacier, measured at $3014^{+48}_{-50}$ m depth, and comparing the strength of the return signal to a direct measurement between the transmitter and the receiver through air. The field attenuation length has been measured to be $\langle L_\alpha \rangle = 947^{+92}_{-85}$ m at 75 MHz averaged over all depths.[25] The power reflection coefficient $R$ is assumed to be 0.3 for the ice-bedrock interface. For a reasonable extrapolation to higher frequencies, this measurement is consistent with measurements in Ref. 39 reported at 150–195 MHz. Accounting for a depth-dependent temperature profile and extrapolating using a frequency dependence of $-0.55$ m/MHz, based on an ensemble of previous measurements, yields a field attenuation length of $\langle L_\alpha \rangle = 1022^{+230}_{-253}$ m over the top 1500 m of ice at 300 MHz.[25]

Another promising site is Moore's Bay on the Ross Ice Shelf, where the average measured depth of the ice is $576 \pm 8$ m.[26] Using a technique where reflection loss (found to be $-1.7$ dB) is separated from attenuation loss by separating the transmitter and receiver by over 500 m at the surface, the measured depth-averaged field attenuation length $\langle L_\alpha \rangle$ between 100 and 850 MHz is $(460 \pm 20) - (180 \pm 40)\nu$ m,

where $\nu$ is the frequency of interest in GHz. This corresponds to a frequency dependence of $-0.18$ m/MHz.[21,26] The shorter attenuation length at Moore's Bay compared to potential deep sites is due primarily to the fact that the ice is warmer at
Moore's Bay.

### 2.2.2. *Attenuation length of rock salt*

Large, naturally-occurring deposits of rock salt have also been investigated as a
possible detection medium for radio emission from neutrino interactions. Investigations at three different locations in the United States (WIPP in New Mexico,
Hockley in Texas, and the Cote Blanche mine operated by the North American Salt
Company in St. Mary Parish, Louisiana)[40,41] indicate that the attenuation length
of rock salt is heavily dependent on the moisture content, layering, and composition
of the salt. The longest attenuation length was measured directly at Cote Blanche
by transmitting impulsive, broadband radio signals directly through the salt from
a transmitter to a receiver with bandwidths spanning 125 to 900 MHz in boreholes
drilled up to 200 ft. into the salt.

The field attenuation length $\langle L_\alpha \rangle$ at 300 MHz was measured to be $63 \pm 3$ m at
Cote Blanche, which is significantly shorter than measurements of glacial ice have
shown. However, the relatively high density of rock salt compared to ice boosts the
neutrino interaction cross section by a factor of about 2.5, which would make up for
some of the volumetric loss due to the shorter radio attenuation length. In addition,
in salt the Cherenkov emission would be broader in solid angle,[42] leading to an
increase in achievable effective volume, and an overburden of bedrock would block
backgrounds from above such as galactic noise and cosmic rays.

### 2.2.3. *Attenuation length of lunar regolith*

The lunar regolith is another possible radio-transparent target for interactions of
highly energetic neutrinos. The radio attenuation length of samples that have been
returned to Earth has been shown to be $\mathcal{O}(20)$ m at 1 GHz and $\mathcal{O}(200)$ m at
100 MHz,[43] comparable to that of rock salt. However, these values were found to
vary across samples, depending on their exact composition. Although estimates of
the depth of the moon's regolith vary widely, and is expected to vary over the lunar
surface, conservative depth estimates are near 10 m.[44] Sensitivity estimates for
the Goldstone Lunar Ultra-high-energy Neutrino Experiment (GLUE) experiment
assume a 10 m regolith depth, while estimates for the NuMoon experiment assume
a depth of up to 500 m.[45,46]

## 3. Experimental strategies

### 3.1. *Assessing the sensitivity of an experiment*

There are a few factors that determine the number of neutrinos expected to be
detected from a given experiment, and these are the factors that go into an

experimental design. We often define an energy-dependent, water-equivalent effective volume × solid angle $[V\Omega]_{\text{eff}}(E)$ for an experiment, where $E$ is the neutrino energy, and it is given by

$$[V\Omega]_{\text{eff}}(E) = V \cdot 4\pi \cdot \varepsilon_V(E) \cdot \rho_{\text{H20}}/\rho_{\text{det}}, \qquad (3)$$

where $V$ is the total volume of the detection medium, $\varepsilon_V(E)$ is the fraction of neutrinos interacting in that volume that pass the trigger, and $\rho_{\text{det}}/\rho_{\text{H20}}$ is the density of the detection medium relative to water. In order to predict the number of neutrino events $N$ detected in an experiment from a given flux model $F(E) = dN/dE/d\Omega/dt$, we need to define an effective area × solid angle $[A\Omega]_{\text{eff}}(E)$ so that

$$N = \int F(E) \cdot [A\Omega]_{\text{eff}}(E) \cdot T \cdot dE, \qquad (4)$$

where $T$ is the livetime of the experiment. If the thin-target approximation is valid, i.e., the dimensions of the detection medium are much smaller than the interaction length $\ell(E)$, then we can take $[A\Omega_{\text{eff}}](E) = [V\Omega]_{\text{eff}}(E)/\ell(E)$.

The effective area typically increases with energy, and there is some energy below which the experiment does not expect to see a significant number of events. This is called the energy threshold, and is not well-defined quantitatively, but is set by the energy-dependent efficiencies at the trigger and analysis stages (analysis efficiencies are not included in the above equations). In turn, the trigger efficiencies are set by the trigger thresholds necessary to maintain the rate of data acquisition that is possible given the level of thermal noise that is seen. Since the high-energy neutrino flux (from GZK models and direct production models) is a falling spectrum at high energies, the event rate can be strongly dependent on the energy threshold.

## 3.2. Experimental approaches

Although most experiments that aim to detect radio emission from high-energy neutrino interactions are similar in their approach, determining the best experimental configuration is a tradeoff among a variety of factors. All experiments that rely on observations of Askaryan emission need to monitor a large volume of radio-transparent dielectric. However, many different approaches have been explored, such as the observation of the Antarctic continent or Greenland from a balloon or satellite, ground-based techniques that deploy instruments directly on the surface or at depth surrounded by large volumes of ice or salt, and observation of the lunar regolith from afar. Other experiments seek the radio emission produced by an air shower resulting from the decay products of a tau lepton produced from a neutrino interaction. All have different benefits and drawbacks.

### 3.2.1. Experiments using the Askaryan technique

We can see the trade-off between balloon-borne experiments and detectors on the ground in terms of the variables defined in Sec. 3.1. Putting a detector farther from the neutrino interactions increases the energy threshold. However, from high altitudes an experiment can view a larger area of ice compared to an experiment on the ground, which increases $[A\Omega]_{\text{eff}}$. In general, detectors that observe large volumes from far distances (from a balloon, satellite, or viewing the moon) are the best probes of the flux of the highest energy neutrinos. We will call these "view-from-a-distance" experiments. Balloon experiments such as ANITA[47] and the proposed ExaVolt Antenna (EVA)[48] fly at ~37 km altitude above the Antarctic ice sheet. ANITA has set the strongest limits above $10^{19.5}$ eV. Telescopes such as the Parkes Lunar Radio Cherenkov Experiment[49] aimed at the moon are seeking Askaryan emission from the sandy regolith at the lunar surface. The PRIDE experiment is even exploring other worlds as possible detection media, such as Enceladus or Europa, icy moons of Saturn and Jupiter, respectively.[50]

Detectors on the ground or embedded in their detection medium, such as the Askaryan Radio Array (ARA) and the Antarctic Ross Ice Shelf Antenna Neutrino Array (ARIANNA), which sit in and on glacial ice, have a lower energy threshold than ones that are situated farther from the interactions. In addition, the livetime can be much longer for an embedded experiment compared to a balloon experiment (years compared to several weeks). The drawback of being so close to the interactions is that a smaller volume of ice can be monitored by each detector element compared to from above, which reduces $[A\Omega]_{\text{eff}}$ for the same number of detectors. Therefore, embedded detectors consist of an array of antennas covering a large area.

View-from-a-distance experiments and in-ice experiments utilizing the Askaryan technique thus have different strengths for probing high-energy astrophysics. For example, experiments that view from far distances have the greatest potential to measure the high-energy cutoff of the astrophysical sources. Experiments on the ground, due to their lower threshold, are necessary to reach the heart of the GZK-induced neutrino spectrum, predicted to be at about $10^{18}$ eV, and if thresholds can be reduced, could measure the astrophysical neutrino spectrum above the ~1 PeV energies corresponding to the highest energy events so far observed by IceCube.

Once neutrino events are observed, embedded detectors also have the potential to achieve better angular resolution for neutrinos. By placing detectors relatively far apart (tens of meters), one can in principle view the Cherenkov cone from different angles, and use measured polarizations to reconstruct the direction of the shower and thus the neutrino direction. This is unlike a balloon-borne experiment whose size is tightly constrained. The energy resolution is also improved with the ability to pinpoint the neutrino interaction location. Although these techniques are not yet mature, their development will become of extreme importance as soon as the first neutrinos are measured using the radio technique.

A few experiments pioneered the radio technique to search for high energy neutrinos by searching for the Askaryan signature in various media. Data from the FORTE[51] satellite was used to search for neutrino interactions in Greenland ice in the late 1990's. The Parkes Lunar Radio Cherenkov Experiment, followed by the GLUE experiment at the Goldstone Observatory in California, both searched for neutrino interactions in the moon's regolith.[45,49,52] RICE was an array of radio antennas that was deployed on strings of AMANDA (the predecessor of IceCube), ran from 1999–2011, and published competitive limits as can be seen in Fig. 1.[14]

### 3.2.2. *Experiments using the air shower technique*

There are also experimental efforts to demonstrate techniques required to detect radio emission from extended air showers induced by neutrino interactions. Tau neutrinos that undergo a charged current interaction in the Earth and are moving at a slight up-going angle or interact in a mountain produce a tau lepton that emerges from the Earth's surface and decays in the atmosphere. The charged particles in the shower(s) resulting from the tau decay generate radio emission in part due to a geomagnetic effect: positively and negatively charged particles are split due to the Earth's magnetic field, yielding geosynchrotron emission. Askaryan radiation is also emitted from the shower. Detection of radio emission from cosmic-ray air showers has been made by a variety of experiments,[53–56] but neutrino-induced air showers are much more rare and have yet to be detected. Auger looks for air showers induced by decays from tau leptons that might emerge from neutrino interactions in the Andes Mountains.[13] The TREND project was designed as a prototype to demonstrate the technique,[55,57] and recently the Giant Radio Array for Neutrino Detection (GRAND) experiment has been proposed to detect radio emission from neutrino-induced extended air showers.[58]

## 4. Balloon experiments

### 4.1. *ANITA: The Antarctic Impulsive Transient Antenna*

The ANITA experiment flies under NASA's Long-Duration Balloon program, lifted to 37 km altitude over the Antarctic continent by a balloon that is ~100 m in diameter at altitude. ANITA searches for radio signals from neutrinos interacting in the ice sheet below. With the horizon at ~700 km distance at float altitude, ANITA uses all of its visible Antarctic ice sheet as its neutrino detection volume. The instrument can view ~$1.5 \times 10^6$ km$^2$ of ice at altitude, and is designed to search for impulsive, broadband signals predominantly in the vertical polarization emerging from the ice below the payload. ANITA uses dual-polarization (horizontal and vertical), quad-ridged horn antennas with 200–1200 MHz bandwidth. For each channel (one polarization of one antenna), the signal at the antenna output is amplified and filtered before being split so that it can be sent through both the trigger and readout chains. Each channel is Nyquist-sampled and read out with fast (>2GSa/sec) digitization, producing a ~100 ns waveform for each channel when a multi-level trigger is satisfied. ANITA-1 was launched in December 2006 and was aloft for 35 days, ANITA-2 flew in 2008–09 for 31 days, and ANITA-3 flew in 2014 for 22 days.

There are two main types of triggered events that need to be rejected in order to search for neutrino candidate events: thermal noise events and human-made noise events. Thermal noise backgrounds are reduced by requiring a high signal-to-noise ratio, a high cross-correlation between waveforms with time delays expected from an incoming plane wave, and a linearly polarized signal. Events that contain modulated, continuous-wave (CW) signals are subject to a notch-filter, which removes power in a narrow band surrounding any peaks in the measured spectrum above a threshold.

ANITA has published results from searches for a diffuse neutrino flux in each of its first two flights, and also reported a targeted search for neutrinos from Gamma Ray Bursts (GRBs) using data from its second flight. ANITA searches for isolated signals from the ice that are not associated with any known bases or with any locations with repeating signals. A blind analysis of data from the second flight of ANITA rejected thermal noise by a factor of $\leq 2.5 \times 10^{-8}$ and yielded one candidate neutrino event on an expected background of $0.97 \pm 0.42$ events. This search produced what is still the world's best limit on the neutrino flux at energies greater than $10^{19.5}$ eV.[15,16] A set of baseline models for cosmogenic neutrino fluxes would predict 0.3–1.0 neutrinos to pass the cuts. A targeted search for neutrinos associated with GRBs was done by narrowing the time window of interest to the 10 minutes surrounding 12 known GRBs that occurred during the second flight of ANITA when the payload was located in quiet regions of Antarctica. The strategy of conducting the search over a reduced time period leads to reduced backgrounds compared to the diffuse search, allowing for the analysis cuts to be loosened. The two hours of data surrounding each GRB, excluding the 10 minute window of interest, was used to estimate the background and set the cuts. The GRB search yielded no candidate events and the best limit on the fluence of neutrinos associated with GRBs at extremely high energies.[59]

ANITA-1 also reported the observation of 16 cosmic rays from geosynchrotron emission. For 14 of the events, the impulses were observed after a reflection from the ice surface, and were flipped in polarity compared to the remaining two events that were observed directly.[53]

ANITA-3 flew 48 antennas compared to 40 in ANITA-2, and triggered on both polarizations (cosmic ray air showers are detected in the horizontal polarization,[53]) with a noise-riding trigger threshold. These improvements led to a factor of $\sim 5$ improvement in expected neutrino event rates compared to ANITA-2. ANITA-3 also included a low-frequency omnidirectional antenna (ALFA), providing additional information for the hundreds of UHE cosmic-ray events that are expected to have been recorded by ANITA-3 via radio emission from cosmic-ray induced atmospheric extended air showers. The left-hand panel of Fig. 3 is a picture of the ANITA-3 payload, fully integrated in August 2014 preparation for its December 2014 flight.

## 4.2. *EVA: The ExaVolt Antenna*

Building on ANITA's strategy of utilizing NASA's long-duration balloons to carry out searches for neutrinos in the UHE regime, EVA[48] is an ambitious project

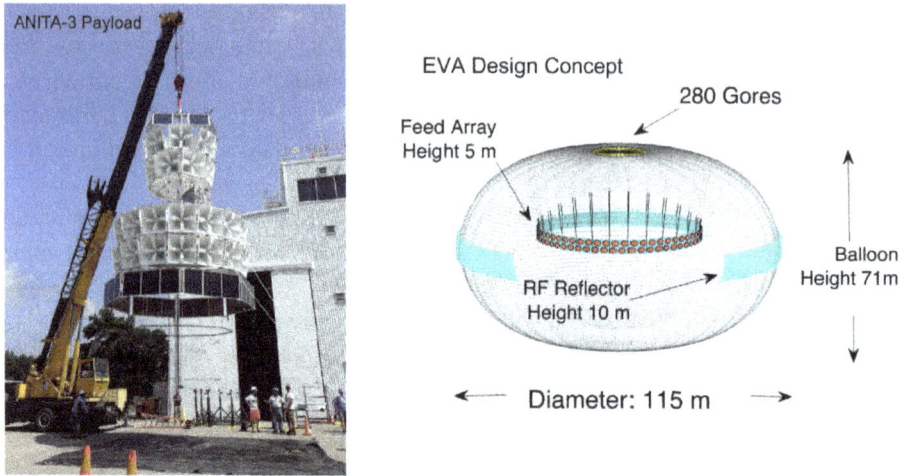

Fig. 3.   Left: A picture of the ANITA-3 payload, fully integrated in August 2014 in preparation for its Antarctic flight in December 2014. ANITA-3 has 96 channels, as well as a low-frequency antenna for cosmic-ray science (visible below the payload in the picture) that was dropped down below the payload after launch. Right: a sketch of the EVA concept.

aiming to transform a 100 m-scale balloon into a high gain, toroidal reflector antenna (see the right-hand panel of Fig. 3). The increase in gain compared to ANITA would mean a $\gtrsim 100$ times reduction in threshold in power, and thus a $\gtrsim 10$ times reduction in threshold in field strength and neutrino energy. What makes the EVA concept feasible is the super-pressure balloon (SPB) technology currently under development by NASA,[60] where the inside of the balloon is kept at a higher pressure than the outside ambient pressure, with differential pressures up to 180 Pa, keeping the shape of the balloon nearly unchanged over a flight. A test flight in 2008 over Antarctica, SPB flight 591NT, reported that the height and diameter of a 7 Mft$^3$ balloon changed by 1% in a 54 day flight.[61]

In September of 2014, the EVA team carried out a test of a prototype instrumented SPB in a hangar at NASA's Wallops Test Facility to demonstrate that a feed array could be deployed inside of a SPB, and that the balloon could be instrumented with an RF reflector-receiver system that could be well understood by modeling.[62] The 1:20 scale prototype was a 5.7 m SPB balloon. Reflector tape was attached to the balloon near its equator and a feed array membrane held the receivers (dual-polarized sinuous patch antennas) and associated electronics over a section of its circumference for the test (see Fig. 4).

An impulsive plane-wave calibration signal was sent to the balloon and was observed at incidence and after reflection. A comparison of the two pulses is consistent with detailed time-domain modeling of the prototype, and the team is currently preparing to propose a flight of a full EVA.

Fig. 4.   Pictures of the EVA prototype test (balloon, toroidal reflector, and dual-polarized sinuous feed antennas) at NASA's Wallops Test Facility.[62]

## 5. Ground-based experiments

### 5.1. *Askaryan effect in ice*

#### 5.1.1. *ARA: The Askaryan Radio Array*

The Askaryan Radio Array (ARA) is a radio array embedded in the ice near the South Pole with the aim of measuring cosmic neutrinos in the energy regime above $\sim 10^{17}$ eV. Searching for Askaryan emission from neutrino interactions in the deep ice near the South Pole, stations of receiver antennas are deployed in dry holes at 200 m depth. A schematic of the ARA design is shown in Fig. 5. Each station consists of 16 antennas, each of which is read out as a separate channel. Each ARA station uses a mixture of antennas designed to measure vertical and horizontal polarizations, deployed along four strings separated by $\sim 10$ m. Calibration pulsers and associated transmitting antennas are also deployed with each station along additional strings.

The trigger and readout chains are similar to ANITA, but the digitized waveforms are $\sim 250$ ns in length. The trigger currently requires that some number of channels in a station (typically 3 out of 16) exceed a power threshold ($\sim 5-6$ times the mean power) within a 110 ns time window, approximately the time it takes a signal to traverse a station.

An initial Testbed prototype station was deployed in the 2010–2011 season at $\sim 30$ m depth, followed by the first full station (A1) in 2011–2012 at 100 m depth. Two other stations, A2 and A3, have been in the ice at 200 m depth since the 2012–2013 season. All three deep stations are currently operational. Two more stations, A4 and A5, will be deployed in the 2017–2018 season.

ARA neutrino searches look for signals that could have originated from neutrino interactions in the ice and reject events that reconstruct to the South Pole Station. Interferometric techniques, with timing based on cross-correlations between waveforms from different antennas, are used to reconstruct directions of signals. ARA has the capability to take into account the bending of the trajectories of signals in

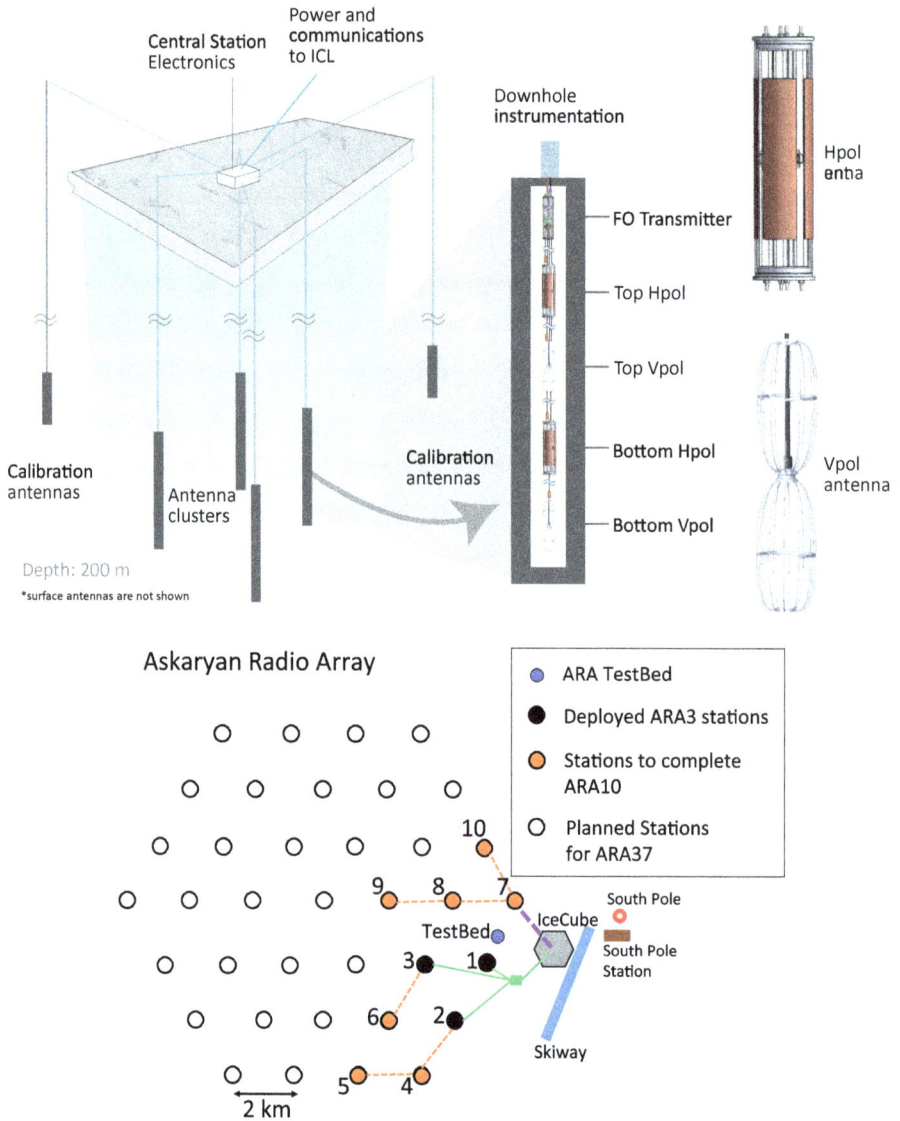

Fig. 5.   Top: the baseline design of an ARA station showing the details of one string and a drawing of the antennas sensitive to each linear polarization, from Ref. 63. Bottom: a map of the proposed locations of ARA stations at the South Pole, from Ref. 64. Currently there are three stations deployed (shown in black), plus a testbed (in blue).

the firn near the surface, where the index of refraction changes with depth. Thermal noise is reduced through requirements that the timing of the signals in the antennas is consistent with a real signal crossing the detector.

Using the ARA Testbed station, ARA carried out a search for cosmic neutrinos from diffuse sources, and then a targeted search for neutrinos from GRBs, with 415 days of livetime using the 2011–2012 data from the ARA Testbed.[64,65] As was done for the ANITA GRB search, neutrinos were sought in a 10 minute time window surrounding the occurrence of each of 57 GRBs. This resulted in the first quasi-diffuse limit on the GRB flux above $10^{16}$ eV. The quasi-diffuse limit is based on the assumption that the 57 GRBs considered for the analysis were representative of all GRBs.

ARA has completed a first search for neutrinos from two of the three deployed deep stations (A2 and A3) using 10 months of data from 2013.[63] No candidates were observed on a background of $0.009 \pm 0.010$ for A2 and $0.011 \pm 0.015$ for A3, resulting in constraints on the cosmic neutrino flux.

### 5.1.2. *ARIANNA: The Antarctic Ross Ice Shelf Antenna Neutrino Array*

ARIANNA is a proposed ground-based array of radio antennas on the Ross Ice Shelf in Antarctica.[66] An ARIANNA array of ~1000 autonomously-powered detectors would achieve a sensitivity similar to a 37-station ARA array. Although a 1000-station wind-powered ARIANNA design has been considered, a 1296 ($36 \times 36$) station array can be powered via solar power and battery backup for the same cost. ARIANNA's shallow design with no drilling needed makes an ARIANNA station simpler to deploy than an ARA station. ARIANNA's proposed $36 \times 36$ station array would span more than 1000 km$^2$ on the ~600 m-deep Ross Ice Shelf compared to an ARA37 that would span ~100 km$^2$ area at the South Pole where the ice is ~2800 m deep. There is a reflecting layer of ocean water at the bottom of the Ross Ice Shelf that acts as a mirror, reflecting radio signals from down-going neutrinos back up to the antennas on the surface of the snow.[67] This means that ARIANNA would have a broader range in visible solid angle compared to ARA due to the ability to see down-going neutrinos, which counteracts the shorter measured radio attenuation length of the ice on the Ross Ice Shelf.

An ARIANNA station, shown in Fig. 6, consists of high-gain log-periodic dipole antennas (LPDAs) with a bandwidth of ~$0.08-1.3$ GHz in snow that are downward-pointing and deployed near the surface of the snow. The HRA (Hexagonal Radio Array) consists of eight prototype stations, most consisting of four LPDAs, and was deployed in the Austral summers of 2012–2013 and 2014–2015. A design station for the full ARIANNA array would contain 6–8 downward-facing LPDAs. When a multi-level trigger is satisfied at a station, waveforms across that station are digitized at 2 GSa/sec. Stations are solar powered, with battery backups that both provide power on cloudy days and extend the life of the stations a few weeks past sundown in low power mode, as discussed in Ref. 68. The achieved livetime is 58% of the year.[69]

ARIANNA performed a first neutrino search using data from the HRA. After searching for events that contain impulsive signals whose arrival times are consistent with a plane wave crossing the station in the absence of any narrow peak in the Fourier spectrum pointing to CW noise, no neutrino candidates were found in four

Fig. 6.  View of one ARIANNA station with solar panels on a tower. The box containing the station's data acquisition box is buried in the ice beneath the tower. Photo credit to Chris Persichilli.

months of operation. From those results, ARIANNA set its first limit on the diffuse neutrino flux above $10^{17}$ eV with three HRA stations.[70] Data from all eight stations of the HRA is currently being analyzed.

### 5.1.3. *GNO: The Greenland Neutrino Observatory*

An effort is ongoing to develop Summit Station, a year-round station run by the National Science Foundation (NSF), as a site for radio detection of UHE neutrinos. The ice at Summit Station is comparable to that at the South Pole in radio clarity (see Sec. 2.2.1) and in thickness (3000 m at Summit Station compared to 2800 at South Pole), but with a shallower firn allowing for shallower deployment of sub-surface antennas. At Summit Station, at 100 m depth the firn density has reached 95% that of deep ice, compared to 140 m at the South Pole.[71,72] A first prototype station for GNO was deployed in the summer of 2015 to monitor the radio-frequency noise environment.

## 5.2. *Lunar experiments*

In 1995, the Parkes Lunar Radio Cherenkov Experiment, using the 64 m Parkes radio telescope, was the first to search for radio emission from neutrino interactions in the moon's regolith, the loose, $\sim$10 m deep layer at the surface of the moon. Due to the expected geometry of neutrino-induced events, lunar radio experiments are expected to be sensitive to neutrinos when viewing the lunar "limb" (the edge of the moon's visible surface).[73] Multiple antennas are beamed for heightened sensitivity to regions along the limb. Radio frequency interference is often reduced in lunar radio experiments by requiring signals in different frequency bands to arrive at the delays expected from dispersion through the ionosphere. GLUE was the next to search for highly energetic neutrino interactions in the moon's regolith in 2000–2003 using the 34 m DSS13 and 70 m DSS14 antennas at the Goldstone Observatory in California for 124 h of observing time, constraining the neutrino flux at energies above $10^{21}$ eV.[45,49] Since then, a series of experiments have followed up with lunar observations with radio telescopes.[52,74–79] The NuMoon experiment observed the moon with the Westerbork Synthesis Radio Telescope array in four frequency bands between 113 and 175 MHz for 47.6 hours, and claims the most competitive constraints on the neutrino flux above $10^{23}$ eV, as seen in Fig. 1.[18] Lunar observations have been proposed with the Low Frequency Array (LOFAR), a recently completed radio antenna array with beaming accomplished electronically rather than mechanically, with a sensitivity projected to improve upon the NuMoon constraints by a factor of 25 as stated in Ref. 18. The Square Kilometer Array (SKA) will be even more sensitive to neutrinos from the moon with this technique due to its wide bandwidth and large collecting area. Construction of Phase 1 of SKA is scheduled to begin in 2018.[80]

## 5.3. *Neutrino-induced air showers*

Auger searches for neutrinos in the UHE regime by distinguishing "young" and "old" showers using their 1600 Surface Detectors, each consisting of a Cherenkov tank with photomultipliers read out with flash analog to digital converters (FADC) with 25 ns resolution.[13] Protons, heavy nuclei and any gamma rays are expected to shower soon after they hit the atmosphere, so by the time their showers hit Auger's Surface Detectors they are old showers, consisting mainly of muons with highly coincident arrival times. Neutrinos however, are equally likely to interact anywhere along their path through the atmosphere, and therefore can be young when they arrive at the Surface Detectors. Also, a tau neutrino can interact and produce a tau lepton in mountains surrounding the observatory and the subsequent tau decay would lead to a young shower seen by Auger. Younger showers would still contain an electromagnetic component when they are observed and would show a broader distribution of arrival times. The greatest separation between young and old showers is expected for events that are highly inclined. Auger sets competitive limits on the tau neutrino diffuse flux only, to which they are the most sensitive. Auger's competitive UHE neutrino flux limit shown in Fig. 1 is a tau neutrino limit multiplied (i.e. weakened) by a factor of 3 for comparison with other experiments whose limits are averaged over all flavors.

The proposed GRAND experiment would consist of an array of antennas in a remote mountainous site to search for air showers induced by the decay of tau leptons in the atmosphere that originate from a charged current interaction of tau neutrinos with the Earth.[58] The GRAND collaboration proposes the deployment of $\mathcal{O}(10^5)$ radio antennas, operating between $30 - 100$ MHz and covering $\mathcal{O}(10^5)$ km$^2$ at a site in the Tianshan mountains in China. The preliminary design includes several sub-arrays, each with area $\mathcal{O}(10^4)$ km$^2$. Initial estimates predict that a $6 \times 10^4$ km$^2$ array would have an effective area several times larger than ARA's effective area between $10^{18} - 10^{19}$ eV, and a $2 \times 10^5$ km$^2$ array would further improve the sensitivity by a factor of $\sim 10$ in the same energy region.

## 6. Goals of future experimental efforts

The most important aim in this field at present is to make a first discovery of neutrinos in the UHE regime. Alongside the firm expectation of the BZ neutrino flux, there is an expected neutrino flux from the astrophysical sources themselves that may be comparable. Once the initial discovery is made, it will open up a unique window to the universe at the highest energies. We will likely be able to say that the highest energy cosmic rays are not purely heavy composition, and that no new physics prohibits a UHE neutrino from reaching us from cosmic distances. We will know the detector effective area that is needed to measure a large enough sample of neutrinos to carry out a particle physics and astrophysics program in this new frontier. In addition, a first discovery will heighten the urgency to develop the tools to reconstruct neutrino directions and measure neutrino energies, which will be necessary as physics priorities are pursued beyond the implications of the first detection.

Both view-from-a-distance and embedded experiments are working to reduce their energy thresholds, which would move their sensitivities to the *left* in Fig. 7, and increase their chance of a discovery with a falling neutrino spectrum. In this figure, one can see the effect of EVA's factor of $\gtrsim 10$ reduction in energy threshold compared to ANITA. Many experiments are working to implement the interferometric techniques that have been successful at the analysis stage to their trigger stage, to enable detection of lower energy events.[81,82] With an interferometric phased array trigger coherently summing hundreds of antennas, in-ice experiments may be able to lower their thresholds enough to reach the higher end of the PeV neutrino spectrum observed by IceCube[9,10] and ascertain whether the spectrum cuts off or continues to higher energies.[82]

Experiments are also working to increase their effective area in order to move their sensitivities *down* in Fig. 7. Ground-based projects ARA and ARIANNA are working to expand their detectors, adding as many stations as funding and logistics allow. The GRAND experiment described in Sec. 5.3 is looking to utilize a vast swath of the atmosphere to seek neutrino-induced extended air showers and will compete with ARA in the same energy range.

The radio technique for neutrino detection allows experimenters to instrument the immense volumes needed for sensitivity to the rare fluxes of neutrinos expected in the UHE regime through a variety of projects that view the detection medium from within, with embedded detectors such as ARA or ARIANNA, or from a distance, with balloon experiments such as ANITA. Radio telescopes pointed at the moon such as NuMoon are the most sensitive at even higher energies. Other projects, such as Auger, search for UHE neutrinos through an air shower signature. Radio techniques provide an opportunity for a long-term UHE astrophysics program that can reach even some of the most pessimistic predictions for the neutrino flux in the UHE regime. The results from the first decade and a half of neutrino searches in this field and the projections for future experiments portend an exciting future.

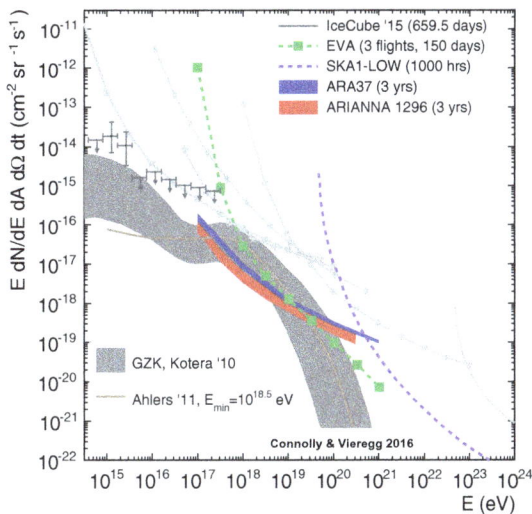

Fig. 7. The sensitivity reach of future radio neutrino experiments at ultra-high energies and above. We show projected limits for ARA37, EVA and SKA Phase 1 with the low frequency antennas.[62,64,80] In light gray we also show the current most competitive limits that were shown in Fig. 1 for comparison.

## Acknowledgements

The authors would like to thank Tom Gaisser and Albrecht Karle for inviting us to write this contribution. We are also grateful to Jordan Hanson, Clancy James, Andrew Romero-Wolf and David Saltzberg for helpful feedback. A. Connolly would also like to thank the National Science Foundation for their support through CAREER award 1255557, BIGDATA Grant 1250720, Grant 1404266 for ARA support, and NASA for their support through Grant NNX12AC55G for EVA development. A. Connolly would also like to thank the United States-Israel Binational Science Foundation for their support through Grant 2012077. A. Connolly and A. Vieregg would like to thank NASA for their support for ANITA through Grant NNX15AC20G. We are grateful to the U.S. National Science Foundation-Office of Polar Programs. A. Vieregg would like to thank the Kavli Institute for Cosmological Physics for their support, and A. Connolly would like to thank the Ohio State University for their support.

## References

1. K. Greisen, End to the cosmic-ray spectrum?, *Phys. Rev. Lett.* **16** (1966) 748–750.
2. G.T. Zatsepin and V.A. Kuz'min, Upper limit of the spectrum of cosmic rays, *JTEP Lett.* **4** (1966) 78.
3. V.S. Berezinsky and G.T. Zatsepin, Cosmic rays at ultra high energies (neutrino?), *Phys. Lett. B.* **28** (1969) 423–424.
4. V.S. Berezinsky and G.T. Zatsepin, *Sov. J. Nucl. Phys.* **11** (1970) 111.
5. A. Connolly, R.S. Thorne, and D. Waters, Calculation of high energy neutrino-nucleon cross sections and uncertainties using the MSTW parton distribution functions and implications for future experiments, *Phys. Rev. D.* **83** (2011) 113009.
6. S.R. Klein and A. Connolly, Neutrino absorption in the Earth, neutrino cross-sections, and new physics. (2013). arXiv:1304.4891.
7. P.W. Gorham, *et al.*, Implications of ultra-high energy neutrino flux constraints for Lorentz-invariance violating cosmogenic neutrinos, *Phys. Rev. D.* **86**(10) (2012) 103006.
8. L.A. Anchordoqui, *et al.*, End of the cosmic neutrino energy spectrum, *Phys. Lett. B.* **739** (2014) 99–101.
9. M.G. Aartsen, *et al.*, First observation of PeV-energy neutrinos with IceCube, *Phys. Rev. Lett.* **111**(2) (2013) 021103.
10. M.G. Aartsen, *et al.*, Observation of high-energy astrophysical neutrinos in three years of IceCube data, *Phys. Rev. Lett.* **113**(10) (2014) 101101.
11. M. Ahlers and F. Halzen, Minimal cosmogenic neutrinos, *Phys. Rev. D.* **86**(8) (2012) 083010.
12. K. Kotera, D. Allard, and A.V. Olinto, Cosmogenic neutrinos: parameter space and detectabilty from PeV to ZeV, *J. Cosm. and Astropart. Phys.* **10** (2010) 013.
13. A. Aab, *et al.*, Improved limit to the diffuse flux of ultra-high energy neutrinos from the Pierre Auger Observatory, *Phys. Rev. D.* **91** (2015) 092008.
14. I. Kravchenko, *et al.*, Updated results from the RICE experiment and future prospects for ultra-high energy neutrino detection at the South Pole, *Phys. Rev. D.* **85**(6) (2012) 062004.

15. P.W. Gorham, *et al.*, Observational constraints on the ultra-high energy cosmic neutrino flux from the second flight of the ANITA experiment, *Phys. Rev. D.* **82**(2) (2010) 022004.

16. P.W. Gorham, *et al.*, Erratum: Observational constraints on the ultra-high energy cosmic neutrino flux from the second flight of the ANITA experiment. (2010). arXiv:1011.5004.

17. M.G. Aartsen, *et al.*, The IceCube Neutrino Observatory — Contributions to ICRC 2015 Part II: Atmospheric and Astrophysical Diffuse Neutrino Searches of All Flavors. (2015). arXiv:1510.05223.

18. S. Buitink, *et al.*, Constraints on the flux of ultra-high energy neutrinos from Westerbork Synthesis Radio Telescope observations, *Astronomy & Astrophysics.* **521** (2010) A47.

19. M. Ackermann, *et al.*, Optical properties of deep glacial ice at the South Pole, *J. Geophys. Res.* **111**(D13) (2006) D13203.

20. M.G. Aartsen, *et al.*, IceCube-Gen2: A vision for the future of neutrino astronomy in antarctica. (2014). arXiv:1412.5106.

21. T. Barrella, S. Barwick, and D. Saltzberg, Ross Ice Shelf *in situ* radio-frequency ice attenuation, *J. Glaciol.* **57** (2011) 61–66.

22. D.Z. Besson, *et al.*, *In situ* radioglaciological measurements near Taylor Dome, Antarctica and implications for ultra-high energy (UHE) neutrino astronomy, *Astropart. Phys.* **29**(2008) 130–157.

23. S. Barwick, *et al.*, South Polar *in situ* radio-frequency ice attenuation, *J. Glaciol.* **51** (2005) 231–238.

24. P. Allison, *et al.*, Design and initial performance of the Askaryan Radio Array prototype EeV neutrino detector at the South Pole, *Astropart. Phys.* **35** (2012) 457–477.

25. J. Avva, *et al.*, An *in situ* measurement of the radio-frequency attenuation in ice at Summit Station, Greenland , *J. Glaciol.* **61** (2015) 1005–1011.

26. J. Hanson, *et al.*, Radar absorption, basal reflection, thickness, and polarization measurements from the Ross Ice Shelf, *J. Glaciol.* **61** (2015) 438–446.

27. Pierre Auger Collaboration, Depth of maximum of air-shower profiles at the pierre auger observatory. II. Composition implications, *Phys. Rev. D.* **90**(12) (2014) 122006.

28. R.U. Abbasi, *et al.*, Study of ultra-high energy cosmic ray composition using Telescope Array's Middle Drum detector and surface array in hybrid mode, *Astropart. Phys.* **64** (2015) 49–62.

29. M. Unger, G.R. Farrar, and L.A. Anchordoqui, Origin of the ankle in the ultra-high energy cosmic ray spectrum, and of the extragalactic protons below it, *Phys. Rev. D.* **92**(12) (2015) 123001.

30. N. Globus, D. Allard, and E. Parizot, A complete model of the cosmic ray spectrum and composition across the galactic to extragalactic transition, *Phys. Rev. D* **92**(2) (2015) 021302.

31. G. Askaryan, Excess negative charge of an electron-photon shower and its coherent radio emission, *Soviet Physics JETP-USSR.* **14**(2) (1962) 441–443.

32. D. Saltzberg, *et al.*, Observation of the Askaryan effect: Coherent microwave cherenkov emission from charge asymmetry in high-energy particle cascades, *Phys. Rev. Lett.* **86** (2001) 2802.

33. P.W. Gorham, *et al.*, Accelerator measurements of the Askaryan effect in rock salt: A roadmap toward teraton underground neutrino detectors, *Phys. Rev. D* **72**(2) (2005) 023002.

34. P.W. Gorham, *et al.*, Observations of the Askaryan effect in ice, *Phys. Rev. Lett.* **99**(17) (2007) 171101.

35. S. Buitink, *et al.* Searching for neutrino radio flashes from the Moon with LOFAR. In eds. R. Lahmann *et al.*, *American Institute of Physics Conference Series*, Vol. 1535, pp. 27–31, (2013).

36. J.D. Bray, *et al.*, LUNASKA neutrino search with the Parkes and ATCA telescopes. In eds. R. Lahmann *et al.*, *American Institute of Physics Conference Series*, Vol. 1535, pp. 21–26, (2013).

37. A. Connolly, *et al.*, Status of SalSA. (2010). arXiv:1010.4347.

38. I.M. Zheleznykh. In ed. R. Prothroe, *Proceedings of the 21st International Cosmic Ray Conference*, Vol. 6, pp. 528–533, (1989).

39. J.A. MagGregor, *et al.*, Radar attenuation and temperature within the Greenland Ice Sheet, *J. Geophys. Res. Earth Surf.* **120** (2015) 983–1008.

40. P.W. Gorham, *et al.*, Measurements of the suitability of large rock salt formations for radio detection of high energy neutrinos, *Nucl. Instrum. Meth. A* **490** (2002) 476–491.

41. A. Connolly, *et al.*, Measurements of radio propagation in rock salt for the detection of high-energy neutrinos, *Nucl. Instrum. Meth. A* **599** (2009) 184–191.

42. J. Alvarez-Muniz, *et al.*, Coherent radio pulses from showers in different media: A unified parameterization, *Phys. Rev. D* **74** (2006) 023007.

43. G.R. Olohoeft and D.W. Strangway, Dielectric properties of the first 100 meters of the moon, *Earth Planet Sci. Lett.* (1975).

44. C. James, The lunar Cherenkov technique — answering the unanswered questions, *Nucl. Instr. Meth.* **662** (2012) S12–S19.

45. P.W. Gorham, *et al.*, Experimental limit on the cosmic diffuse ultra-high energy neutrino flux, *Phys. Rev. Lett.* **93**(4) (2004) 041101.

46. O. Scholten, *et al.*, Optimal radio window for the detection of ultra-high energy cosmic rays and neutrinos off the moon, *Astroparticle Physics.* **26** (2006) 219.

47. P.W. Gorham, *et al.*, The Antarctic Impulsive Transient Antenna Ultra-high Energy Neutrino Detector design, performance, and sensitivity for 2006-2007 balloon flight, *Astropart. Phys.* **32** (2009) 10–41.

48. P.W. Gorham, *et al.*, The ExaVolt Antenna: A large-aperture, balloon-embedded antenna for ultra-high energy particle detection, *Astropart. Phys.* **35** (2011) 242–256.

49. C.W. James, *et al.*, Limit on UHE neutrino flux from the Parkes Lunar Radio Cherenkov Experiment, *Mon. Not. Roy. Astron. Soc.* **379** (2007) 1037–1041.

50. T. Miller, R. Schaefer, and H.B. Sequeira, PRIDE (Passive Radio [frequency] Ice Depth Experiment): An instrument to passively measure ice depth from a Europan orbiter using neutrinos, *Icarus.* **220** (2012) 877–888.

51. N.G. Lehtinen, *et al.*, FORTE satellite constraints on ultra-high energy cosmic particle fluxes, *Phys. Rev. D.* **69** (2004) 013008.

52. J.D. Bray, *et al.*, A lunar radio experiment with the Parkes radio telescope for the LUNASKA project, *Astropart. Phys.* **65** (2014) 22–39.

53. S. Hoover, *et al.*, Observation of ultra-high-energy cosmic rays with the ANITA balloon-borne radio interferometer, *Phys. Rev. Lett.* **105**(15) (2010) 151101.

54. P. Schellart, *et al.*, Detecting cosmic rays with the LOFAR radio telescope, *Astron. Astrophys.* **560** (2013) A98.

55. O. Martineau-Huynh, *et al.*, Status of the TREND project. (2012). arXiv:1204.1559.

56. A. Aab, *et al.*, Energy estimation of cosmic rays with the engineering radio array of the Pierre Auger Observatory. (2015). arXiv:1508.04267.

57. O. Brusova, *et al.*, Radio detection of neutrinos from behind a mountain. In *Proceedings of the 30th International Cosmic Ray Conference*, Vol. 5, pp. 1585–1588, (2008).

58. O. Martineau-Huynh, *et al.* The Giant Radio Array for Neutrino Detection. In *European Physical Journal Web of Conferences*, Vol. 116, p. 03005, (2016).

59. A. G. Vieregg, *et al.*, The first limits on the ultra-high energy neutrino fluence from gamma-ray bursts, *Astrophys. J.* **736** (2011) 50.

60. H. Cathey, The NASA super pressure balloon — A path to flight, *Advances in Space Research.* **44** (2009) 23–38.

61. F.B.K. Baginski, Estimating the deployment pressure in pumpkin balloons, *AIAA Journal of Aircraft.* **48** (2011) 235–247.

62. A. Romero-Wolf, *et al.*, The ExaVolt Antenna mission concept and technology developments. In *Proceedings of the 34th International Cosmic Ray Conference*, (2015).

63. P. Allison, *et al.*, Performance of two Askaryan Radio Array stations and first results in the search for ultra-high energy neutrinos. (2015). arXiv:1507.08991.

64. P. Allison, *et al.*, First constraints on the ultra-high energy neutrino flux from a prototype station of the Askaryan Radio Array, *Astropart. Phys.* **70** (2015) 62–80.

65. P. Allison, *et al.*, Constraints on the ultra-high energy neutrino flux from gamma-ray bursts from a prototype station of the Askaryan Radio Array. (2015). arXiv:1507.00100.

66. S.W. Barwick. *et al.*, Design and performance of the ARIANNA Hexagonal Radio Array Systems. (2014). arXiv:1410.7369.

67. F. Wu. Using ANITA-I to constrain ultra high energy neutrino-nucleon cross section. PhD thesis, University of California, Irvine, (2009).

68. S.A. Kleinfelder, Design and performance of the autonomous data acquisition system for the ARIANNA high energy neutrino detector, *IEEE Trans. Nucl. Sci.* **60**(2) (2013) 612–618.

69. S.W. Barwick, *et al.*, Livetime and sensitivity of the ARIANNA Hexagonal Radio Array. In *Proceedings of the 34th International Cosmic Ray Conference*, (2015).

70. S.W. Barwick, *et al.*, A first search for cosmogenic neutrinos with the ARIANNA Hexagonal Radio Array, *Astropart. Phys.* **70** (2015) 12–26.

71. R.B. Alley and B.R. Koci, Ice-core analysis at site A, Greenland: Preliminary results, *Ann. Glaciol.* **10** (1988) 1–4.

72. K.C. Kuivinen, A 237-meter ice core from South Pole Station, *Antarctic Journal of the United States.* **18**(5) (1983) 113–114.

73. J.D. Bray, The sensitivity of past and near-future lunar radio experiments to ultra-high-energy cosmic rays and neutrinos, *Astropart. Phys.* **77** (2016) 1–20.

74. A.R. Beresnyak, *et al.*, Limits on the flux of ultra high-energy neutrinos from radio astronomical observations, *Astronomy Reports.* **49** (2005) 127–133.

75. C. James, *et al.*, LUNASKA experiments using the Australia Telescope Compact Array to search for ultra-high energy neutrinos and develop technology for the lunar Cherenkov technique, *Phys. Rev. D.* **81** (2010) 042003.

76. C. James. Ultra-high energy particle detection with the lunar Cherenkov technique. PhD thesis, The University of Adelaide, (2009).

77. T.R. Jaeger, R.L. Mutel, and K.G. Gayley, Project RESUN, a radio EVLA search for UHE neutrinos, *Astropart. Phys.* **34** (2010) 293–303.

78. R.E. Spencer, *et al.*, La Luna: Lovell Attempts LUnar Neutrino Acquisition. In *10th European VLBI Network Symposium and EVN Users Meeting: VLBI and the New Generation of Radio Arrays*, p. 97, (2010).

79. J.D. Bray, *et al.*, Limit on the ultra-high-energy neutrino flux from lunar observations with the Parkes radio telescope, *Phys. Rev. D.* (2015).

80. J.D. Bray, *et al.*, Lunar detection of ultra-high-energy cosmic rays and neutrinos with the Square Kilometre Array. (2015). arXiv:1408.6069.

81. A. Romero-Wolf, *et al.*, An interferometric analysis method for radio impulses from ultra-high energy particle showers, *Astropart. Phys.* **60** (2015) 72–85.
82. A.G. Vieregg, K. Bechtol, and A. Romero-Wolf, A technique for detection of PeV neutrinos using a phased radio array, *J. Cosm. and Astropart. Phys.* **1602**(02) (2016) 005.

# Index